D1731978

Aerospace Thermal Structures and Materials for a New Era

Edited by
Earl A. Thornton
University of Virginia
Charlottesville, VA

Volume 168
PROGRESS IN ASTRONAUTICS AND AERONAUTICS

Paul Zarchan, Editor-in-Chief
Charles Stark Draper Laboratory
Cambridge, Massachusetts

Technical papers selected from the Second University of Virginia Thermal Structures Conference,Charlottesville, Virginia, October 18–20, 1994, and subsequently revised for this volume.

Published by the American Institute of Aeronautics and Astronautics, Inc.
370 L'Enfant Promenade, SW, Washington, DC 20024-2518

ISBN 1-56347-182-5

Preface

In November 1990, the First University of Virginia Thermal Structures Conference was held in Charlottesville, Virginia. A collection of selected papers from the conference, *Thermal Structures and Materials for High Speed Flight*, was published as Volume 140 of the AIAA Progress in Astronautics and Aeronautics Series. The emphasis of the papers was on the technologies needed for the interdisciplinary design and development of thermal structures for high-speed flight.

Since 1990, there have been significant changes in the world that have altered aerospace research programs throughout the United States, Europe, and the Far East. In the United States, research and development for the National Aerospace Plane and the Space Exploration Initiative has ended. Furthermore, research on high-performance military aircraft has been reduced due to the end of the Cold War. Now the nation is placing increased emphasis on civil aeronautics and on space missions oriented toward Earth. In these latter missions, thermal structural problems remain important because challenging problems occur for long-duration supersonic flight in the atmosphere, in advanced high-temperature propulsion systems, and in the transient environment of extreme hot and cold encountered in Earth orbit.

This volume presents a collection of papers presented at the Second University of Virginia Thermal Structures Conference held in Charlottesville, Virginia, October 18–20, 1994. The fundamental goal of the conference was to expose participants to important problems and emerging technologies for the thermal structures encountered in new aeronautics and astronautics missions. This volume presents selected papers from the conference organized in four chapters treating analytical and experimental studies of thermal structures, analysis of high-temperature composites, and performance of aircraft materials.

Chapter 1, Analysis of Thermal Structures, deals with analytical studies of thermal structures, and a diverse set of modern problems is presented. The chapter begins with two interdisciplinary analyses of cryogenic fuel tanks. The development of reusable flight-weight cryogenic fuel tanks is one of the most difficult challenges in the design of advanced hypersonic aircraft. In the first paper, *Greer* presents a computational fluid dynamics analysis for the flow in a tank subject to a rapid drain rate and a high heat flux. The fluid motion within the cryogenic liquid is predicted as the first step leading to a structural analysis for the thermal stresses in the tank wall. In a closely related analysis, *Ko* presents a finite element structural analysis of the same tank. The effects of the liquid fill level on tank thermal stresses and critical buckling temperatures are investigated. The two papers demonstrate the multidisciplinary character of advanced thermal-structural analysis. Another multidisciplinary problem encountered on hypersonic vehicles is the response of heated structures to the high acoustic loads from advanced propulsion systems. In the next paper, *Lee* describes an analysis of a thermally buckled plate experiencing large-amplitude random vibrations. The relationships of RMS displacements, strains, and stresses to temperature and acoustic loads are established.

The last two papers of Chapter 1 describe problems encountered in thermally induced behavior of structures in space. The paper by *Malla and Ghoshal* presents a survey of temperature effects on space structures. The paper observes that the extreme temperature gradients possible in space give host to a number of static and dynamic structural problems not encountered in a terrestrial environment. Thermally induced vibration effects in large light-weight structures are among the problems discussed. In the final paper in Chapter 1, *Kim and McManus* present analyses of an insulated experimental package in the thermal environment of the space shuttle payload bay. The thermal analysis is complicated by radiation and shadowing effects. The effects of the thermal environment and quasi-static structural deformations on the performance of the experiment are evaluated.

Chapter 2, Experimental Studies of Thermal Structures, presents three papers devoted to experimental studies of thermal structures. In the first paper, *Blosser* describes the challenge of providing appropriate boundary conditions for aerospace thermal-structural tests. Test boundary conditions are difficult to design because of inherent conflicts between providing structural restraint and prescribing known heat transfer conditions. The conflicts are described, and several thermal-structural tests of aerospace structures are illustrated. In a related paper, *Polesky and Shideler* consider an inverse boundary condition problem. A structure has been subjected to thermal loads, and the data indicates the test fixture provides an unknown elastic restraint. The applicability of least square techniques for determining the elastic restraint from deflection measurements is investigated. The paper demonstrates how inverse data reduction techniques aid in planning thermal-structural tests. In the final paper in the chapter, *Foster and Thornton* describe an experimental study of thermally induced vibrations of a simulated spacecraft boom. Both stable and unstable oscillations are demonstrated, and three-dimensional dynamic response effects are described.

Taken together, the eight papers presented in Chapters 1 and 2 describe a significant number of contemporary thermal-structural problems. The papers show that analysts are addressing a variety of multidisciplinary, nonlinear thermal-structural problems with modern computational techniques. At the same time, experimentalists are trying to conduct well-defined tests to validate models and nonlinear analyses as well as to identify new phenomena. In spite of major advances in computational technology over the last several decades, the papers demonstrate that the interplay of analysis and experiment remains important, particularly in new multidisciplinary aerospace applications.

Chapter 3, Analysis of High Temperature Composites, is the longest chapter in the book with eight papers. The number and diversity of papers on high-temperature composites reflects the growing role of these materials in elevated temperature applications. The chapter begins with a description of recent advances in a new form of composite materials—functionally graded composites. The idea of functionally graded composites involves varying the microstructural details through nonuniform distribution of the reinforcement phase. *Pindera* et al. describe the advantages of the approach and summarize recent developments of a theory for the thermomechanical response. Another new form of advanced composites are woven textile-based materials. Textile composites, in comparison to laminated composites, have significantly greater damage tolerance and resistance to delamination. *Glaessgen and Griffin* present a three-dimensional, finite-element micromechanical analysis of the thermal response of a textile-based composite. The effects of geometric and material parameters on the distributions of displacement, strain, and stress are studied.

The next five papers treat thermally related problems encountered with traditional laminated composites in advanced applications. As laminated composites are being accepted for aircraft applications, researchers are developing sophisticated computational techniques to optimize designs. *Noor* describes advances in the development of computational models for sensitivity analysis of the thermomechanical buckling and postbuckling responses of composite panels. In the analyses, sensitivity derivatives are computed that predict the effect of various design parameters on a panel's response. Numerical studies illustrate the approach for composite panels with circular cutouts and prescribed temperature changes. In the next paper treating laminated composites, *Couick* describes an analysis simulating laser-induced damage in a composite structure. The analysis employs laminated beam models and provides closed-form analytical solutions for thermal stresses due to complicated temperature distributions.

An important consideration in the successful application of laminated composites with polymer-matrices to aircraft structures is their long-term behavior when subject to environmental cycles of temperature and moisture. The traditional approach to evaluating these long-term effects is a large number of tests. *Chamis and Singhal* describe an approach for evaluating the long term behavior by using computational simulation with probabilistic mechanics and a few experiments. The approach is used to quantify uncertainties in the long term behavior of the composites. *Wetherhold and Wang* address the problem of minimizing deformations in a laminated composite exposed to through-the-thickness temperature gradients. The problem is of fundamental interest for applications where thermally induced bending or extensional deformations degrade the performance of a structure. The paper investigates analytical techniques for designing beam laminations to eliminate thermal bending or to match a desired in-plane thermal expansion. For applications of laminated composites in a dynamic environment with varying temperatures, the understanding of thermoelastic waves is important. In the last paper in this series on laminated composites, *Hawwa and Nayfeh* study the propagation of harmonic waves in anisotropic laminated composites. A novel feature of their analysis is the use of a non-Fourier heat conduction law so that temperatures are propagated with finite speeds. This type of model has been studied in the past for isotropic materials, but it has received little attention for anisotropic materials. In the last paper in the chapter on high-temperature composites, *Warren and Wadley* discuss metallurgical aspects of the superplastic deformation of a titanium alloy that may be controlled by advanced manufacturing processes. The work is of interest for the manufacture of metal matrix composites.

In the final chapter of the book, Performance of Aircraft Materials, three papers assess the performance of aircraft materials. In the first paper, *Staley* reviews the current use of aluminum alloys in subsonic airframe structures and predicts future trends. The relationship between alloy properties and performance requirements are described. Emphasis is placed on recent shifts from performance-driven material development to an emphasis on costs. In a closely related paper, *Williams* describes the role of material performance on aircraft engines. The emphasis is placed on commercial instead of military engines, and the requirements for future products are discussed. The role of materials in meeting the challenges of competitive performance and costs is described. As a formal way of assessing the effect of technology changes on engine performance, *Generazio* describes a systems approach to perform a benefits analysis. The approach may be used to identify high-payoff research areas and show the link between advanced concepts and the accrued benefits of the research. Particular attention is given to the role of advanced composite materials and manufacturing process variations that affect the system's benefits.

The editor gratefully acknowledges the contributions of the organizers of the technical sessions of the Second Thermal Structures Conference at which these papers were presented. At the University of Virginia, the organizers include Professors John A. Wert, Edgar A. Starke Jr., Ahmed K. Noor, Carl T. Herakovich, and Dana M. Elzey. We also appreciate the support of Dr. Michael F. Card of NASA Langley Research Center as well as Dr. Donald B. Paul and Mr. Christopher L. Clay of the Wright Laboratories at Wright-Patterson Air Force Base. The editor is also appreciative of the contributions of Jeanne Godette and Paul Zarchan, Editor-in-Chief of the AIAA Progress in Astronautics and Aeronautics Series. Finally, the contributors to this volume are generously thanked for their patience, cooperation, and care in preparation of their papers.

Earl A. Thornton
February 1995

Table of Contents

Preface

Chapter 1. Analysis of Thermal Structures

Numerical Modeling of a Cryogenic Fluid within a Fuel Tank............3
D. S. Greer, *NASA Dryden Flight Research Center, Edwards, California*

Thermocryogenic Buckling and Stress Analyses of a Partially Filled Cryogenic Tank Subjected to Cylindrical Strip Heating..............22
W. L. Ko, *NASA Dryden Flight Research Center, Edwards, California*

Random Vibration of Thermally Buckled Plates.......................41
J. Lee, *Wright Laboratory, Wright-Patterson AFB, Ohio*

On Thermally-Induced Vibrations of Structures in Space 68
R. B. Malla and A. Ghoshal, *University of Connecticut, Storrs, Connecticut*

Transient Thermal-Structural Response of a Space Structure with Thermal Control Materials 96
Y. A. Kim and H. L. McManus, *Massachusetts Institute of Technology, Cambridge, Massachusetts*

Chapter 2. Experimental Studies of Thermal Structures

Boundary Conditions for Aerospace Thermal-Structural Tests 119
M. Blosser, *NASA Langley Research Center, Hampton, Virginia*

Inverse Analysis for Structural Boundary Condition Characterization of a Panel Test Fixture 145
S. P. Polesky and J. L. Shideler, *NASA Langley Research Center, Hampton, Virginia*

An Experimental Investigation of Thermally Induced Vibrations of Spacecraft Structures 163
R. S. Foster and E. A. Thornton, *University of Virginia, Charlottesville, Virginia*

Chapter 3. Analysis of High Temperature Compsosites

Recent Advances in the Mechanics of Functionally Graded Composites...181
M.-J. Pindera, *University of Virginia, Charlottesville, Virginia*, J. Aboudi, *Tel-Aviv University, Ramat-Aviv, Israel*, and S. M. Arnold, *NASA Lewis Reserach Center, Cleveland, Ohio*

Micromechanical Analysis of Thermal Response in Textile-Based Composites ... 204
E. H. Glaessgen and O. H. Griffin Jr., *Virginia Polytechnic and State University, Blacksburg, Virginia*

Recent Advances in the Sensitivity Analysis for the Thermomechanical Postbuckling of Composites Panels ... 218
A. K. Noor, *University of Virginia, NASA Langley Research Center, Hampton, Virginia*

Laser Induced Thermal Stresses in Composite Materials ... 240
J. R. Couick, *Wright Laboratory, Wright-Patterson AFB, Ohio*

Quantification of Uncertainties of Hot-Wet Composite Long Term Behavior ... 259
C. C. Chamis, *NASA Lewis Research Center, Cleveland, Ohio,* and S. N. Singhal *NYMA Inc., Brook Park, Ohio*

Minimizing Thermal Deformation by Using Layered Structures ... 273
R. C. Wetherhold and J. Wang, *State University of New York, Buffalo, New York*

Harmonic Generalized Thermoelastic Waves in Anisotropic Laminated Composites ... 293
M. A. Hawwa and A. H. Nayfeh, *University of Cincinnati, Ohio*

The Superplastic Deformation Behavior of Physical Vapor Deposited Ti-6Al-4V ... 323
J. Warren amd H.N.G. Wadley, *University of Virginia, Charlottesville, Virginia*

Chapter 4. Performance of Aircraft Materials

Aluminum Alloys for Subsonic Aircraft ... 343
J. T. Staley, *Aluminum Company of America, Alcoa Center, Pennsylvania*

Materials Requirements for Aircraft Engines ... 359
J. C. Williams, *GE Aircraft Engines, Cincinnati, Ohio*

Benefits Estimation of New Engine Technology Insertion ... 385
E. R. Generazio and C. C. Chamis, *NASA Lewis Reserach Center, Cleveland, Ohio*

Author Index for Volume ... 401

List of Series Volumes ... 403

Chapter 1. Analysis of Thermal Structures

Numerical Modeling Of A Cryogenic Fluid Within A Fuel Tank

Donald S. Greer
NASA Dryden Flight Research Center,
Edwards, California 93523-0273

Abstract

The computational method developed to study the cryogenic fluid characteristics inside a fuel tank in a hypersonic aircraft is presented. The model simulates a rapid draining of the tank by modeling the ullage vapor and the cryogenic liquid with a moving interface. A mathematical transformation was developed and applied to the Navier-Stokes equations to account for the moving interface. The formulation of the numerical method is a transient hybrid explicit–implicit technique where the pressure term in the momentum equations is approximated to first order in time by combining the continuity equation with an ideal equation of state.

Nomenclature

a	= Van der Waals constant, Pa m^6
b	= Van der Waals constant, m3
c	= speed of sound, m/s
FI	= flux, radial direction
FJ	= flux, circumferential direction
g	= gravity, m/s^2
h	= enthalpy, J/kg

*Aerodynamics Branch Research Engineer, P. O. Box 273.

k = conductivity, J/m s K

MSF = Mach scaling factor, $c_{\text{real}}/c_{\text{numerical}}$

P = pressure, Pa

q = heat flux, J/m^2 sec

R = radius relative to interface, $f(\theta, t)$, m

$\boldsymbol{R}$ = fluid constant, J/kg K

$\boldsymbol{R}^*$ = ideal fluid constant, J/kg K

r = radial coordinate, m

$\mathscr{r}$ = transformed radius coordinate, r/R

T = temperature, K

t = time, s

V = velocity, m/s

α = thermal diffusivity, m^2/s

$\Delta\mathscr{r}$ = distance between radial grid points, m

$\Delta\vartheta$ = distance between circumferential grid points, rad

θ = circumferential coordinate, rad

ϑ = transformed circumferential coordinate, rad

μ = viscosity, kg/m s

v = specific volume, m^3/kg

ρ = density, kg/m^3

Superscripts

n = time step index

Subscripts

i,j = spatial coordinate indices

$i+, i$ = spatial coordinate at $i+\frac{1}{2}, i-\frac{1}{2}$

$j+, j$ = spatial coordinate at $j+\frac{1}{2}, j-\frac{1}{2}$

$r, \mathscr{r}$ = radial direction

s = constant entropy

θ, ϑ = circumferential direction

Introduction

Liquid hydrogen has long been considered one of the most advantageous fuels for hypersonic aircraft. The National Advisory Committee on Aeronautics (NACA), Lewis Flight Propulsion Laboratory, Cleveland, Ohio, began to study hydrogen as a fuel source in the 1950s. The NACA study included a flight test program with a U.S. Air Force B-57 which had one engine modified for operation with hydrogen fuel.[1] The test program was successful, and the engine typically operated at an altitude of 50,000 ft and Mach 0.75 for 17 min with hydrogen fuel. Since this initial test program, there have been several studies to investigate the aspects of liquid hydrogen-fueled aircraft. An excellent review is presented by Brewer.[2]

Developing reusable flight-weight cryogenic fuel tanks is one of the most difficult challenges in designing advanced hypersonic aircraft and the next generation of spacecraft. The most recent cryogenic fuel tank studies were conducted for the National Aero-Space Plane. McDonnell Douglas Corporation, St. Louis, Missouri, constructed and tested with liquid hydrogen a 3600-liter, graphite-reinforced, epoxy thermoset composite tank. The General Dynamics Space Systems Division, San Diego, California, also constructed and tested a 2600-liter graphite and epoxy liquid hydrogen tank. Presently, NASA Dryden Flight Research Center, Edwards, California, is involved in a research program to continue the study and to increase understanding of liquid hydrogen cryogenic fuel tanks. The initial phase of the program includes the design of a test facility and the development of a generic research cryogenic tank (Fig. 1) for initial testing and verification of analytical models. Testing of the tank will simulate the severe

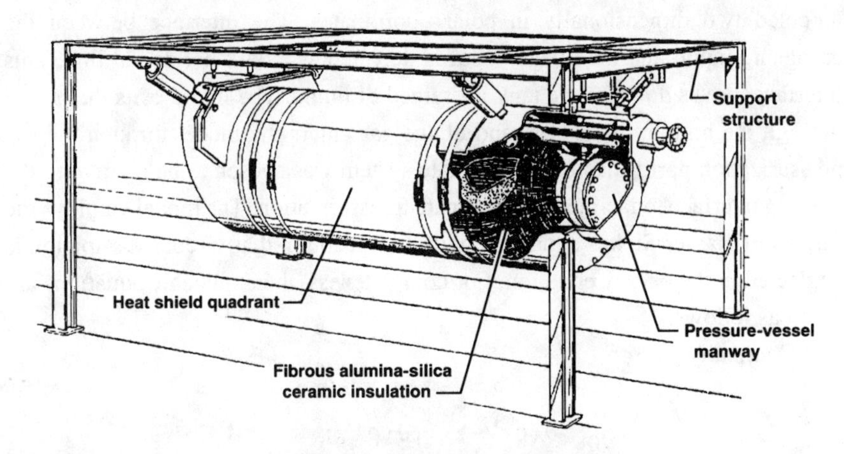

Fig. 1 Generic research cryogenic tank.

aeroheating effects of hypersonic flight by subjecting the tank exterior to large heat fluxes. Details of the generic research cryogenic tank and test setup have been reported.[3]

These hypersonic cryogenic fuel tanks must be able to survive the thermal gradients of exterior gaseous aeroheating temperatures as high as 7000 K and internal cryogenic temperatures as low as 20 K. See Ref. 4 for a detailed discussion on hypersonic aerodynamic heating. The understanding of the cryogenic fluid dynamic motion inside the tank is essential to predict and analyze the thermal stress gradients though the tank wall. The fluid dynamic model will predict cryogenic fluid temperatures, heat-transfer coefficients on the inner tank wall, and stratification effects in the ullage gas. An analysis using SINDA'85/FLUINT showed that these are the relevant interior parameters in determining the fuel tank thermal stresses.[3]

The generic research cryogenic tank which this model was developed to simulate has met with budgetary constrains, and test data for verification of the model are unavailable. Comparison of the model with test data will be the subject of future efforts. This paper presents the computational fluid dynamic modeling development effort to simulate the generic research cryogenic tank when subjected to high external heat flux and rapid draining. The numerical methods were developed with emphasis on increasing the time step to reduce computer resources.

Model Description

Figure 2 shows the cross-sectional area of the generic research cryogenic tank modeled two dimensionally in polar coordinates. The interface between the cryogenic liquid and the ullage vapor is depicted as a dark horizontal line. This interface moves down as the tank is drained of liquid. The liquid exits the model through the bottom drain port, and ullage gas enters the model through the top pressurization port. The polar coordinate system was chosen to be relative to the moving interface to facilitate a coordinate transformation. The model employs the full set of Navier-Stokes equations with the exception that viscous dissipation is neglected in the energy equation. For completeness, the applicable equations are given as follows.

Continuity

$$\frac{\partial \rho}{\partial t} + \frac{\partial (\rho r V_r)}{\partial r} + \frac{1}{r}\frac{\partial (\rho V_\theta)}{\partial \theta} = 0 \tag{1}$$

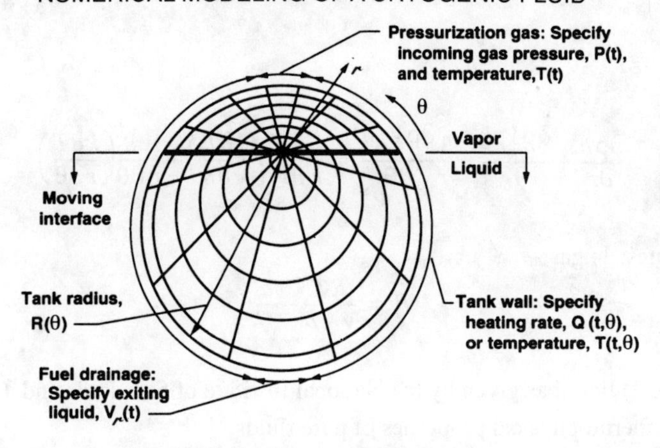

Fig. 2 Modeled cross-sectional area.

Radial momentum

$$\rho\left(\frac{\partial V_r}{\partial t}+V_r\frac{\partial V_r}{\partial r}+\frac{V_\theta}{r}\frac{\partial V_r}{\partial \theta}-\frac{V_\theta^2}{r}\right) \tag{2}$$

$$=\frac{1}{r}\frac{\partial}{\partial r}\left(r\mu\left\{2\frac{\partial V_r}{\partial r}-\frac{2}{3}\left[\frac{1}{r}\frac{\partial (rV_r)}{\partial r}+\frac{1}{r}\frac{\partial V_\theta}{\partial \theta}\right]\right\}\right)$$

$$+\frac{1}{r}\frac{\partial}{\partial \theta}\left\{\mu\left[r\frac{\partial}{\partial r}\left(\frac{V_\theta}{r}\right)\right]+\frac{1}{r}\frac{\partial V_r}{\partial \theta}\right\}$$

$$-\frac{\mu}{r}\left(2\left[\frac{1}{r}\frac{\partial V_\theta}{\partial \theta}+\frac{V_r}{r}\right]-\frac{2}{3}\left[\frac{1}{r}\frac{\partial (rV_r)}{\partial r}+\frac{1}{r}\frac{\partial V_\theta}{\partial \theta}\right]\right)$$

$$-\frac{\partial P}{\partial r}+\rho g_r$$

Circumferential momentum

$$\rho\left(\frac{\partial V_\theta}{\partial t}+V_r\frac{\partial V_\theta}{\partial r}+\frac{V_\theta}{r}\frac{\partial V_\theta}{\partial \theta}+\frac{V_r V_\theta}{r}\right) \tag{3}$$

$$=\frac{1}{r^2}\frac{\partial}{\partial r}\{r^2\mu\left[r\frac{\partial}{\partial r}\left(\frac{V_\theta}{r}\right)+\frac{1}{r}\frac{\partial V_r}{\partial \theta}\right]\}$$

$$+\frac{1}{r}\frac{\partial}{\partial \theta}\left\{\mu\left[\frac{2}{r}\frac{\partial V_\theta}{\partial \theta}+\frac{2V_r}{r}-\frac{2}{3r}\frac{\partial (rV_r)}{\partial r}-\frac{2}{3r}\frac{\partial V_\theta}{\partial \theta}\right]\right\}$$

$$-\frac{1}{r}\frac{\partial P}{\partial \theta}+\rho g_\theta$$

Energy

$$\frac{\partial \rho h}{\partial t} + \frac{\partial \rho V_r h}{\partial r} + \frac{1}{r}\frac{\partial \rho V_\theta h}{\partial \theta} = \frac{1}{r}\frac{\partial}{\partial r}\left(r k \frac{\partial T}{\partial r}\right) + \frac{1}{r}\frac{\partial}{\partial \theta}\left(\frac{k \partial T}{r \partial \theta}\right) \tag{4}$$

States: 1) gas

$$P = \frac{\boldsymbol{R}T}{v - b} - \frac{a}{v^2} \tag{5}$$

and 2) liquid as given by the National Institute of Standards and Technology (NIST) thermophysical properties of pure fluids.[5]

Boundary conditions for the model are divided into three regions: tank wall, ullage pressurant gas port, and liquid drain port. The wall boundary conditions are defined as the no-slip condition, and the heat flux is specified as a function of circumference and time. These conditions are expressed as follows:

$$V_r = V_\theta = 0 \tag{6}$$

$$q = f(\theta, t) \tag{7}$$

The ullage pressurant gas port boundary is specified at constant pressure and temperature with no heat flux

$$P = P_{\text{gas}} \tag{8}$$

$$T = T_{\text{gas}} \tag{9}$$

$$q = 0 \tag{10}$$

The liquid drain port boundary is specified as a free boundary with no circumferential velocity, at constant pressure, and with no heat flux

$$V_\theta = 0 \tag{11}$$

$$P = P_{\text{liquid}} \tag{12}$$

$$q = 0 \tag{10}$$

The interface movement is calculated from the liquid drain rate and the geometry. Interface conditions are equal temperature and equal velocity,

$$T_{gas} = T_{liquid} \tag{13}$$

$$V_{\theta_{gas}} = V_{\theta_{liquid}} \tag{14}$$

Because of the numerical difficulties associated with moving boundaries, a transformation is defined which allows the transformed computational space to be invariant to the moving vapor and liquid interface. This transformation is graphically shown in Fig. 3 and is defined as follows:

$$\vartheta = \theta \tag{15}$$

$$\mathscr{r} = r/R \tag{16}$$

$$\frac{\partial}{\partial r} = \frac{1}{R}\frac{\partial}{\partial \mathscr{r}} \tag{17}$$

$$\frac{\partial^2}{\partial r^2} = \frac{1}{R^2}\frac{\partial^2}{\partial \mathscr{r}^2} \tag{18}$$

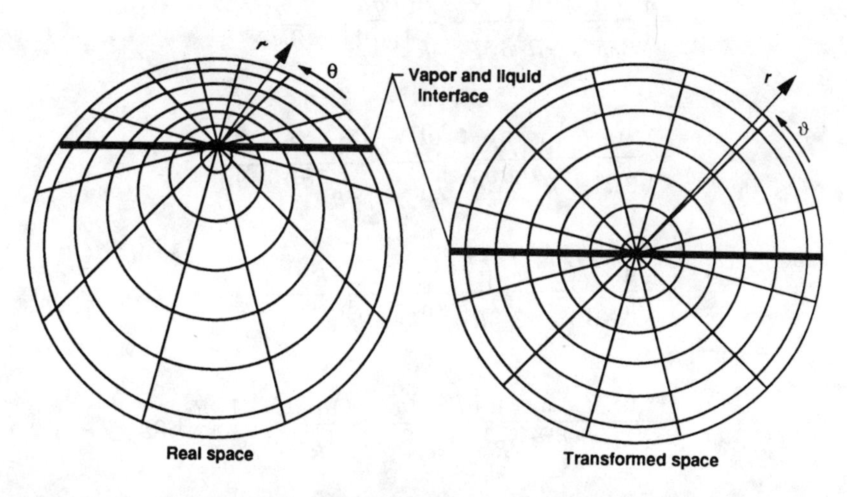

Fig. 3 Coordinate transformation relative to the vapor and liquid interface.

$$\frac{\partial}{\partial\theta} = \frac{\partial}{\partial\vartheta} - \frac{\mathscr{r}}{R}\frac{\partial R}{\partial\vartheta}\frac{\partial}{\partial\mathscr{r}} \tag{19}$$

$$\frac{\partial^2}{\partial\theta^2} = \frac{\partial^2}{\partial\vartheta^2} + \left[-\frac{\mathscr{r}}{R}\frac{\partial^2 R}{\partial\vartheta^2} + \frac{\mathscr{r}^2}{R^2}\frac{\partial R}{\partial\vartheta}\frac{\partial^2 R}{\partial\mathscr{r}\,\partial\vartheta} + \frac{2\mathscr{r}}{R^2}\left(\frac{\partial R}{\partial\vartheta}\right)^2\right]\frac{\partial}{\partial\mathscr{r}} \tag{20}$$
$$-\frac{2\mathscr{r}}{R}\frac{\partial R}{\partial\vartheta}\frac{\partial^2}{\partial\mathscr{r}\,\partial\vartheta} + \frac{\mathscr{r}^2}{R^2}\left(\frac{\partial R}{\partial\vartheta}\right)^2\frac{\partial^2}{\partial\mathscr{r}^2}$$

This transformation is now applied to the model equations to obtain the applicable Navier-Stokes equations for the computational space. These equations are presented as follows.

Continuity

$$\frac{\partial\rho}{\partial t} + \frac{1}{\mathscr{r}R}\frac{\partial(\rho\mathscr{r}V_{\mathscr{r}})}{\partial\mathscr{r}} + \frac{1}{\mathscr{r}R}\frac{\partial(\rho V_\vartheta)}{\partial\vartheta} - \frac{1}{R^2}\frac{\partial R}{\partial\vartheta}\frac{\partial(\rho V_\vartheta)}{\partial\mathscr{r}} = 0 \tag{21}$$

Radial momentum

$$\rho\left(\frac{\partial V_{\mathscr{r}}}{\partial t} + \left[\frac{V_{\mathscr{r}}}{R} - \frac{V_\vartheta}{R^2}\frac{\partial R}{\partial\vartheta}\right]\frac{\partial V_{\mathscr{r}}}{\partial\mathscr{r}} + \frac{V_\vartheta}{\mathscr{r}R}\frac{\partial V_{\mathscr{r}}}{\partial\vartheta} - \frac{V_\vartheta^2}{\mathscr{r}}\right) \tag{22}$$

$$= \left(\frac{4\mu}{3\mathscr{r}R^2} - \frac{\mu}{\mathscr{r}R^3}\frac{\partial^2 R}{\partial\vartheta^2} + \frac{2\mu}{\mathscr{r}R^4}\left[\frac{\partial R}{\partial\vartheta}\right]^2\right)\frac{\partial V_{\mathscr{r}}}{\partial\mathscr{r}}$$

$$+\left(\frac{4\mu}{3R^2} + \frac{\mu}{R^4}\left[\frac{\partial R}{\partial\vartheta}\right]^2\right)\frac{\partial^2 V_r}{\partial\mathscr{r}^2} - \frac{2\mu}{\mathscr{r}R^3}\frac{\partial R}{\partial\vartheta}\frac{\partial^2 V_{\mathscr{r}}}{\partial\mathscr{r}\,\partial\vartheta}$$

$$+\frac{\mu}{\mathscr{r}^2R^2}\frac{\partial^2 V_{\mathscr{r}}}{\partial\vartheta^2} - \frac{7\mu}{3\mathscr{r}^2R^2}\frac{\partial V_\vartheta}{\partial\vartheta} - \frac{2\mu}{\mathscr{r}R^3}\frac{\partial R}{\partial\vartheta}\frac{\partial V_\vartheta}{\partial\mathscr{r}}$$

$$+\frac{\mu}{3\mathscr{r}R^2}\frac{\partial^2 V_\vartheta}{\partial\mathscr{r}\,\partial\vartheta} - \frac{\mu}{R^3}\frac{\partial R}{\partial\vartheta}\frac{\partial^2 V_\vartheta}{\partial\mathscr{r}^2} - \frac{4\mu}{3\mathscr{r}^2R^2}V_{\mathscr{r}} - \frac{1}{R}\frac{\partial P}{\partial\mathscr{r}} + \rho g_{\mathscr{r}}$$

Circumferential momentum

$$\rho\left(\frac{\partial V_\vartheta}{\partial t}+\left[\frac{V_r}{R}-\frac{V_\vartheta \partial R}{R^2 \partial \vartheta}\right]\frac{\partial V_\vartheta}{\partial r}+\frac{V_\vartheta}{rR}\frac{\partial V_\vartheta}{\partial \vartheta}+\frac{V_r V_\vartheta}{rR}\right) \tag{23}$$

$$=\frac{7\mu}{3r^2R^2}\frac{\partial V_r}{\partial \vartheta}-\frac{8\mu}{3rR^3}\frac{\partial R}{\partial \vartheta}\frac{\partial V_r}{\partial r}+\frac{\mu}{3rR^2}\frac{\partial^2 V_r}{\partial r\,\partial \vartheta}$$

$$-\frac{\mu}{3rR^3}\frac{\partial R}{\partial \vartheta}\frac{\partial^2 V_r}{\partial r^2}$$

$$+\left(\frac{\mu}{rR^2}+\frac{8\mu}{3rR^4}\left[\frac{\partial R}{\partial \vartheta}\right]^2-\frac{4\mu}{3rR^3}\frac{\partial^2 R}{\partial \vartheta^2}\right)\frac{\partial V_\vartheta}{\partial r}$$

$$+\frac{4\mu}{3r^2R^2}\frac{\partial^2 V_\vartheta}{\partial \vartheta^2}-\frac{8\mu}{3rR^3}\frac{\partial R}{\partial \vartheta}\frac{\partial^2 V_\vartheta}{\partial r\,\partial \vartheta}$$

$$+\left(\frac{\mu}{R^2}+\frac{4\mu}{3R^3}\left[\frac{\partial R}{\partial \vartheta}\right]^2\right)\frac{\partial^2 V_\vartheta}{\partial r^2}-\frac{\mu V_\vartheta}{r^2R^2}$$

$$-\frac{1}{rR}\frac{\partial P}{\partial \vartheta}+\frac{1}{R^2}\frac{\partial R}{\partial \vartheta}\frac{\partial P}{\partial r}+\rho g_\vartheta$$

Energy

$$\frac{\partial \rho h}{\partial t}+\frac{1}{R}\frac{\partial \rho V_r h}{\partial r}+\frac{1}{rR}\frac{\partial \rho V_\vartheta h}{\partial \vartheta}-\frac{1}{R^2}\frac{\partial R}{\partial \vartheta}\frac{\partial \rho V_\vartheta h}{\partial r} \tag{24}$$

$$=\left(\frac{2k}{rR^4}\left[\frac{\partial R}{\partial \vartheta}\right]^2-\frac{k}{rR^3}\frac{\partial^2 R}{\partial \vartheta^2}\right)\frac{\partial T}{\partial r}+\frac{k}{rR^2}\frac{\partial}{\partial r}\left(r\frac{\partial T}{\partial r}\right)$$

$$+\frac{k}{R^4}\left[\frac{\partial R}{\partial \vartheta}\right]^2\frac{\partial^2 T}{\partial r^2}-\frac{2k}{rR^3}\frac{\partial R}{\partial \vartheta}\frac{\partial^2 T}{\partial r\,\partial \vartheta}+\frac{k}{r^2R^2}\frac{\partial^2 T}{\partial \vartheta^2}$$

Gas and liquid states remain unchanged. Boundary conditions at tank wall

$$V_r = V_\vartheta = 0 \tag{25}$$

$$q = f(\vartheta, t) \tag{26}$$

The boundary conditions at the ullage pressurant and liquid drain ports remain unchanged. Interface conditions also remain unchanged.

Note that the transformed equations reduce to the standard Navier-Stokes equations for cylindrical coordinates when R = const and $\partial R/\partial\vartheta = \partial^2 R/\partial\vartheta^2 = 0$.

Numerical Method

Finite difference equations were specifically developed for this model using a volume integral method technique which is discussed in lecture notes by Lick.[6] The development of the continuity and energy equations is purely explicit, time-marching algorithms. The development of the momentum equations is longitudinally implicit, transversely explicit, time-marching algorithms. Longitudinally implicit means, for example, that in solving for V_r in the radial momentum equation all terms that are discretized in the radial direction are implicit. Conversely, terms that are discretized in the theta direction are explicit (reverse for the theta momentum equation).

The longitudinally implicit approach to the momentum equations was chosen to increase the stability of the solution to the nonlinear Navier-Stokes equations. This approach improved Reynolds cell instability associated with purely explicit time-marching techniques. The finite difference equations for the transformed Navier-Stokes equations are second-order accurate in space and first-order accurate in time and are presented as follows.

Continuity

$$\begin{aligned}\rho_{i,j}^{n+1} &= \rho_{i,j}^{n} \\ &+ \left(\frac{r_{i-}}{r_i}FI_{i-,j}^1 - \frac{r_{i+}}{r_i}FI_{i+,j}^1 + FI_{i-,j}^2 - FI_{i+,j}^2\right)\frac{\Delta t}{\Delta r} \\ &+ \frac{1}{R_{i,j}^n}(FJ_{i,j-} - FJ_{i,j+})\frac{\Delta t}{\Delta r}\end{aligned} \tag{27}$$

where

$$FI_{i+,j}^1 = \left(\frac{\rho V_r}{R}\right)_{i+,j}^n \tag{28}$$

$$FI_{i+,j}^2 = \left(\frac{\rho V_\vartheta}{R^2}\frac{\partial R}{\partial\vartheta}\right)_{i+,j}^n \tag{29}$$

$$FJ_{i,j+} = \left(\frac{\rho V_{\vartheta}}{r}\right)^{n}_{i,j+} \qquad (30)$$

Radial momentum

$$V^{n+1}_{r_{i,j}} = V^{n}_{r_{i,j}} + \left\{ -\frac{V^{n}_{r_{i,j}}}{R^{n}_{j}} + \frac{V^{n}_{\vartheta_{i,j}}}{R^{n2}_{j}}\frac{\partial R^{n}_{j}}{\partial\vartheta} + \frac{\mu_{i,j}}{\rho^{n}_{i,j}R^{n3}_{j}} \right. \qquad (3\ )$$

$$\left. \left[\frac{2}{r_{i,j}R^{n4}_{j}}\left(\frac{\partial R^{n}_{j}}{\partial\vartheta}\right)^{2} - \frac{1}{r_{i}}\frac{\partial^{2}R^{n}_{j}}{\partial\vartheta^{2}}\right]\right\}$$

$$\left(V^{n+1}_{r_{i+1,j}} - V^{n+1}_{r_{i+1,j}}\right)\frac{\Delta t}{2\Delta r} + \left[\frac{4\mu_{i,j}}{3\rho^{n}_{i,j}R^{n2}_{j}} + \frac{\mu_{i,j}}{\rho^{n}_{i,j}R^{n4}_{j}}\left(\frac{\partial R^{n}_{j}}{\partial\vartheta}\right)^{2}\right]$$

$$\left(\frac{r_{i+,j}}{r_{i,j}}V^{n+1}_{r_{i+1,j}} - 2V^{n+1}_{r_{i,j}} + \frac{r_{i-,j}}{r_{i,j}}V^{n+1}_{r_{i-1,j}}\right)\frac{\Delta t}{\Delta r^{2}}$$

$$-\left(\frac{V^{n}_{r_{i,j}}}{r_{i,j}R^{n}_{j}} + \frac{7\mu_{i,j}}{3\rho^{n}_{i,j}r^{2}_{i,j}R^{n2}_{j}}\right)\left(V^{n}_{\vartheta_{i,j+}} - V^{n}_{\vartheta_{i,j-}}\right)\frac{\Delta t}{\Delta\vartheta},$$

$$+\frac{\mu_{i,j}}{\rho^{n}_{i,j}r^{2}_{i,j}R^{n2}_{j}}\left(V^{n+1}_{r_{i,j+1}} - 2V^{n+1}_{r_{i,j}} + V^{n+1}_{r_{i,j-1}}\right)\frac{\Delta t}{\Delta\vartheta^{2}}$$

$$+\frac{4\mu_{i,j}}{3\rho^{n}_{i,j}r_{i,j}R^{n4}_{j}}\frac{\partial R^{n}_{j}}{\partial\vartheta}\left(V^{n}_{\vartheta_{i+,j}} - V^{n}_{\vartheta_{i-,j}}\right)\frac{\Delta t}{\Delta r}$$

$$+\frac{\mu_{i,j}}{3\rho^{n}_{i,j}R^{n3}_{j}}\frac{\partial R^{n}_{j}}{\partial\vartheta}\left(V^{n}_{\vartheta_{i+1,j}} - 2V^{n}_{\vartheta_{i,j}} + V^{n}_{\vartheta_{i-1,j}}\right)\frac{\Delta t}{\Delta r^{2}}$$

$$-\frac{4\mu_{i,j}}{3\rho^{n}_{i,j}r^{2}_{i,j}R^{n4}_{j}}V^{n+1}_{r_{i,j}}\Delta t + \frac{V^{n2}_{\vartheta_{i,j}}}{r_{i}R^{n}_{j}}\Delta t + g_{r_{i,j}}\Delta t$$

$$-\frac{(P^{n+1}_{i+,j} - P^{n+1}_{i-,j})}{R^{n}_{j}\rho^{n}_{i,j}}\frac{\Delta t}{\Delta r}$$

where the pressure term P^{n+1} in the radial momentum equation is approximated to first order in time by combining the continuity equation with an ideal equation of state. The result is as follows:

$$P^{n+1}_{i,j} = P^{n}_{i,j} + \Delta P^{n \rightarrow n+1}_{i,j} \tag{32}$$

$$\begin{aligned} P^{n+1}_{i+,j} &= F^{n}_{i+,j} \\ &+ \frac{\rho^{n}_{i+,j} \boldsymbol{R} T_{i+,j}}{R^{n}_{j}} \left[\left(V^{n}_{r_{i,j}} - V^{n}_{r_{i+1,j}} \right) \frac{\Delta t}{\Delta r} \right. \\ &\left. + \left(V^{n}_{\vartheta_{i,j-1}} - V^{n}_{\vartheta_{i,j}} \right) \frac{\Delta t}{r \Delta \vartheta} \right] \end{aligned} \tag{33}$$

Circumferential momentum

$$\begin{aligned} V^{n+1}_{\vartheta_{i,j}} &= V^{n}_{\vartheta_{i,j}} - \frac{V^{n}_{\vartheta_{i,j}}}{2 r_{i,j} R^{n}_{j}} \left(V^{n+1}_{\vartheta_{i,j+1}} - V^{n+1}_{\vartheta_{i,j-1}} \right) \frac{\Delta t}{\Delta \vartheta} \\ &+ \frac{4 \mu_{i,j}}{3 \rho^{n}_{i,j} r^{2}_{i,j} R^{n2}_{j}} \left(V^{n+1}_{\vartheta_{i,j+1}} - 2 V^{n+1}_{\vartheta_{i,j}} + V^{n+1}_{\vartheta_{i,j-1}} \right) \frac{\Delta t}{\Delta \vartheta^{2}} \\ &+ \frac{7 \mu_{i,j}}{3 \rho^{n}_{i,j} r^{2}_{i,j} R^{n2}_{j}} \left(V^{n}_{r_{i,j+}} - V^{n}_{r_{i,j-}} \right) \frac{\Delta t}{\Delta \vartheta} \\ &+ \frac{8 \mu_{i,j}}{3 \rho^{n}_{i,j} r_{i,j} R^{n3}_{j}} \frac{\partial R^{n}_{j}}{\partial \vartheta} \left(V^{n}_{r_{i+,j}} - V^{n}_{r_{i-,j}} \right) \frac{\Delta t}{\Delta r} \\ &- \frac{\mu_{i,j}}{3 \rho^{n}_{i,j} r_{i,j} R^{n3}_{j}} \frac{\partial R^{n}_{j}}{\partial \vartheta} \left(V^{n}_{r_{i+1,j}} - 2 V^{n}_{r_{i,j}} + V^{n}_{r_{i-1,j}} \right) \frac{\Delta t}{\Delta r^{2}} \\ &+ \left\{ \frac{\mu_{i,j}}{\rho^{n}_{i,j} R^{n2}_{j}} \left[\frac{1}{r_{i,j}} + \frac{8}{3 r_{i,j} R^{n2}_{j}} \left(\frac{\partial R^{n}_{j}}{\partial \vartheta} \right)^{2} - \frac{4}{3 r_{i} R^{n}_{j}} \frac{\partial^{2} R^{n}_{j}}{\partial \vartheta^{2}} \right] \right. \\ &\left. - \frac{V^{n}_{r_{i,j}}}{R^{n}_{j}} + \frac{V^{n}_{\vartheta_{i,j}}}{R^{n2}_{j}} \frac{\partial R^{n}_{j}}{\partial \vartheta} \right\} \left(V^{n}_{\vartheta_{i+1,j}} - V^{n}_{\vartheta_{i-1,j}} \right) \frac{\Delta t}{2 \Delta r} \end{aligned} \tag{34}$$

$$+\frac{\mu_{i,j}}{\rho^n_{i,j}R^{n2}_j}\left[1+\frac{4}{3R^{n2}_j}\left(\frac{\partial R^n_j}{\partial\vartheta}\right)^2\right]$$

$$\left(V^n_{\vartheta_{i+1,j}}-2V^n_{\vartheta_{i,j}}+V^n_{\vartheta_{i-1,j}}\right)\frac{\Delta t}{\Delta r^2}$$

$$-\left(\frac{\mu_{i,j}}{\rho^n_{i,j}r^2_1R^{n2}_j}+\frac{V^n_{r_{i,j}}}{r_iR^n_j}\right)V^{n+1}_{\vartheta_{i,j}}\Delta t$$

$$+\frac{1}{\rho^n_{i,j}R^{n2}_j}\frac{\partial R^n_j}{\partial\vartheta}\left(P^n_{i+,j}-P^n_{i+,j}\right)\frac{\Delta t}{\Delta r}+g_{\vartheta_{i,j}}\Delta t$$

$$-\frac{\left(P^n_{i,j+}-P^n_{i,j-}\right)}{\rho^n_{i,j}r_iR^n_j}\frac{\Delta t}{\Delta\vartheta}$$

$$-\frac{\boldsymbol{R}T_{i+,j}}{\rho^n_{i,j}r^2_1R^{n2}_j}\left[\rho^n_{i,j+}\left(V^{n+1}_{\vartheta_{i,j}}-V^{n+1}_{\vartheta_{i,j+1}}\right)\right.$$

$$\left.-\rho^n_{i,j-}\left(V^{n+1}_{\vartheta_{i,j-1}}-V^{n+1}_{\vartheta_{i,j}}\right)\right]\frac{\Delta t^2}{\Delta\vartheta^2}$$

$$-\frac{\boldsymbol{R}T_{i+,j}}{\rho^n_{i,j}r^2_1R^{n2}_j}\left[\rho^n_{i,j+}\left(\frac{r_{i+}}{r_i}V^n_{r_{i-j}}-\frac{r_{i+}}{r_i}V^n_{r_{i+j+}}\right)\right.$$

$$\left.-\rho^n_{i,j-}\left(\frac{r_{i-}}{r_i}V^n_{r_{i-,j-}}-\frac{r_{i+}}{r_i}V^n_{r_{i+j-}}\right)\right]\frac{\Delta t^2}{\Delta r\Delta\vartheta}$$

Energy

$$h^{n+1}_{i,j}=h^n_{i,j}+\frac{1}{r_i}\left(FI_{i+,j}-FI_{i-,j}\right)+\left(FJ_{i,j+}-FJ_{i,j-}\right)\tag{35}$$

where

$$FI_{i+j}=\tag{36}$$

$$\left(\frac{-r_{i+}V^n_{r_{i+,j}}}{R^n_j}+\frac{r_{i+}V^n_{r_{i+,j}}}{R^{n2}_j}\frac{\partial R^n_j}{\partial\vartheta}\right)\rho^n_{i+,j}h^n_{i+,j}$$

$$+ k^n_{i+,j}\left[\frac{1}{R^{n3}_j}\frac{\partial^2 R^n_j}{\partial \vartheta^2} - \frac{2}{R^{n4}_j}\left(\frac{\partial R^n_j}{\partial \vartheta}\right)^2\right]T^n_{i+,j}$$

$$+ k^n_{i+,j}\left[\frac{r_{i+}}{R^{n2}_j} + \frac{r_{i+}}{R^{n4}_j}\left(\frac{\partial R^n_j}{\partial \vartheta}\right)^2\right]\frac{(T^n_{i+1,j} T^n_{i,j})}{\Delta r}$$

$$FJ_{i,j+} = k^n_{i,j+}\frac{1}{r^2_i R^{n2}_{j+}}\frac{(T^n_{i,j+1} - T^n_{i,j})}{\Delta\vartheta} - \frac{V^n_{\vartheta_{i+,j}}}{r_i R^n_{j+}}\rho^n_{i,j+}h^n_{i,j+} \tag{37}$$

The boundary and interface finite difference equations are derived directly from the volume integral method technique and are very similar to these equations. Because of the similarity, they are not shown. These equations form a complete set of time-marching, finite difference equations. The first step in the numerical procedure is to solve the momentum equations for the velocities at the next time step, V_r^{n+1} and V_ϑ^{n+1}. The formulation of the momentum difference equations is such that a tridiagonal matrix is solved for each ray of radial velocities. In addition, a tridiagonal matrix is solved for each circlet of circumferential velocities. The average velocity over the time step, $V^{avg} = (V^{n+1} + V^n)/2$, is then used in the continuity and energy equations to obtain density and temperature for the next time step, $\rho^{n+1}_{i,j}$ and $T^{n+1}_{i,j}$. The pressure $P^{n+1}_{i,j}$ is then calculated from the state equation as a function of $\rho^{n+1}_{i,j}$ and $T^{n+1}_{i,j}$.

A staggered grid was used to develop the circumferential and radial velocities (that is, velocity nodes centered between pressure nodes). The staggered grid was selected to avoid special consideration of the no-slip radial boundary condition. This special consideration arises because longitudinal pressure disturbances or acoustical waves traveling in the radial direction are not correctly modeled in a standard grid by assuming the wall velocity to be zero. The fluid molecules in the vicinity near the wall have velocity when disturbed by radial longitudinal acoustical waves. Thus, specifying a zero velocity for a wall boundary node does not represent the total average velocity for the boundary node when using a standard grid. This difficulty is avoided by using a staggered grid and a no-flux boundary condition at the wall because the radial velocities are not calculated at the boundary in the staggered grid but are calculated at a location of $\Delta r/2$ from the boundary. In addition to the staggered grid, variable node spacing was incorporated to model the tank wall boundary layer with more densely populated nodes.

The stiffness often associated with the Navier-Stokes equations can present numerical difficulties. Stiffness is a term which characterizes a system which has two vastly different time scales. In this problem, the stiffness results from the disparity between the numerical time step associated with the speed of sound in the fluid, of the order of 0.00001 s, and the drain time of the tank, of the order of 1000 s. The numerical solution would require an order of 100×10^6 time steps. One method of dealing with the difficulties associated with stiffness is to adjust the characteristic time scales of the variables.[7, 8] The stiffness in this problem is reduced by adjusting the Mach number of the fluid by artificially decreasing the speed of sound in the computation. The Mach number of the flow within the cryogenic tank is expected to be of the order of 0.0001 which is nearly incompressible. Because compressible flows with Mach numbers of 0.1 or less can be approximated as incompressible flows with less than 1% error,[9] it is reasonable to expect that the Mach number of this problem can be increased from 0.0001 to 0.1 with less than 1% error. The Mach number is increased in the computation by scaling the pressure variable as it relates to the speed of sound through the relation,

$$c^2 = \left(\frac{\partial P}{\partial \rho}\right)_s \tag{38}$$

The numerical time step associated with the scaled pressure variable can be increased to the order of 0.01 s, thus reducing the required time steps for solution to 100,000.

Results and Discussion

These numerical calculations were conducted on small grids (0.01-m radius with 280 nodes and 0.05-m radius with 345 nodes) to save computer resources during model development. Modeling of the generic research cryogenic tank will require a larger grid (0.75-m radius with approximately 5,000 – 10,000 nodes). This grid will substantially increase computer resources. Full-scale calculations were considered an unwarranted use of computer resources because testing of the generic research cryogenic tank has not begun. Data for model comparison are currently not available.

Figures 4 and 5 show the numerical effect of scaling the Mach number. These figures present time histories of velocity at the center of the gas region (Fig. 4) and at the center of the liquid region (Fig. 5) for various Mach scaling factors [MSF ($c_{real}/c_{numerical}$)]. The calculations were performed assuming a tank radius of 0.01 m with a 10- by 28-node grid. Initially, the gas and liquid regions are at rest.

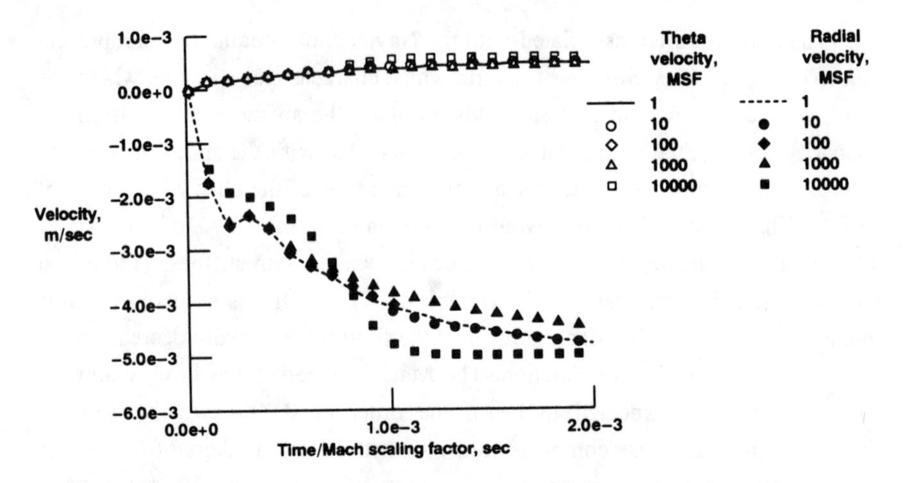

Fig. 4 Numerical effect of scaling the Mach number in the gas region.

The model is begun by applying a constant impulsive drain rate velocity at the gas and liquid interface. These figures represent the time response to an impulsively applied drain rate. Excellent agreement for Mach scaling factors up to 100 and fair agreement for Mach scaling factors up to 1000 are also shown. The time step required for stability in these calculations were 1.e-9 s (MSF = 1), 1.e-7 s (MSF = 100), and 1.e-6 s (MSF = 1000). Thus by using a Mach scaling factor of 1000, computational time is decreased by three orders of magnitude in this test case.

Figures 4 and 5 also show the physical modeling effect of scaling the Mach numbers. Note that the time axis is a ratio of the Mach scaling factor. Thus, when an impulsive disturbance is applied to the model, scaling the Mach number

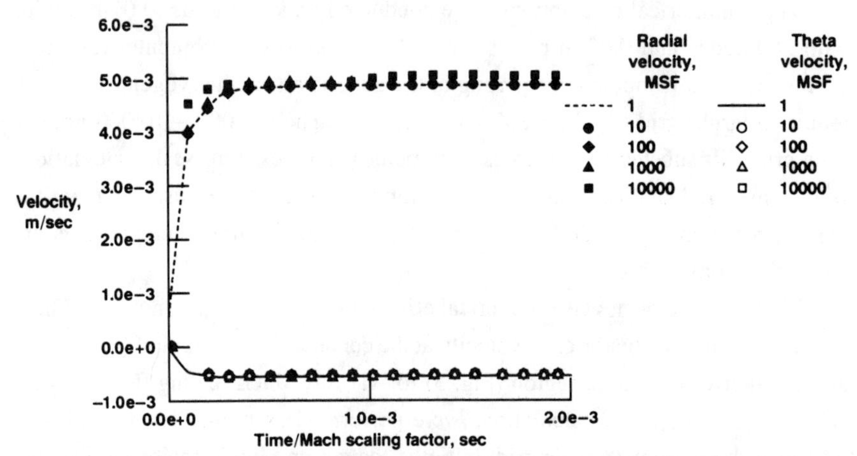

Fig. 5 Numerical effect of scaling the Mach number in the liquid region.

increases the time required for the modeled fluid to respond to the disturbance. The final effect of the disturbance, or the velocities, remain unchanged. In this model, the interest is in fluid characteristics which occur over large time scales (minutes). As a result, characteristic responses which occur in time scales of microseconds can be increased to seconds as long as they do not affect the final result.

Figures 6 and 7 show the numerical results of applying a first-order correction to the pressure term as shown in eqs. 32 and 33 for the ullage gas and liquid regions. In these figures, ***R*/*R***** is the ratio of the fluid constant used in the numerical calculations to the ideal fluid constant. Thus, an ***R*/*R***** = 0 would imply no first-order correction. An ***R*/*R***** = 1 would imply an ideal first-order correction. The deviation is defined as the maximum velocity difference between a particular test case and a standard test case calculated with a time step of 1.e-6 s and ***R*/*R***** set at zero.

These calculations were performed assuming a tank radius of 0.01 m with a 10- by 28-node grid and a Mach scaling factor of 100. As expected, the first-order correction term increased the stability of the program and allowed larger time steps in the calculations. However, the large deviations with increasing ***R*/*R***** for the liquid (17% at ***R*/*R***** = 1, Δt = 0.5e-5) were unanticipated. The large difference in deviation between the gas and liquid regions may result from the difference in Reynolds cell number. The average Reynolds cell number in the gas region is 0.65 compared to 8.5 in the liquid region. Reynolds cell numbers greater than 2 are

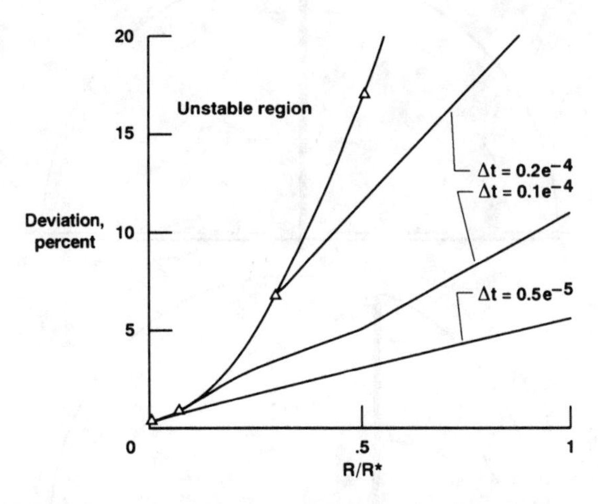

Fig. 6 Numerical deviation of first-order correction to the pressure term in the gas region; *R*/*R = 0 implies no first-order correction, *R*/*R** = 1 implies an ideal first-order correction.**

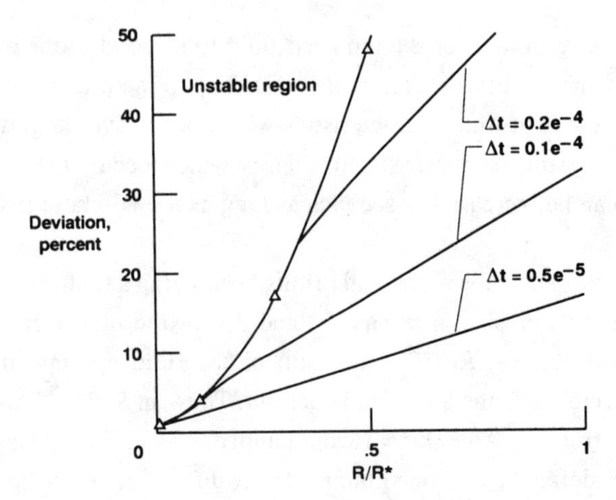

Fig. 7 Numerical deviation of first-order correction to the pressure term in the liquid region.

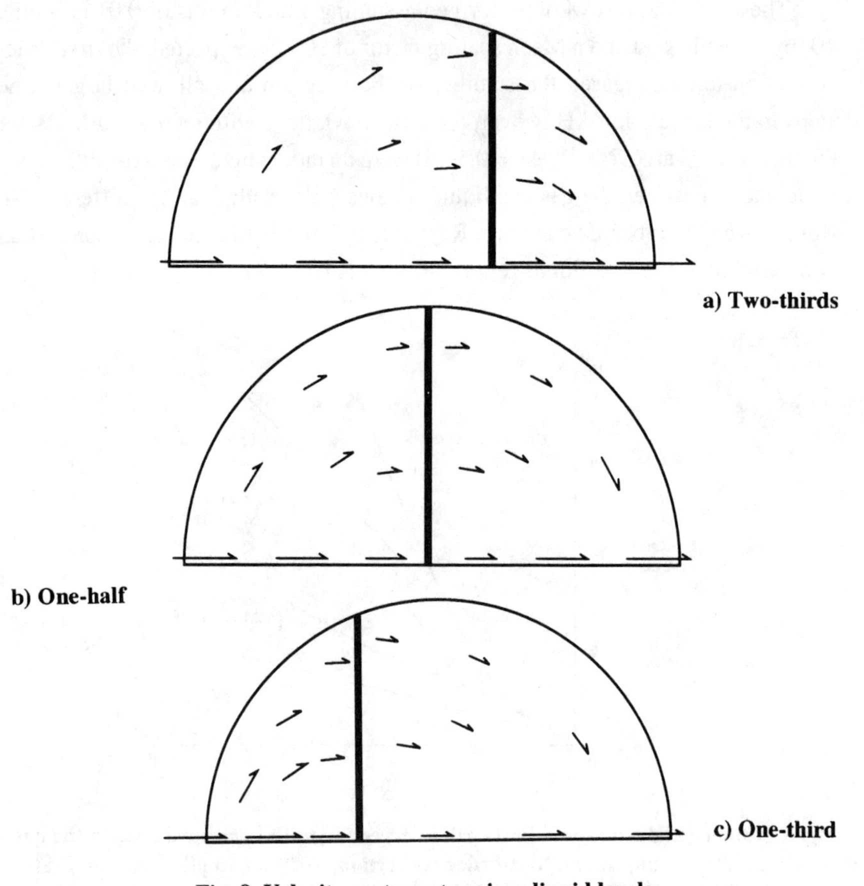

Fig. 8 Velocity vectors at various liquid levels.

known to cause instabilities in the finite difference technique. An $\boldsymbol{R/R^*} = 0.25$ appears to be the optimal value for this test case. As a result, the computational time could be reduced by half (by doubling time step) with a loss in numerical accuracy of less than 3% in the gas region and 10% in the liquid region.

Figures 8a, 8b, and 8c show the effect of the moving interface. These figures present velocity vectors at two-thirds, one-half, and one-third of the gas and liquid interface heights. The calculations were conducted on one side of the tank as symmetry is assumed, a tank radius of 0.05 m, a 15×23 node grid, a Mach scaling factor of 100, an $\boldsymbol{R/R^*}$ set at 0.25, and a drain rate of 0.002 m^3/s.

Concluding Remarks

A computational fluid dynamic model has been developed to model a cryogenic fluid tank while subjected to rapid drain rate and high heat flux. The model incorporates a moving interface, a transient hybrid explicit–implicit finite difference technique, a Mach scaling factor, and a first-order correction to the pressure term in the momentum equations. The Mach scaling factor decreases the computational time of the model without a significant loss of accuracy. The first-order correction to the pressure term increases stability and decreases computational time with some loss of accuracy.

The generic research cryogenic tank which this model was developed to simulate has met with budgetary constrains, and test data for verification of the model are unavailable. Comparison of the model with test data will be the subject of future efforts.

References

[1]Fenn, D. B., Acker, L. W., and Algranti, J. S., "Flight Operation of a Pump-Fed, Liquid Hydrogen Fuel System," NASA TM-252, April 1960.

[2]Brewer, G. D., *Hydrogen Aircraft Technology*, CRC Press, Inc., Boca Raton, FL, 1991.

[3]Stephens, C. A., Hanna, G. J., and Gong, L., "Thermal–Fluid Analysis of the Fill and Drain Operations of a Cryogenic Fuel Tank," NASA TM-104273, Dec. 1993.

[4]Anderson, J. D., Jr., *Hypersonic and High Temperature Gas Dynamics*, McGraw Hill, New York, 1989.

[5]McCarty, R. D., Arp, V., and Friend, D. G., "NIST Thermophysical Properties of Pure Fluids, Version 3.0, Users Guide: NIST Standard Reference Database 12," National Institute of Standards and Technology, Gaithersburg, MD, 1992.

[6]Lick, W. J., *Difference Equations from Differential Equations, Engineering Services.*, Vol. 41, Springer-Verlag, New York, 1989.

[7]Chorin, A. J., "A Numerical Method for Solving Incompressible Viscous Flow Problems," *Journal of Computational Physics*, Vol. 2, 1967, pp. 12–26.

[8]Bussing, T. R. A. and Murman, E. M., "Finite-Volume Method for the Calculation of Compressible Chemically Reacting Flows," *AIAA Journal*, Vol. 26, No. 9, 1988, pp. 1070–1078.

[9]Schlichting, H., *Boundary-Layer Theory*, McGraw Hill, New York, 1979, pp. 9–10, 19.

Thermocryogenic Buckling and Stress Analyses of a Partially Filled Cryogenic Tank

William L. Ko*

NASA Dryden Flight Research Center, Edwards, California 93523-0273

Abstract

Thermocryogenic buckling and stress analyses were conducted on a horizontally oriented cryogenic tank using the finite element method. The tank is a finite length circular cylindrical shell with its two ends capped with hemispherical shells. The tank is subjected to cylindrical strip heating in the region above the liquid-cryogen fill level and to cryogenic cooling below the fill level (i.e., under thermocryogenic loading). The effects of cryogen fill level on the buckling temperature and thermocryogenic stress field were investigated in detail. Both the buckling temperature and stress magnitudes were relatively insensitive to the cryogen fill level. The buckling temperature, however, was quite sensitive to the radius-to-thickness ratio. Deformed shapes of the cryogenic tank under different thermocryogenic loading conditions are shown, and high-stress domains were mapped on the tank wall for the strain-gauge installations. The accuracies of solutions from different finite element models were compared.

Introduction

A finite length circular cylindrical shell with its two ends capped with hemispherical shells or hemispheroidal shells of revolution is a popular geometry for pressure vessels. These shapes of vessels also are used commonly as cryogenic fuel tanks for liquid-propellant rockets. A pressure vessel is loaded under uniform internal pressure, and the stress field generated in the tank wall is axisymmetric and can be calculated relatively easily.[1] When used as a cryogenic fuel tank for a

---

*Senior Scientist. Aerostructures Branch, P. O. Box 273. AIAA Member.

liquid-propellant rocket motor for vertical liftoff, the tank axis is oriented vertically; therefore, the stress field generated in the tank wall is also axisymmetric, but the induced stresses vary with axial location (governed by the fill level and the end effect). When used as a cryogenic fuel tank for a hypersonic flight vehicle for horizontal takeoff (e.g., space plane for single-stage horizontal takeoff to space), the insulated tank is carried inside the fuselage of the vehicle (or could form part of the fuselage of the vehicle), and its axis is oriented horizontally.

During one mission, the liquid-cryogen fill level starts from empty to full (at takeoff), then gradually comes down as the fuel is consumed during flight, and finally reaches empty at the end of the mission. Thus, the tank wall goes through a history of being cooled longitudinally below the liquid-cryogen fill level (which is constantly decreasing during the mission), and also being heated longitudinally (as the result of aerodynamic heating, even through insulations) in the region above the fill level. During a flight, the tank is subjected to the fill-level-dependent thermocryogenic loading, and the thermocryogenic stress field induced in the tank wall is no longer axisymmetric. The stresses in the tank wall then change with both the fill level and axial location because of the end effect, which is magnified by the shortness of the tank. Because of simultaneous heating and cryogenic cooling (changing with fill level) in different regions of the horizontally oriented cryogenic tank, thermocryogenic buckling could take place in certain high-compression zones of the heated zone if the thermal loading is too severe.

In addition to the thermocryogenic loading, the tank also is subjected to cryogen liquid-pressure loading, internal pressure loading, and tank-wall inertia loading. Severe liquid sloshing inside a large fuel tank (dynamic loading) could disturb the control of the flight vehicle.

A circular cylindrical shell with hemispherical bulkheads has been considered as a potential candidate cryogenic tank geometry for hypersonic flight vehicles. Thus, studies on the thermocryogenic performance of this type of tank are required. Some results of the thermal response of this horizontally oriented cryogenic tank subjected to simulated aerodynamic heating profiles were reported by Stephens and Hanna.[2,3] The results of their studies could be used as the basis to conduct thermocryogenic buckling and stress analyses of a cryogenic tank of this geometry.

Some existing closed-form solutions were obtained by Hill,[4] Abir and Nardo,[5] and Bushnell, Smith[6] and Thornton[7] for calculating thermal buckling temperatures and thermal stresses in a thin circular cylindrical shell heated along a narrow axial strip. These solution equations, however, might not give accurate results for the case of a finite length circular cylindrical tank with hemispherical bulkheads, because the cylindrical shell considered by those investigators was long enough so that the end effect could be neglected.

This paper concerns thermocryogenic buckling and stress analyses of a horizontally oriented cryogenic tank subjected to cylindrical strip heating. The tank is of relatively short circular cylindrical shell with its two ends capped with hemispherical shells. Because the cryogenic tank under consideration is relatively short, the end effect could be felt in most of the cylindrical section area. Therefore, the finite element method is used in the present thermocryogenic buckling and stress analyses. The results presented show how the thermocryogenic buckling temperatures and stress field in the tank wall change with the liquid-cryogen fill level and also with tank-wall thickness.

Description of Problem

Figure 1 shows the geometry of the cryogenic fuel tank and the coordinate systems. The circular cylindrical section of length l, and the hemispherical bulkheads have radius R and wall thickness t. The xyz system is the global coordinate system for the tank, and the $x'y'z'$ system is the local coordinate system used in the finite element mesh generation of a hemispherical bulkhead.

The cryogenic fuel tank is partially filled and subjected to thermocryogenic loading with temperature distribution shown in Fig. 2. The hot region will be heated under constant temperature differential of $\Delta T = 1$ °F (i.e., temperature of hot region minus temperature of cold region), except for the end regions of the temperature profile where the temperature decreases linearly over a 12 deg arc down to zero at the liquid-cryogen fill level. The ramp zone of the temperature profile is about 15 deg according to the heat transfer analysis.[2,3] Because of the finite element sizing chosen, however, a 12 deg temperature ramp was used. For the present analysis, the temperature was assumed constant through the tank-wall thickness. The tank is supported at two end points. One end point is fixed, and the other end point can move freely only in the z direction.

Finite Element Models

The structural performance and resizing (SPAR) finite element program was used in setting up two finite element models. Because of the symmetry

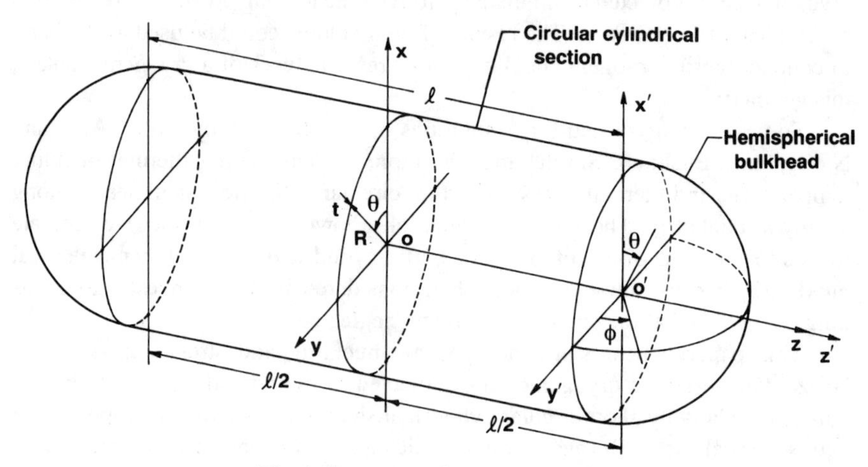

Fig. 1 Geometry of cryogenic tank.

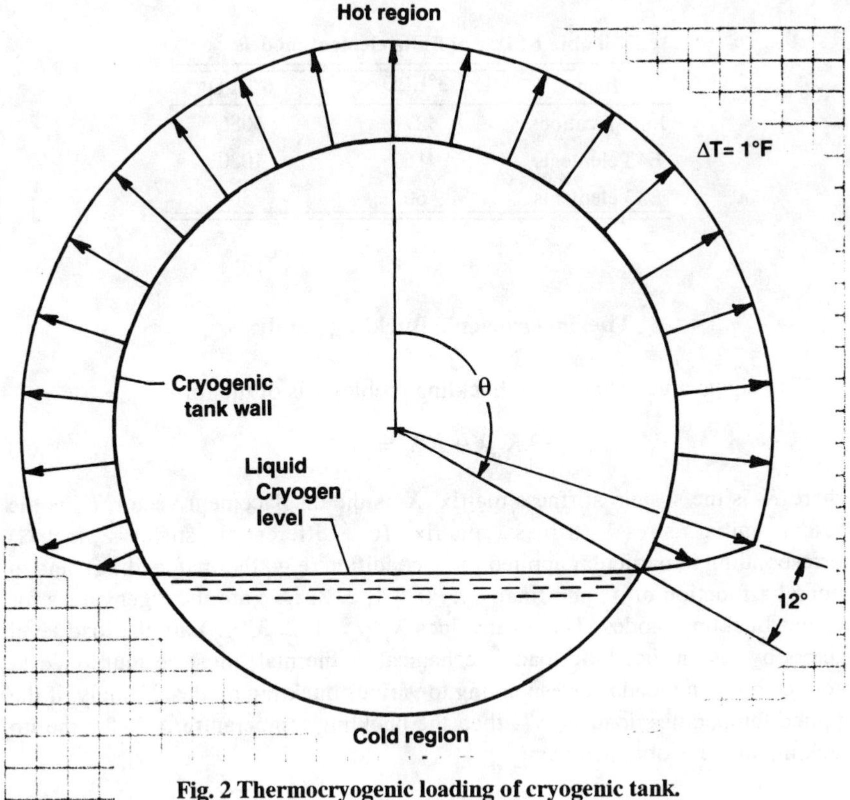

Fig. 2 Thermocryogenic loading of cryogenic tank.

with respect to the *xy* and *xz* planes, only a quarter of the tank lying in the region $0 \leq \theta \leq 180$ deg, $z \geq 0$, was modeled (Fig. 1). The SPAR commands SYMMETRY PLANE = 2 and SYMMETRY PLANE = 3 then were used in the constraint definition to generate the whole tank for the stress computations. Figures 3 and 4 show the finer model named 3°ELD (ELement Definition) and a coarser model named 6°ELD set up for the tank, respectively. In these graphical displays, the SPAR command SYMMETRY PLANE = 3 was used in generating the model's mirror image with respect to the *xy* plane. The elements in the circular cylindrical section were generated based on the *xyz*-coordinate system. In the mesh generation of the hemispherical bulkhead, the $x'y'z'$-coordinate system was used so that the horizontal grid lines would match the fuel fill levels. The purpose of using two models was to study the finite element solution convergency and to define mesh density required to obtain adequate buckling and stress solutions. The finer model 3°ELD was needed for obtaining satisfactory buckling solutions and smooth buckling mode shapes. For the buckling analysis, past experience showed that the element length-to-thickness ratio must be about 5 to 1. The 3°ELD model was set up according to this criterion. Table 1 compares the sizes of the two finite element models. As will be seen later, the 3°ELD model requires roughly 10 times the computer time needed to run the 6°ELD model in the eigenvalue extractions.

Table 1 Sizes of finite element models

Item	3°ELD	6°ELD
Joint locations	4271	1086
*E*43 elements	4160	1020
*E*33 elements	60	30

Thermocryogenic Buckling Analysis

The eigenvalue equation for buckling problems is of the form

$$\lambda K_g X + KX = 0 \tag{1}$$

where K is the system stiffness matrix, X is the displacement vector, K_g is the system initial stress stiffness matrix (or differential stiffness matrix) corresponding to particular applied force condition (e.g., thermal loading), and in general a function of X, and finally λ_i (i = 1, 2, 3, ...) are the eigenvalues for various buckling modes. The eigenvalues λ_i (i = 1, 2, 3, ...) are the load level factors by which the static load (mechanical or thermal) must be multiplied to produce buckling loads corresponding to various buckling modes. Namely, if the applied temperature load is ΔT, then the buckling temperature ΔT_{cr} for the ith buckling mode is obtained from

$$\Delta T_{cr} = \lambda_i \Delta T \tag{2}$$

Equation (1) will give the eigenvalues (either positive or negative) in the neighborhood of zero. If one desires to find the eigenvalue in the neighborhood of c, then the following "shifted" eigenvalue equation may be used:

$$(\lambda - c) K_g X + (K + cK_g) X = 0 \tag{3}$$

As will be seen later, using Eq. (3), the number of eigenvalue iterations could be greatly reduced (i.e., fast eigenvalue convergency).

In the eigenvalue extractions, the SPAR program uses an iterative process consisting of a Stodola matrix iteration procedure, followed by a Rayleigh-Ritz procedure, and then followed by a second Stodola procedure. This process results in successively refined approximations of m eigenvectors associated with the m eigenvalues of Eq. (1) closest to zero. Whetstone[8] describes the details of this process.

Thermocryogenic Stress Analyses

The thermocryogenic stress solutions (for different fill levels) are the byproducts of thermocryogenic buckling analysis, because the static stress

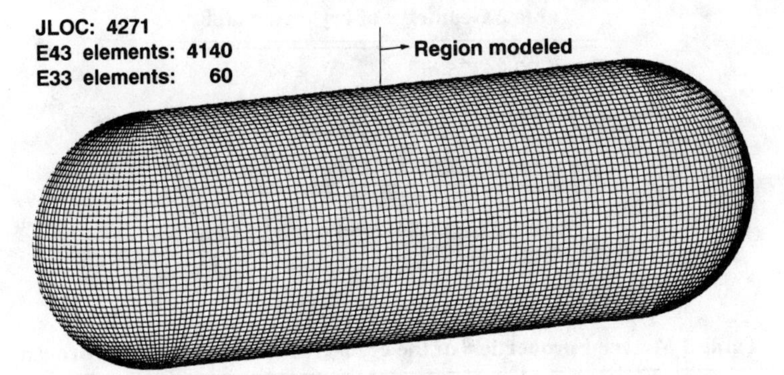

Fig. 3 Finite element model 3°ELD set up for cryogenic tank.

analysis must be performed before the eigenvalue solutions could be obtained. The thermocryogenic stress solutions are based on the unit temperature load $\Delta T = 1$ °F (Fig. 2) and are the influence function type stress solutions. For other ΔT, which could be as high as 300 °F (Refs. 2 and 3), those stress solutions must be multiplied by the actual value of ΔT to obtain actual stresses under linear elasticity.

Numerical Results

The numerical results of the thermocryogenic buckling and stress analyses presented in this article are based on the following physical properties of the stainless-steel cryogenic tank carrying liquid hydrogen as cryogen.

Physical Properties

The geometry and material properties of the cryogenic tank and the cryogen are shown in Tables 2 and 3.

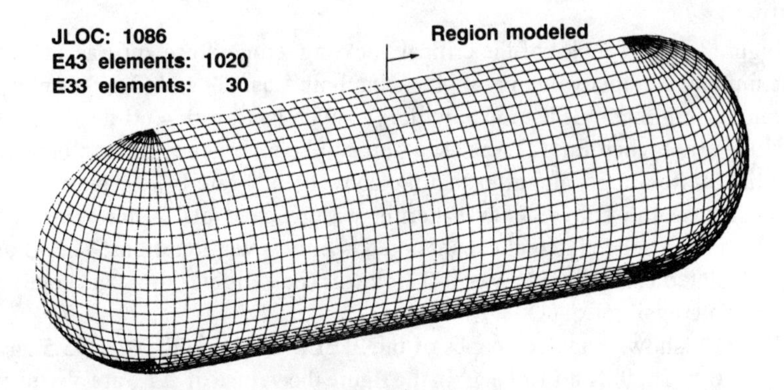

Fig. 4 Finite element model 6°ELD set up for cryogenic tank.

Table 2 Geometry of cryogenic tank

R = 29.84375 in.[a]
l = 120 in.[a]
R/t = 95.5,[a] 150, 200, 250, 300, 350, 400

[a]Geometry of NASA generic research cryogenic tank.

Table 3 Material properties[a] of the cryogenic tank and liquid hydrogen

	Low temperature	High temperature
E, lb/in.^{2}	27.9×10^6	27.9×10^6
ν	0.28	0.28
ρ, lb/in.^{3}	0.29	0.29
α, in./in.-°F	9.0×10^{-6}	7.8×10^{-6}
ρ_H, lb/in.^{3}	0.002685	n.a.

[a]Stainless-steel SA 240 type 304.

Thermocryogenic Buckling

In the eigenvalue extractions, the maximum number of iterations was set to 100. However, for most cases in which the nonshifted eigenvalue equation (1) was used, the convergency criterion $\{|(\lambda_i - \lambda_{i-1})/\lambda_i| < 10^{-4}\}$ for eigenvalue iterations could be reached in fewer than 100 iterations. If shifted eigenvalue equation (3) was used, the number of eigenvalue iterations could be reduced greatly.

Figure 5 shows a plot of the critical buckling temperature ΔT_{cr} as a function of the number of eigenvalue iterations using both Eqs. (1) and (3). This plot was generated using the 3°ELD model, with R/t = 95.5 having θ = 60-deg fill level.

Notice that when Eq. (1) was used, the rapid convergency rate occurs during the initial 30 iterations, and after that the convergency rate is very slow. For this particular case the convergency criterion was reached at 72 iterations. When Eq. (3) was used, the eigenvalue converged at only seven iterations. The value of ΔT_{cr} calculated using Eq. (3) was 1–3 °F lower (i.e., slightly more accurate) than that calculated using Eq. (1).

Figure 6 shows buckled shapes of the 3°ELD model with R/t = 95.5 having fill levels θ = 30, 90, and 150 deg. In the figure the values of ΔT_{cr} are also shown. Notice that the buckling is local in nature and occurs in a small, central region of

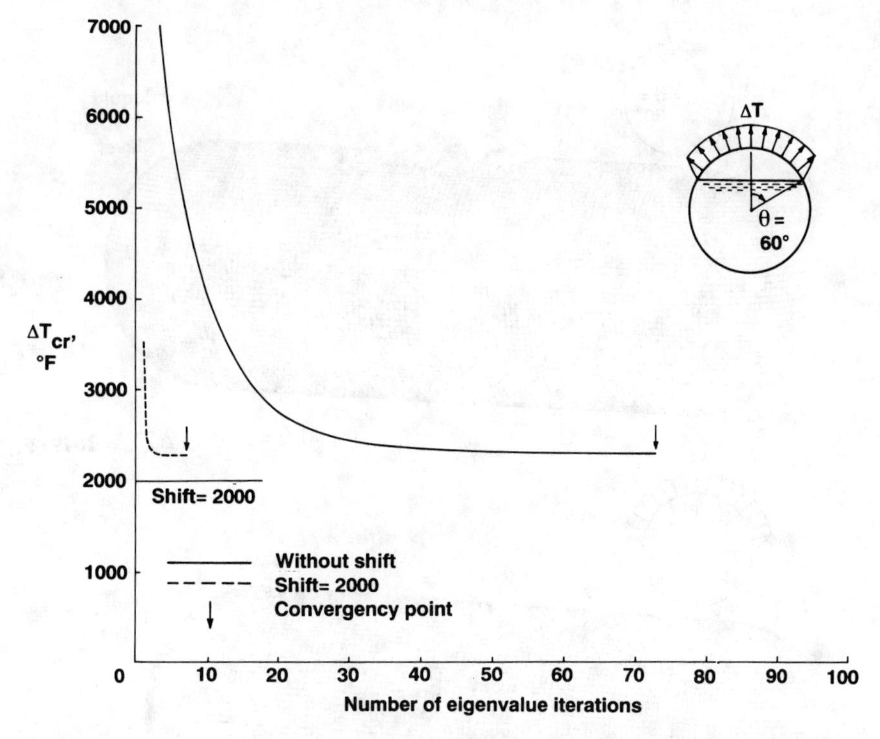

Fig. 5 Convergency curves of eigenvalue solutions; θ = 60 deg; 3°ELD finite element model.

the tank slightly above the fill level, where the peak axial compressive stress lies. Table 4 summarizes the thermocryogenic buckling temperature ΔT_{cr} of a cryogenic tank having $R/t = 95.5$ under different fill levels.

In Table 4, for the fill levels of θ = 30, 90, 120, and 168 deg, the lowest eigenvalues were found to be negative and, therefore, the eigenshifting method was used to search the lowest positive eigenvalues. The negative ΔT_{cr} implies that the heated zone (Fig. 2) turned out to be a cold zone (i.e., tank turns upside down). The 6°ELD model underpredicts the buckling temperatures by more than 200 °F, and the percent solution difference is practically insensitive to the change of the fill level.

Figure 7 shows the buckled shapes of 3°ELD with θ = 60-deg fill level but different R/t (i.e., R/t = 95.5 and 350). The number of buckles increases slightly as the value of R/t increases. In those figures the values of ΔT_{cr} and R/t are indicated.

Table 5 summarizes the thermocryogenic buckling temperature ΔT_{cr} of cryogenic tanks with different R/t ratios having identical fill levels of θ = 60 deg.

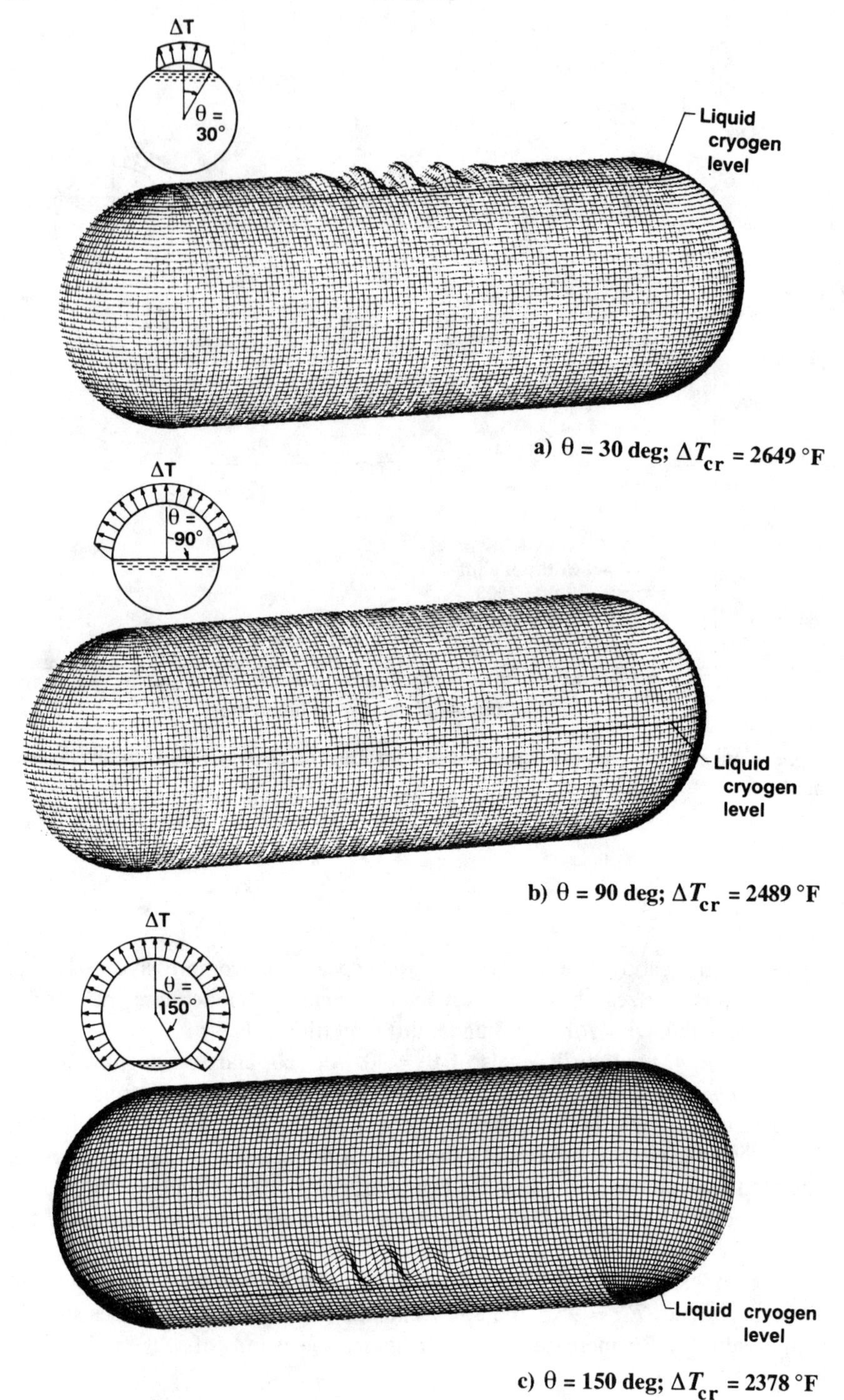

a) θ = 30 deg; ΔT_{cr} = 2649 °F

b) θ = 90 deg; ΔT_{cr} = 2489 °F

c) θ = 150 deg; ΔT_{cr} = 2378 °F

Fig. 6 Buckled shape of partially filled cryogenic tank subjected to cylindrical strip heating; 3°ELD finite element model; R/t = 95.5.

Table 4 Thermocryogenic buckling temperatures for different fill levels (*R*/*t* = 95.5)

	ΔT_{cr}, °F			
θ, deg	3°ELD	6°ELD	$\Delta(\Delta T_{cr})$, °F	Solution difference, %
30	2649	2323	326	12
60	2292	2063	229	10
90	2489	2241	248	10
120	2426	2182	244	10
150	2378	2114	264	11
168	2007	1801	206	10

Notice from Table 5 that the 6°ELD model gives lower values of ΔT_{cr} than those calculated using the 3°ELD model. The discrepancy of the eigenvalue solutions between the two models averages slightly more than 200 °F, and the percent solution difference increases with the increasing R/t.

The present study confirmed that to obtain satisfactory buckling solutions with smooth buckling shapes, the element density of the finite element model must be such that the length-to-thickness ratio is about 5. Thus, for thinner tanks, finer mesh may be required. However, because of excessive computer time problems, finer mesh was not used for high R/t tanks.

Figure 8 shows plots of ΔT_{cr} as a function of the fill level for the two finite element models using data shown in Table 4. The buckling temperature is seen to be relatively insensitive to the change of the fill level.

Figure 9 shows plots of ΔT_{cr} as a function of R/t for the $\theta = 60$ deg fill-level case using data presented in Table 5. The buckling temperature is quite sensitive to the change of R/t. The rate of decrease of ΔT_{cr} is faster in the lower R/t region and becomes slower as R/t increases.

The CPU time required to extract eigenvalues using the 3°ELD model is almost 10 times that required when using the 6°ELD model. For each eigenvalue

Table 5 Thermocryogenic buckling temperatures for different *R*/*t* (θ = 60 deg fill level)

		R/*t*						
		95.5	150	200	250	300	350	400
ΔT_{cr}, °F	3°ELD	2292	1415	1060	849	707	604	525
ΔT_{cr}, °F	6°ELD	2063	1198	832	609	451	347	286
$\Delta(\Delta T_{cr})$, °F		229	217	228	240	256	257	239
Solution difference, %		10	15	22	28	36	43	46

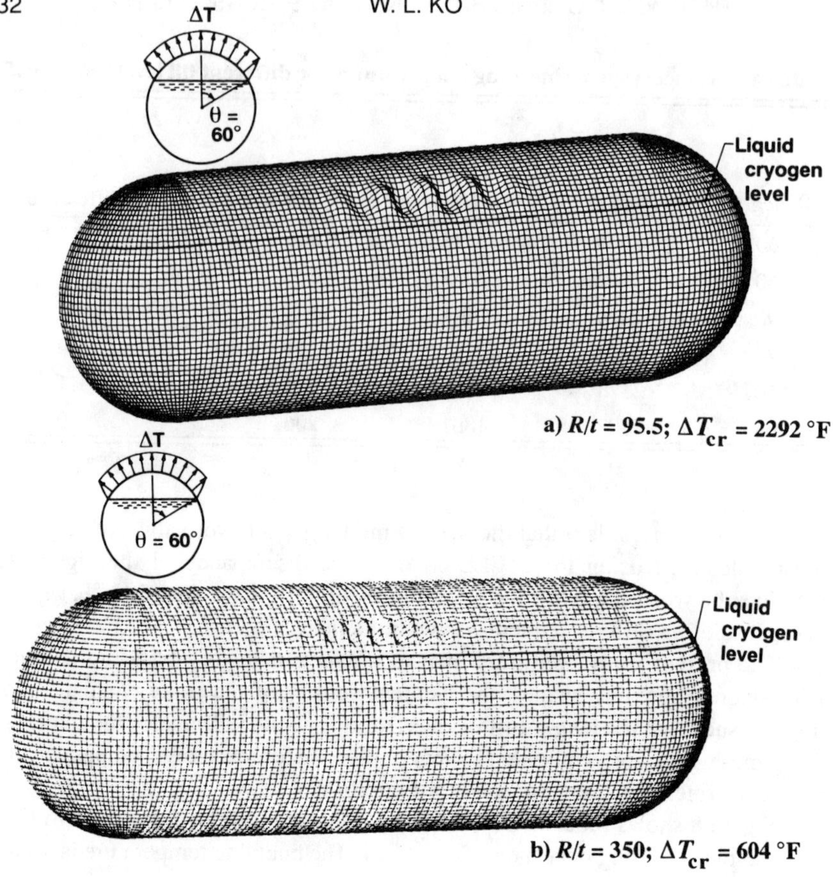

Fig. 7 Buckled shape of partially filled cryogenic tank subjected to cylindrical strip heating; 3°ELD finite element model; θ = 60 deg.

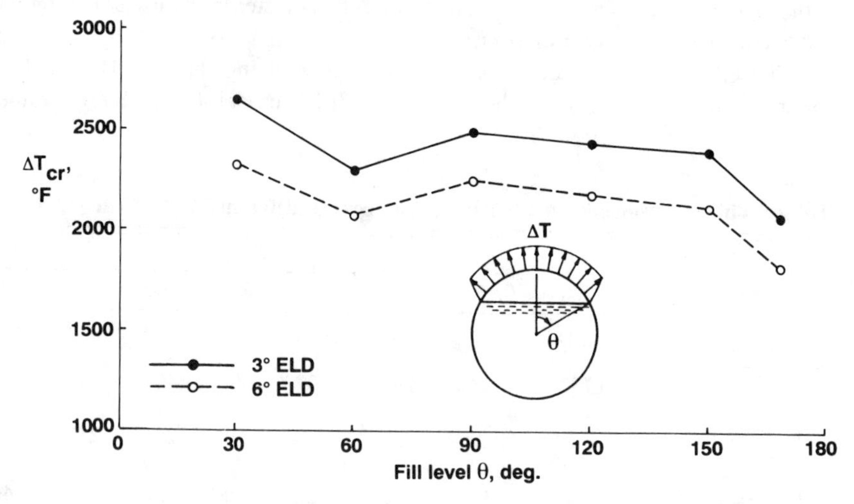

Fig. 8 Variation of buckling temperature ΔT_{cr} with liquid-cryogen fill level θ, R/t = 95.5.

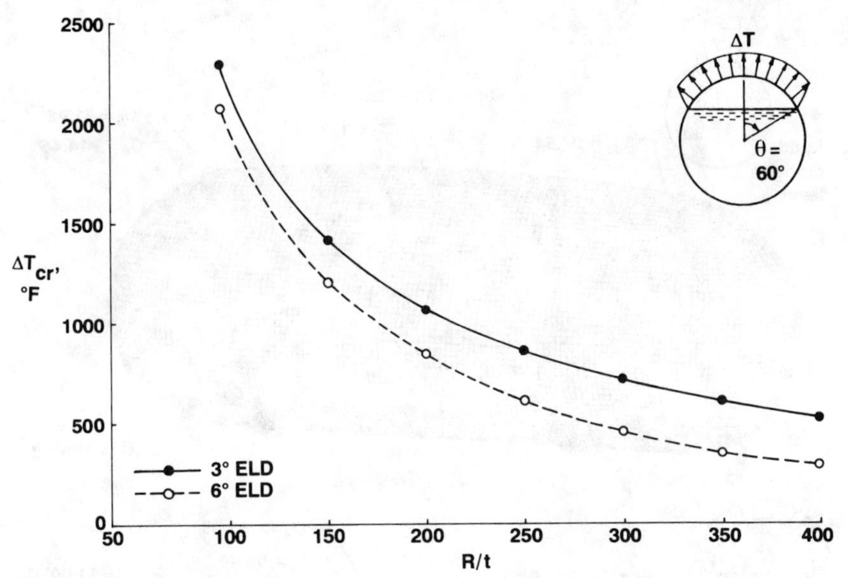

Fig. 9 Plot of buckling temperature ΔT_{cr} as a function of radius-to-thickness ratio *R*/*t*, θ = 60 deg, *R* = 29.84375 in.

iteration, the 3°ELD and 6°ELD models required about 5.23 and 0.69 min of CPU time, respectively. As will be seen later, for the stress analysis, the 6°ELD model, which requires much shorter CPU time, gave fairly good stress solutions.

Induced Stresses and Deformations

As mentioned before, the thermocryogenic stress solutions are generated during the thermocryogenic buckling analysis. Figure 10 shows the deformed shapes of the 3°ELD model with fill levels $\theta = 30$, 90, and 150 deg subjected to thermal loading of $\Delta T = 1$ °F. At higher fill levels (i.e., $\theta = 30$ deg), the hot region bulged upward, and the cold region caved in slightly in the region below the fill-level line. The deformed shape at fill level $\theta = 90$ deg is the most interesting. The top and fill-line regions of the tank caved in, and the central region of the tank wall bulged out at three angular locations to form three lobes. At low fill levels ($\theta =$ 150 deg), the bottom of the tank caved in severely. Figure 10 also shows the locations of peak (positive or negative) stress points and stress magnitudes at those points. Notice that the peak tension and peak compression of the axial stresses σ_z are at the midsection of the tank. The peak axial tensile stress is always at the fill level, and the peak axial compressive stress is slightly above the fill level where the potential thermocryogenic buckling could occur. The peak values of shear stress $\tau_{\theta z}$ are at the cylinder–hemisphere junctures and are slightly above the fill level. The peak tension and peak compression of tangential stresses σ_θ are near the peak shear stress points but are lying in the hemispherical bulkhead regions.

Table 6 summarizes the stress magnitudes at high-stress points for different fill levels calculated from the two finite element models. The 6°ELD model gives

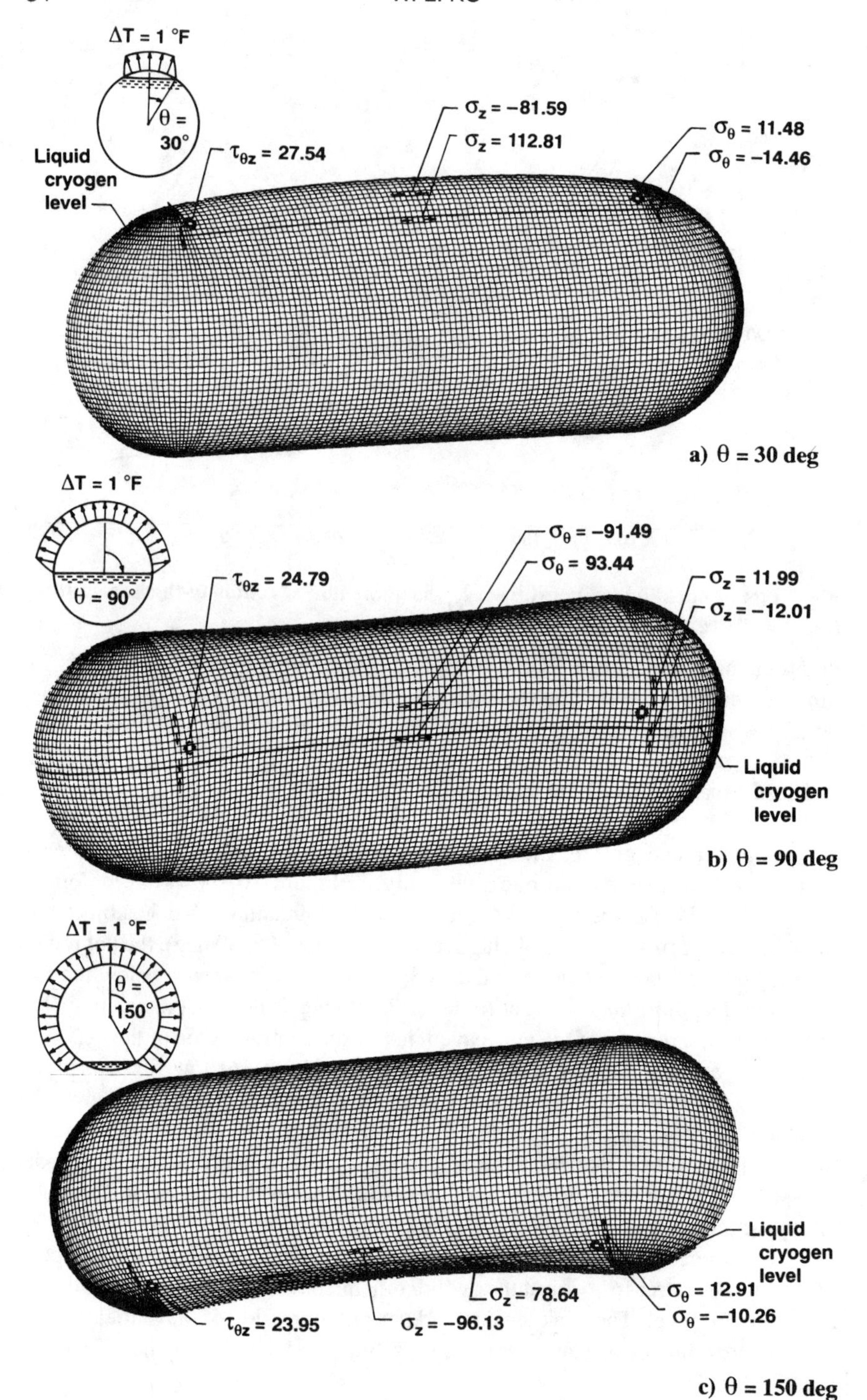

Fig. 10 Deformed shapes of cryogenic tank and locations of high-stress points, ΔT = 1 °F, stresses expressed in pounds per square inch; 3°ELD finite element model.

Table 6 Stresses at high-stress points; thermocryogenic loading, ΔT = 1 °F

θ, deg	Model	σ_θ, lb/in.^{2}		σ_z, lb/in.^{2}		$\tau_{\theta z}$, lb/in.^{2}
30	3°ELD	11.48	–14.46	112.81	–81.59	27.54
30	6°ELD	7.79	–13.32	106.03	–77.74	24.57
60	3°ELD	11.89	–11.41	88.56	–98.58	24.26
60	6°ELD	10.33	–8.97	82.52	–92.07	20.99
90	3°ELD	11.99	–12.01	93.44	–91.49	24.79
90	6°ELD	8.67	–8.62	86.96	–85.13	21.42
120	3°ELD	11.82	–12.49	95.53	–93.32	25.15
120	6°ELD	8.74	–10.04	89.54	–87.18	21.80
150	3°ELD	12.91	–10.26	78.64	–96.13	23.95
150	6°ELD	11.52	–9.29	70.94	–88.76	20.71
168	3°ELD	15.82	–14.08	107.74	–108.14	30.11
168	6°ELD	12.95	–9.74	106.03	–103.14	27.34

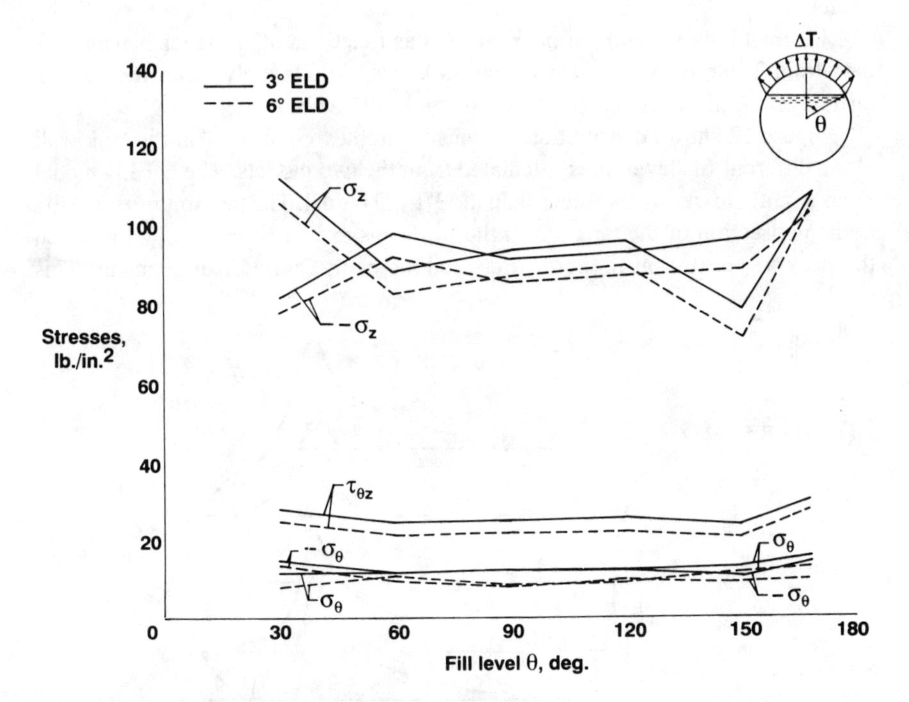

Fig. 11 Plots of stresses at high-stress points as functions of liquid-cryogen fill level θ, thermocryogenic loading, ΔT = 1 °F.

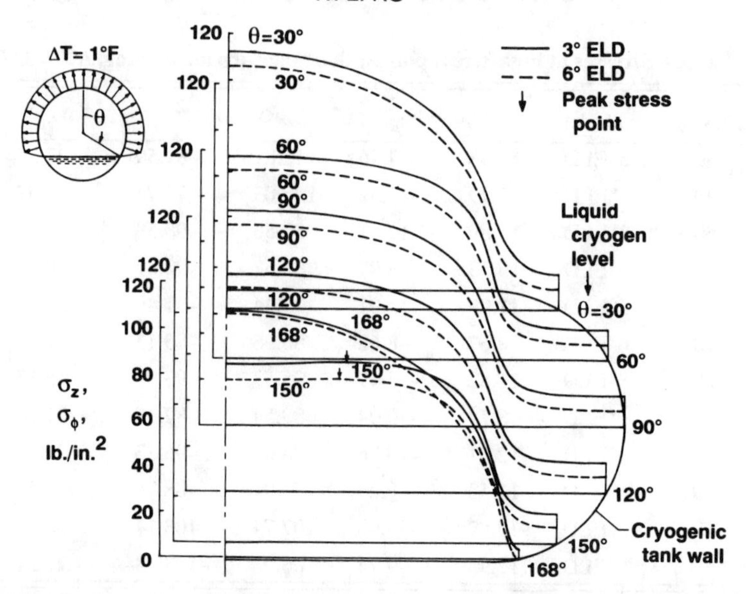

Fig. 12 Distributions of tensile stresses σ_z and σ_ϕ in the cryogenic tank wall along liquid-cryogen fill-level line, ΔT = 1 °F.

slightly lower stress intensities as compared with those calculated from the 3°ELD model.

Figure 11 shows plots of peak stresses as functions of fill level θ using the data from Table 6. Notice that the stress levels are relatively insensitive to the change of the fill level except at low and high fill levels.

Figure 12 shows distributions of tensile stresses σ_z and σ_ϕ in the tank wall along different fill-level lines calculated from the two models. The 6°ELD model gives slightly lower stress values than the 3°ELD model. The maximum σ_z occurs at the midsection of the tank for all the fill levels except θ = 150 deg for which the peak σ_z occurs near the quarter-sections of the cylindrical segment. This

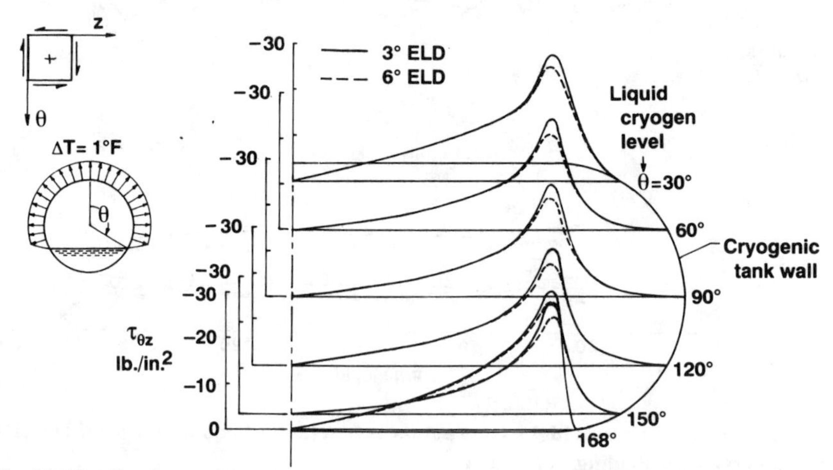

Fig. 13 Distributions of shear stresses $\tau_{\theta z}$ in the cryogenic tank wall along liquid-cryogen fill-level lines, ΔT = 1 °F.

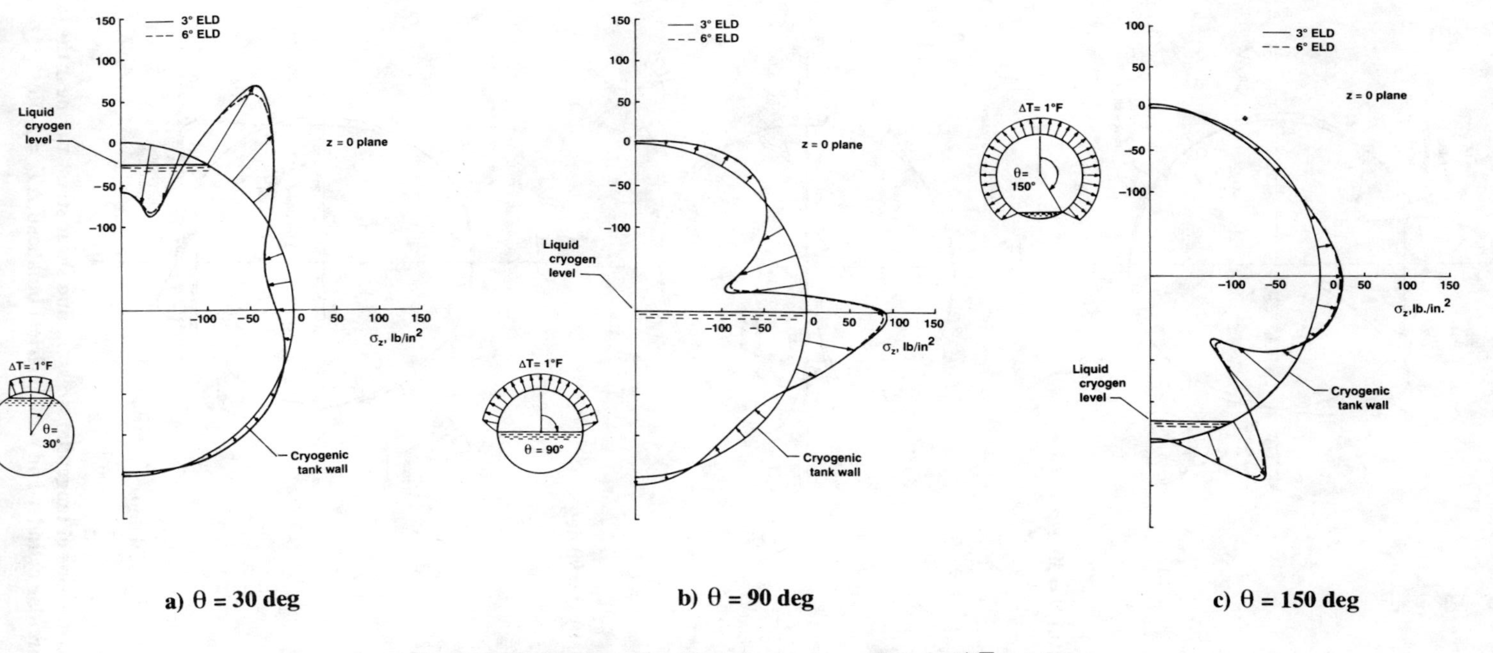

Fig. 14 Distributions of axial stress σ_z in $z = 0$ plane, $\Delta T = 1$ °F.

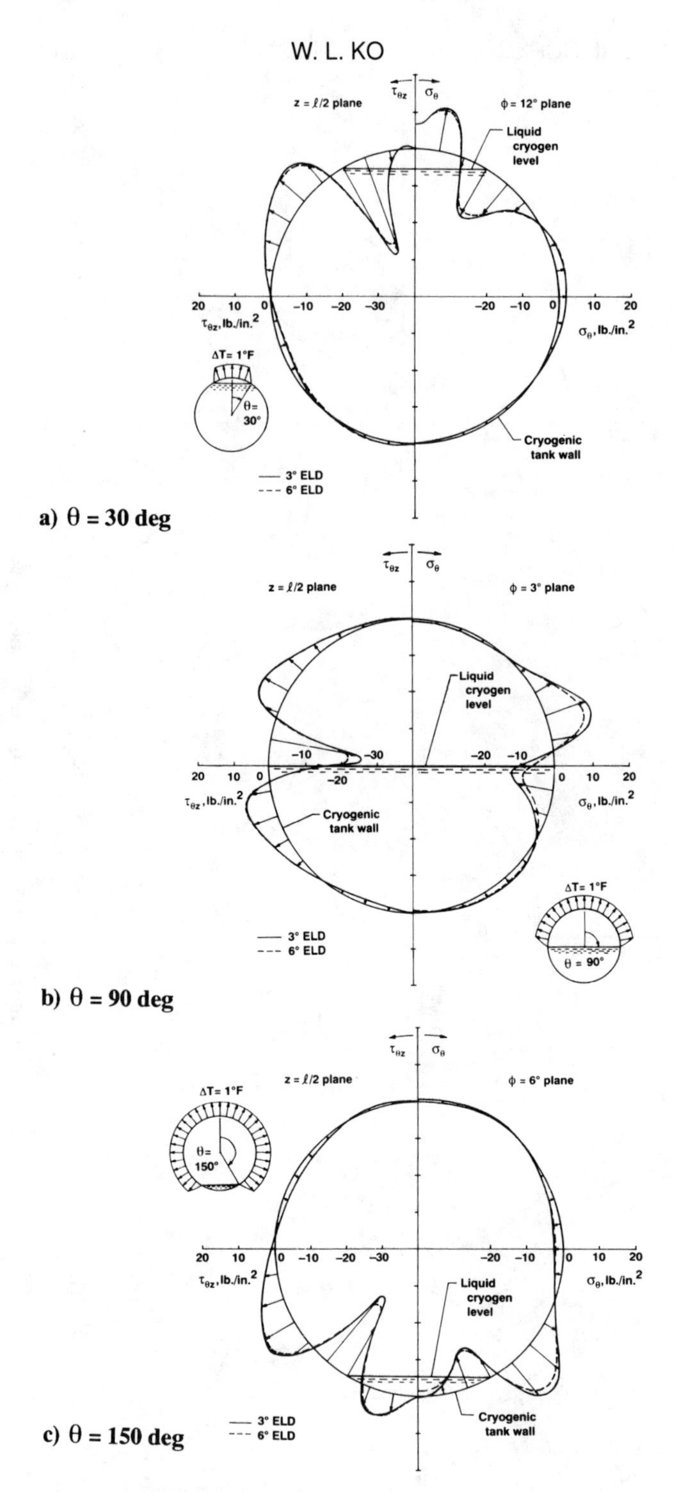

Fig. 15 Distributions of tangential stress σ_θ and shear stress $\tau_{\theta z}$ near the junction region of circular cylinder and semispherical bulkhead, ΔT = 1 °F.

figure shows that the two-dimensional analysis is not applicable for the present short-tank case.

Figure 13 shows distributions of the shear stress $\tau_{\theta z}$ in the tank wall along the different fill-level lines. The shear stress concentration occurs at the cylinder–hemisphere junction zone, and its intensity increases slightly at very high and very low fill levels (c.f., Table 6). Again, the 6°ELD model gives slightly lower stress concentrations than those given by the 3°ELD model.

Figure 14 shows circumferential distributions of σ_z in the $z = 0$ plane (i.e., tank central cross section) for fill levels $\theta = 30, 90$, and 150 deg. The two models give almost identical stress distributions except at the high-stress (tension or compression) zones, where the 6°ELD model consistently gives slightly lower stress values than those given by the 3°ELD model. The locations of the stress concentration points are always in the vicinities of the fill-level line, and they migrate with the changing fill-level line.

Figure 15 shows the circumferential distributions of tangential stress σ_θ in the meridian plane (i.e., ϕ = constant plane) and shear stress $\tau_{\theta z}$ in the $z = l/2$ plane for fill levels $\theta = 30, 90$, and 150 deg. The two models give quite close stress solutions except at high-stress regions. Like the previous case, the stresses calculated from the 6°ELD model are slightly lower than those calculated from the 3°ELD model in the high-stress regions. Again, the stress concentration points lie in the neighborhood of the fill-level line and move together with the fill-level line.

High-Stress Domains

Based on the preceding thermocryogenic stress analyses, the domains of high-stress points may be mapped on the cryogenic tank as shown in Fig. 16. In the experimental measurement of stresses, the high-stress domains are the areas where strain gauges should be installed for obtaining the highest data outputs. Ko[9] provides the detailed version of this paper, which also includes stress analysis of the cryogenic tank under 1) cryogenic liquid-pressure loading, 2) internal-pressure loading, and 3) tank-wall inertia loading.

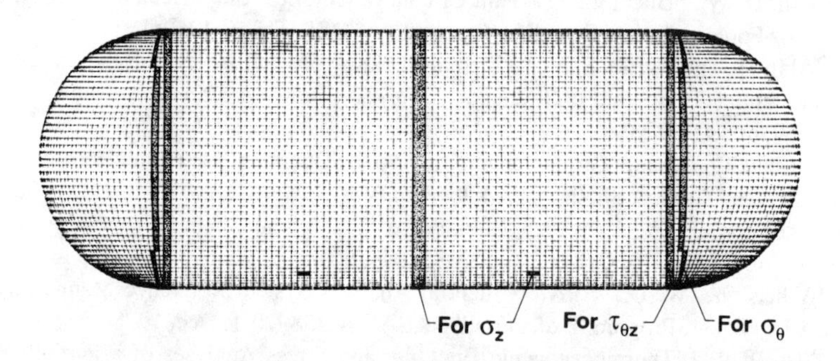

Fig. 16 High-stress domains for strain-gauge installations.

Concluding Remarks

Two finite element models of a cryogenic tank were set up for thermocryogenic buckling analysis and thermocryogenic stress analysis. The results of the analyses are summarized in the following.

1) The thermocryogenic buckling temperature ΔT_{cr} was insensitive to the liquid-cryogen fill level; however, it was sensitive to the radius-to-thickness ratio R/t and decreased with the increase of R/t.

2) In thermocryogenic buckling analysis, the 6°ELD model gave lower values of ΔT_{cr} and could not give as smoothly buckled shapes as those the 3°ELD model gave. For thermocryogenic buckling analysis, the finite element density should be such that the element length is about five times the element thickness.

3) In the stress analysis, both 3°ELD and 6°ELD models gave very close stress solutions. Therefore, the 6°ELD model is adequate for reasonably accurate stress solutions, because it requires about 1/10 of the computer central processing unit time than that required to run the 3°ELD model.

4) Thermocryogenic loading induced the peak tangential stresses in the region bounded by the ϕ = 3–12 deg meridian planes; the peak axial stresses occurred in the $z = 0$ plane for all the fill levels except the θ = 150 deg fill-level case, for which the peak axial tensile stress occurred near the $z = \pm l/4$ planes. The peak shear stresses occurred in the $z = \pm l/2$ planes. The locations of those high-stress points moved with the changing fill level.

5) High-stress domains were mapped on the cryogenic tank wall for experimental strain-gauge installations.

References

[1]Timoshenko, S. P. and Krieger, S. W., *Theory of Plates and Shells*, McGraw-Hill, New York, 1959, pp. 481–485.

[2]Stephens, C. A. and Hanna, G. J., "Thermal Modeling and Analysis of a Cryogenic Tank Design Exposed to Extreme Heating Profiles," NASA CR-186012, June 1991.

[3]Hanna, G. J., and Stephens, C. A., "Predicted Thermal Response of a Cryogenic Fuel Tank Exposed to Simulated Aerodynamic Heating Profiles With Different Cryogens and Fill Levels," NASA CR-4395, Sept. 1991.

[4]Hill, D. W., "Buckling of a Thin Circular Cylindrical Shell Heated Along an Axial Strip," Air Force Cffice of Scientific Research, AFOSR-TN-59-1250, Dec. 1959.

[5]Abir, D., and Nardo, S. V., "Thermal Buckling of Circular Cylindrical Shells Under Circumferential Temperature Gradients," *Journal of the Aero/Space Sciences*, Vol. 29, Dec. 1959, pp. 803–808.

[6]Bushnell, D. and Smith, S., "Stress and Buckling of Nonuniformly Heated Cylindrical and Conical Shells," *AIAA Journal*, Vol. 9, No. 12, 1972, pp. 2314–2321.

[7]Thornton, E. A., "Thermal Buckling of Plates and Shells," *Applied Mechanics Review*, Vol. 46, No. 10, Oct. 1993, pp. 485–506.

[8]Whetstone, W. D., "SPAR Structural Analysis System Reference Manual, System Level 13A, Vol. 1, Program Execution," NASA CR-158970-1, Dec. 1978.

[9]Ko, W. L., "Thermocryogenic Buckling and Stress Analyses of a Partially Filled Cryogenic Tank Subjected to Cylindrical Strip Heating," NASA TM-4579, Sept. 1994.

Random Vibration of Thermally Buckled Plates:
I Zero Temperature Gradient across the Plate Thickness

Jon Lee*

Wright Laboratory,
Wright-Patterson Air Force Base, Ohio 45433

Abstract

In modal representation the large-amplitude vibration of a thermally buckling/buckled plate can be modeled by coupling Duffing's oscillators for prebuckled modes and Holmes' buckled-beam oscillators for postbuckled modes. Whereas stochastic Duffing's oscillator has long been studied as a prototype model for structural nonlinearity, the random vibration of Holmes' oscillator is new in structural applications. It is shown that the negative thermal stiffness engenders a bimodal Fokker–Planck distribution, and the thermal moment induced by a temperature gradient across the plate thickness renders the Fokker–Planck distribution asymmetric. Consequently, the equivalent linearization technique which works quite well for Duffing's oscillator cannot be extended to Holmes' oscillator in a straightforward fashion. Thus we have carried out successive cumulant analyses for the second- and fourth-order moments which are needed to compute the root mean square strain and stress on a simply supported and clamped plate. In this paper we restrict ourselves to the case of zero temperature gradient across the plate thickness. At high plate temperature the asymptotic form of moments is governed by the static snap-through displacement alone, independent of the random forcing. This has been confirmed by the recent numerical simulation of Vaicaitis.

I. Problem Statement

Recently, Lee[1] has derived the modal equations for the transverse displacement of an isotropic plate undergoing thermal buckling due to the immov-

---

* Research Scientist, Structural Dynamics Branch (WL/FIBG).

able edge constraint imposed on all plate edges. The heating of the plate is depicted by three thermal terms; (1) a uniform plate temperature above the room temperature, (2) a temperature variation over the midplane, and (3) a temperature gradient across the plate thickness when the top and bottom sides of plate are held at different temperatures. It has been shown that the thermal expansion due to terms (1) and (2) plays the role of thermal stiffness and, hence, can be lumped into a combined stiffness together with the structural stiffness. Since thermal stiffness has a negative sign, the combined stiffness first decreases and then vanishes at the critical buckling temperature. It becomes increasingly more negative as the magnitudes of terms (1) and (2) are raised. On the other hand, term (3) gives rise to thermal moment which contributes to combined forcing together with the external forcing. All this can be illustrated mostly succinctly by a typical single-mode equation for displacement q

$$\ddot{q} + \beta \dot{q} + k_o(1-s)q + \alpha q^3 = f_o + f(t) \tag{1}$$

which is Eq. (46) in Ref. 1. Here, $\beta\dot{q}$ is the viscous damping, $k_o(1-s)q$ the combined stiffness, αq^3 the cubic stiffness, f_o the thermal moment due to temperature gradient, and $f(t)$ the external random forcing. Appendix A gives α, β, f_o, k_o, and s for the simply supported and clamped plates.

First of all, the combined stiffness consists of structural $k_o q$ and thermal $-sk_o q$ contributions, where s reflects the strength of thermal terms (1) and (2). We have $s=0$ in the non-thermal case, and $s=1$ corresponds to the critical buckling temperature. For the prebuckled plate ($s<1$) the combined stiffness is positive so that Eq. (1) has the form of Duffing's oscillator which has long been investigated as a prototype nonlinear model in stochastic dynamics.[2–4] On the other hand, Eq. (1) reduces to the buckled-beam equation of Holmes[5] for $s>1$ with negative combined stiffness. Because of this sign change, the closure approximation developed for Duffing's oscillator cannot be applied to the random vibration of Holmes' oscillator.

Furthermore, in the absence of f_o, the zero-mean forcing can give rise to zero-mean displacement. However, when $f_o \neq 0$ the displacement would have moments of all orders. Now, to eliminate f_o from the right-hand side of Eq. (1), we define the static displacement by

$$\alpha Q^3 + k_o(1-s)Q - f_o = 0 \tag{2}$$

For $f_o = 0$ we have the symmetric snap-through displacement

$$Q = \begin{cases} Q_1 = 0 & \text{for } s \leq 1 \\ Q_{2,n} = (-1)^n \sqrt{(s-1)k_o/\alpha} \quad (n=0,1) & \text{for } s>1 \end{cases} \tag{3}$$

However, when $f_o \neq 0$ we have

$$Q = \begin{cases} Q_3 = A^{1/3} - B^{1/3} & \text{for } s \leq s^* \\ Q_{4,n} = 2\sqrt{\dfrac{k_o(s-1)}{3\alpha}} \cos\left(\dfrac{\upsilon + 2\pi n}{3}\right) \ (n=0,1) & \text{for } s > s^* \end{cases} \tag{4}$$

where

$$(A, B) = \pm \frac{f_o}{2\alpha} + \sqrt{\left(\frac{f_o}{2\alpha}\right)^2 + \left(\frac{k_o(1-s)}{3\alpha}\right)^3}$$

$$\upsilon = \arccos\sqrt{\left(\frac{f_o}{2\alpha}\right)^2 \Big/ \left(\frac{k_o(s-1)}{3\alpha}\right)^3} \quad \text{and} \quad s^* = 1 + \left(\frac{3\alpha}{k_o}\right)\left(\frac{f_o}{2\alpha}\right)^{2/3}$$

Note that Eq. (4) reduces to Eq. (3) when $f_o = 0$. For a simply supported and clamped plate, we have reproduced in Fig. 1 the root loci (3) and (4) already presented in Ref. 1. Note that Q is the trough of potential (strain) energy $U = k_o(1-s)q^2/2 + \alpha q^4/4 - f_o q$. When $f_o = 0$ the potential energy is an even function $U(q) = U(-q)$. Hence, we see that f_o destroys the symmetry property of Duffing's oscillator as defined in Ref. 6. An asymmetric potential energy has also been observed in an electrical power system model[7] and imperfect structures model.[8] Here, Q_1 is the trough of a single-well potential ($s < 1$), and $Q_{2,0}$ and $Q_{2,1}$ are troughs of the double-well potential ($s > 1$). Note that Q_3 and $Q_{4,n}$ are the asymmetric distortion of Q_1 and $Q_{2,n}$. Then, splitting the total displacement q into static displacement Q and fluctuation x

$$q = Q + x \tag{5}$$

we obtain the local equation of motion about Q[9]

$$\ddot{x} + \beta\dot{x} + Kx + 3\alpha Q x^2 + \alpha x^3 = f(t) \tag{6}$$

where $K = k_o(1-s) + 3\alpha Q^2$. This is the starting point of the stochastic dynamic analysis.

The investigation of this paper was initiated by an attempt to validate the equivalent linearization technique used in Ref. 1. It is first necessary to establish a reference statistics by the Fokker–Planck distribution (Sec. II). An extension of equivalent linearization is sought in the hierarchical cumulant formulation in Sec. III. The second-order cumulant analysis represents a natural equivalent linearization when the third-order moment is nonzero, and successive cumulant analyses up to the eighth-order are carried out in Sec. III.B. In contrast, applying equivalent linearization in the usual manner calls for certain Gaussian assumptions that cannot be justified other than by the end (Sec. IV). Besides the variance, the root mean square (rms) strain and stress requires the knowledge of the fourth-order moment (Sec. V) that the equivalent linearization procedure cannot provide explicitly.

II. Stationary Fokker–Planck Distribution

Under the restriction of $f_o = 0$ assumed throughout this paper, the potential energy is symmetric and so is its trough (Fig. 1). Note that this symmetry will carry over to the distribution function. Using the conjugate variables $x_1 = x$ and $x_2 = \dot{x}$, Eq. (6) can be put in the Ito stochastic differential equation

$$\begin{pmatrix} \mathrm{d}x_1 \\ \mathrm{d}x_2 \end{pmatrix} = \begin{pmatrix} x_2 \\ -\beta x_2 - K x_1 - 3\alpha Q x_1^2 - \alpha x_1^3 \end{pmatrix} + \begin{pmatrix} 0 \\ \mathrm{d}W(t) \end{pmatrix} \tag{7}$$

together with the increments of the Wiener process $W(t)$ obeying $<\mathrm{d}W(t)> = 0$ and $<\mathrm{d}W(t)\mathrm{d}W(s)> = 2D\delta(t-s)$, where $< >$ denotes ensemble average and D is the constant power input. The Fokker–Planck equation for distribution function $P(x_1, x_2)$ is given in standard texts[10,11]

$$\frac{\partial P}{\partial t} = -\frac{\partial (x_2 P)}{\partial x_1} + \frac{\partial}{\partial x_2}(\beta x_2 + K x_1 + 3\alpha Q x_1^2 + \alpha x_1^3)P + D\frac{\partial^2 P}{\partial x_2^2} \tag{8}$$

It is known that stationary solution ($\partial P/\partial t = 0$) has the form $P = \exp(-\beta H/D)$, where the Hamiltonian H is sum of the kinetic and potential energies.[12] Since x_1 and x_2 are independent, one may integrate out x_2 and write $P = \exp(-\beta U/D)$, where the potential energy $U = Kx^2/2 + \alpha Q x^3 + \alpha x^4/4$ is written by dropping the subscript.

With such a Fokker–Planck distribution, we can compute moments $M_n = \int_{-\infty}^{\infty} P(x)\, x^n \, \mathrm{d}x$ of all order n, which can be expressed in a dimensionless integral form by $x = y(2D/\alpha\beta)^{1/4}$

$$M_n = \left(\frac{2D}{\alpha\beta}\right)^{\frac{n+1}{4}} \int_{-\infty}^{\infty} P(y)\, y^n \, \mathrm{d}y, \quad (n \geq 0) \tag{9}$$

where $P(y) = \exp\{-(zy^2 + \sqrt{2}z y^3 + y^4/2)\}$. Here, the parameter $z = (s-1)\sqrt{2}k_o/\sigma_o^2\alpha$ is formed by $K = 2(s-1)k_o$, $Q = Q_{2,0}$, and $\sigma_o^2 = D/\beta k_o$. Being proportional to $(s-1)$, it measures plate temperature above the critical buckling temperature. At the critical buckling temperature ($z = 0$) the Fokker–Planck distribution has a single peak at $y = 0$ (Fig 2). However, for $z > 0$ the Fokker–Planck distribution becomes bimodal; one peak at $y = 0$ and the other peak at $y = -\sqrt{2}z$. Note that the midpoint of two peaks at $y = -\sqrt{z/2}$ moves away from $y = 0$ as z increases. In fact, transformation (5) is superfluous in the present case of $f_o = 0$, so that one may formulate the Fokker–Planck equation directly from Eq. (1). It is interesting to point out that the emergence of symmetric bimodal Fokker–Planck distribution due to the negative linear stiffness was first discussed by Andronov, Vitt and Pontryagin[13] in 1933, and the reciprocal relationship between the potential energy and Fokker–Planck distribution has been depicted in Fig. 6.6 of Haken's book.[14]

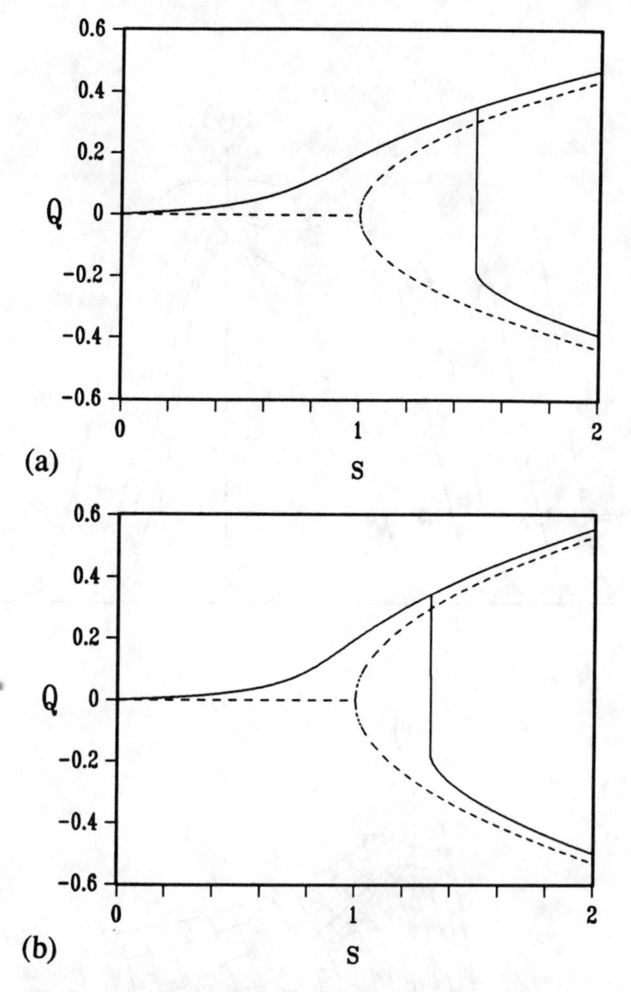

Fig. 1 Trough of potential energy surface, – – – – Q_1 and Q_2, and ——— Q_3 and Q_4: a) simply supported plate, and b) clamped plate.

After shifting to the midpoint of two peaks by $y = r - \sqrt{z/2}$, we obtain symmetric moment integral expression

$$M_n = \left(\frac{2D}{\alpha\beta}\right)^{\frac{n+1}{4}} \int_{-\infty}^{\infty} P(r)\left(r - \sqrt{\frac{z}{2}}\right)^n \mathrm{d}r, \quad (n \geq 0) \tag{10}$$

where $P(r) = \exp\{-\frac{1}{2}(r^2 - \frac{z}{2})\}$. By denoting $I_n = \int_{-\infty}^{\infty} P(r)\, r^n\, \mathrm{d}r$, the mean $m_1 = M_1/M_0$ is given by

$$m_1 = -\left(\frac{2\sigma_o^2 k_o}{\alpha}\right)^{1/4} \sqrt{\frac{z}{2}} \tag{11}$$

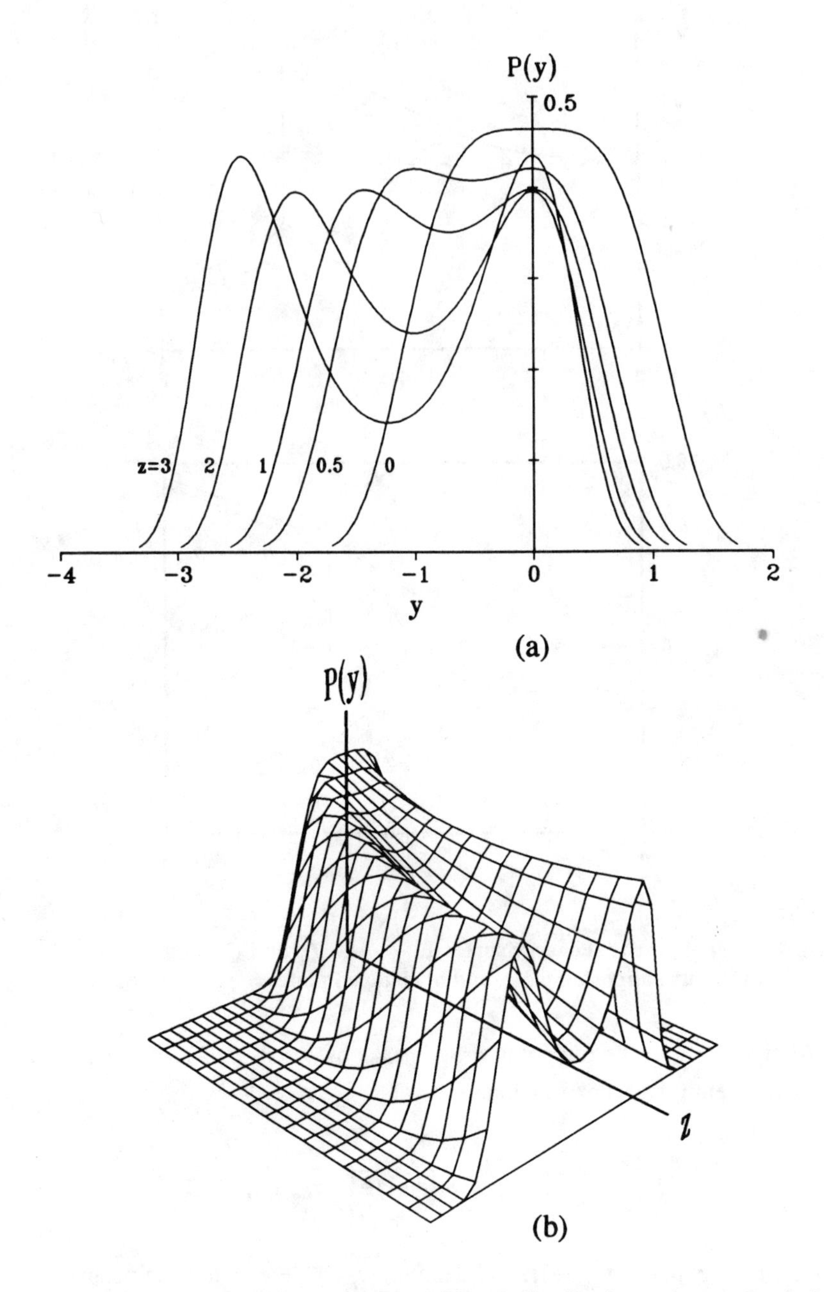

Fig. 2 Development of bimodal Fokker–Planck distributions: a) some typical z values, and b) surface plot for z=(0, 3).

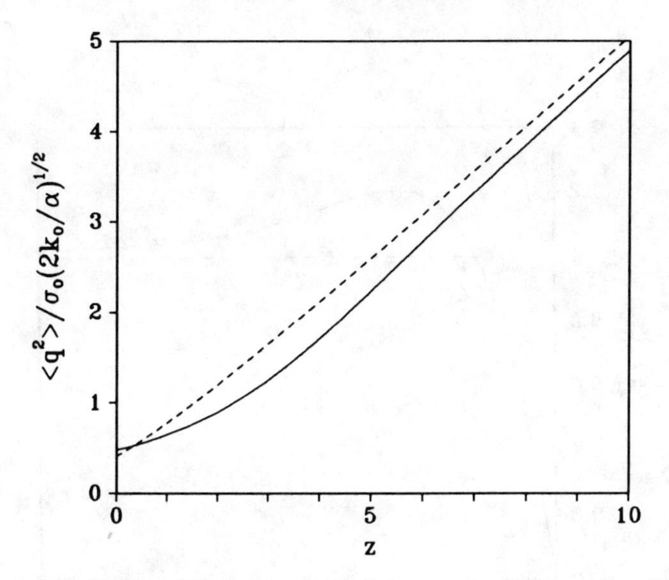

Fig. 3 Dimensionless second-order moment, ——— Fokker–Planck, and - - - - equivalent linearization.

in view of $I_1/I_0 = -\sqrt{z/2}$ and $2D/\alpha\beta = 2\sigma_o^2 k_o/\alpha$. Similarly, we have for moments $m_n = M_n/M_0$ of any order n

$$
\begin{aligned}
m_2 &= J_2 + m_1^2 \\
m_3 &= 3m_1 J_2 + m_1^3 \\
m_4 &= J_4 + 6m_1^2 J_2 + m_1^4, \ldots
\end{aligned}
\tag{12}
$$

where $J_n = (2\sigma_o^2 k_o/\alpha)^{n/4} I_n/I_0$. Now, using the relationships between m_n and the central moments μ_n (e.g., Eq (15.4.4) of Cramer[15])

$$
\begin{aligned}
\mu_2 &= m_2 - m_1^2 \\
\mu_3 &= m_3 - 3m_1 m_2 + 2m_1^3 \\
\mu_4 &= m_4 - 4m_1 m_3 + 6m_1^2 m_2 - 3m_1^4, \ \ldots
\end{aligned}
\tag{13}
$$

we find that the odd μ_n $(n \geq 3)$ are identically zero and the even moments are $\mu_n = J_n$ $(n \geq 2)$.

Although we now have the moments of Eq. (6), our original problem is to compute the total moments by averaging both sides of Eq. (5)

$$
\begin{aligned}
<q> &= Q + m_1 \\
<q^2> &= Q^2 + 2Qm_1 + m_2
\end{aligned}
$$

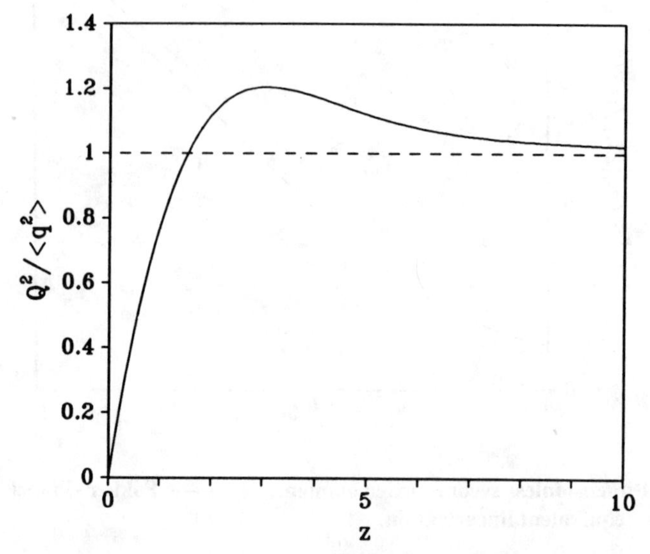

Fig. 4 Ratio $Q^2/<q^2>$ approaching unity.

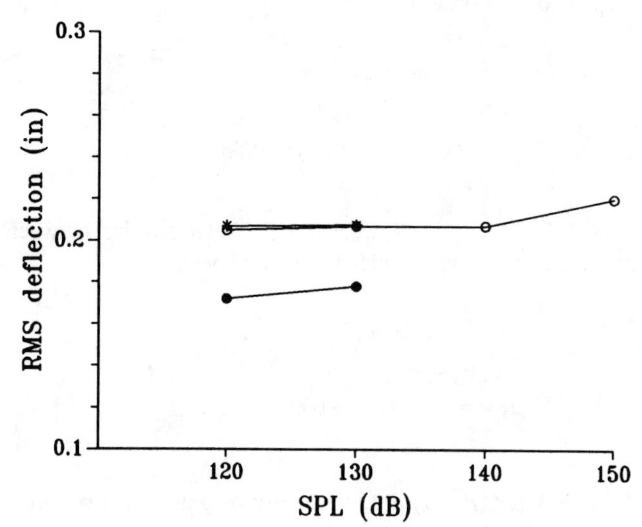

Fig. 5 Dependence of rms displacements on the sound pressure level: • h=0.0416 in., * h=0.0624 in., and o h=0.0832 in.; taken from Figs. 10–11, 15, and 33 of Ref. 16 for DT=1000°F.

$$<q^3> = Q^3 + 3Q^2 m_1 + 3Qm_2 + m_3$$
$$<q^4> = Q^4 + 4Q^3 m_1 + 6Q^2 m_2 + 4Qm_3 + m_4 \qquad (14)$$

In terms of z, we see at once that $Q = -m_1$. In words, the snap-through displacement Q is given by $-m_1$, which is the midpoint of two peaks in Fig. 2. Consequently, we have $<q> = 0$ from the first of Eq. (14) and the second of Eq. (14) gives rise to $<q^2> = J_2$. Similarly,

$$<q^n> = J_n \text{ (even } n\text{)}, \; <q^n> = 0 \text{ (odd } n\text{)} \qquad (15)$$

Here, J_n are obtained by the numerical evaluation of I_n.

Because of $Q = -m_1$, the dimensionless snap-through displacement is simply given by $Q/(2\sigma_o^2 k_o/\alpha)^{1/4} = \sqrt{z/2}$. We present in Fig. 3 the dimensionless variance $<q^2>/(2\sigma_o^2 k_o/\alpha)^{1/2}$ which increases gradually for small z but then increases linearly like z for $z > 8$. We, therefore, infer that $<q^2>/(2\sigma_o^2 k_o/\alpha)^{1/2} \approx Q^2/(2\sigma_o^2 k_o/\alpha)^{1/2} \propto z/2$ for large z, as also substantiated in Fig. 4 by the ratio $Q^2/<q^2>$ approaching unity as $z \to \infty$. In view of $z = (s-1)\sqrt{2k_o/\sigma_o^2\alpha}$, we then conclude for large z that

$$<q^2> \approx Q^2 \approx (s-1)(k_o/\alpha) \qquad (16)$$

In other words, $<q^2>$ and Q^2 are not only identical for large z but, more important, independent of the random forcing σ_o^2. This asymptotic independence-upon-forcing can be checked out by the numerical simulation of Vaicaitis[16] who has computed rms displacements of randomly excited composite plates over a wide range of plate temperatures. Compiled in Fig. 5 are the rms deflections of three composite plates with different thicknesses at ΔT=1000°F, where ΔT is uniform plate temperature rise above the room temperature. Using the critical buckling temperatures estimated in Ref. 16, ΔT=1000°F corresponds to $s \approx 123$, 45, and 24 for the three composite plates of thicknesses h= 0.0416, 0.0624, and 0.0832 in., respectively. At any rate, these s values are large enough to validate asymptotic high-temperature relation (16), as evidenced by the rms deflections of the time histories which are indeed insensitive to the forcing sound pressure level in Fig. 5. Though qualitative, such a validation is quite remarkable in that Eq. (16) was deduced from a single-mode analysis. In contrast, the numerical simulation of Vaicaitis[16] involves 12, 8, and 6 modes for the composite plates of h=0.0416, 0.0624, and 0.0832 in., respectively.

For a further comparison, we first show in Fig. 6 the dimensionless $<q^2>^{1/2}$ which increases as $\sqrt{z}$ for large z. We have reproduced in Fig. 7 for the convenience of readers, the rms displacement of the composite plate of h=0.0632 in. over ΔT up to 1000°F. Note that the z in Fig. 6 includes both the structure

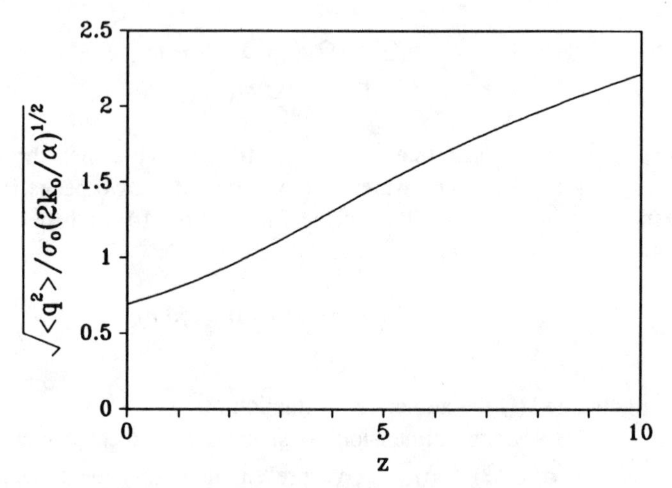

Fig. 6 Square root of Fokker–Planck variance.

parameter $\sqrt{2k_o/\sigma_o^2\alpha}$ and temperature rise $(s - 1)$, whereas the ΔT in Fig. 7 refers only to plate temperature rise. Hence, Figs. 6 and 7 cannot be compared quantitatively. Although it appears in Fig. 6 that $<q^2>^{1/2}$ increases roughly linearly for small z, we know that $<q^2>^{1/2} \sim \sqrt{z}$ for large z. Hence, the rms deflection in Fig. 7 should increase as $\sqrt{\Delta T}$ in the limit as $\Delta T \to \infty$, and not as linearly as one might infer from the four sampling points in the figure.

III. Cumulant Analyses based on the Abridged Edgeworth Series

The practical goal of stochastic dynamic analysis is to estimate the lowest order moments of, say, the mean and variance directly from the equation of

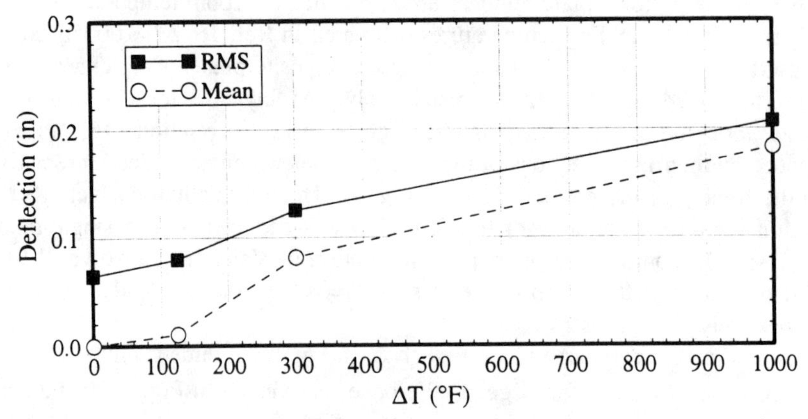

Fig. 7 Root mean square displacements of the composite plate of thickness *h*= 0.0632 in.; reproduction of Fig. 15 for SPL=120 dB in Ref. 16.

motion. One such method is the equivalent linearization technique to be discussed in Sec. IV, the success of which is hinged on departure from the Gaussian condition. Since the departure can best be quantified by the cumulants, it is natural to consider the Gram–Charlier or Edgeworth series expansion.[17-19]

We begin with the Hermite polynomial expansion of an arbitrary distribution[15,20]

$$\wp(x) = \left\{1 + \sum_{n=3}^{\infty} \frac{c_n}{n!} H_n\left(\frac{x-m}{\sigma}\right)\right\}\Phi\left(\frac{x-m}{\sigma}\right) \tag{17}$$

where

$$\Phi\left(\frac{x-m}{\sigma}\right) = \frac{1}{\sigma\sqrt{2\pi}} \exp\left\{-\tfrac{1}{2}\left(\frac{x-m}{\sigma}\right)^2\right\}$$

is the Gaussian distribution with mean m and variance σ^2, and the Hermite polynomials are $H_0 = 1$, $H_1 = x$, $H_2 = x^2 - 1$, $H_3 = x^3 - 3x$, $H_4 = x^4 - 6x^2 + 3, \ldots$ (Note that the use of x here is not related to Sec. II). Expansion (17) with the coefficients c_n expressed in central moments is the Gram-Charlier expansion (Eq. (6.32) of Kendall and Stuart[20]). On the other hand, let us express c_n in cumulants κ_n

$$\begin{gathered} c_n = \kappa_n / \sigma^n \ (n = 3,\ 4,\ 5), \quad c_6 = (\kappa_6 + 10\kappa_3^2)/\sigma^6 \\ c_7 = (\kappa_7 + 10\kappa_7\kappa_3)/\sigma^7, \quad c_8 = (\kappa_8 + 10\kappa_5\kappa_3 + 35\kappa_4^2)/\sigma^8, \ldots \end{gathered} \tag{18}$$

(Eq (6.41) of Kendall and Stuart[20]). Then, expansion (17) with cumulant coefficients (18) is the Edgeworth series. Note that the abridgment proposed in Ref. 4 omits the quadratic cumulant terms in Eq. (18). Furthermore, since the odd cumulants vanish identically for a symmetric distribution, the abridged Edgeworth series for the present problem is

$$\mathrm{P}(x) = \left\{1 + \sum_{n=4,6,8,\ldots}^{\infty} \frac{\kappa_n}{n!\sigma^n} H_n\left(\frac{x-m}{\sigma}\right)\right\}\Phi\left(\frac{x-m}{\sigma}\right) \tag{19}$$

Since $\wp(x)$ and $\mathrm{P}(x)$ are identical up to the sixth-order term, the abridgment shows up first in the eighth-order term for the present problem. We now compute the first several moments of $\mathrm{P}(x)$

$$\begin{gathered} m_1 = m, \quad m_2 = \sigma^2 + m^2, \quad m_3 = 3\sigma^2 m + m^3 \\ m_4 = \kappa_4 + 3\sigma^4 + 6\sigma^2 m^2 + m^4 \\ m_5 = 5\kappa_4 m + 15\sigma^4 m + 10\sigma^2 m^3 + m^5 \\ m_6 = \kappa_6 + 15\sigma^6 + 15\kappa_4(\sigma^2 + m^2) + 45\sigma^4 m^2 + 15\sigma^2 m^4 + m^6, \ldots \end{gathered} \tag{20}$$

Noting that $m = \kappa_1$ and $\sigma^2 = \kappa_2$, Eq. (20) expresses moments in terms of the cumulants, according to abridged Edgeworth series (19).

A. Stationary Moment Equations

From Eq. (7) we have the hierarchical moment equations as derived in Appendix B

$$
\begin{aligned}
-Km_1 - 3\alpha Q m_2 - \alpha m_3 &= 0, \\
D/\beta - Km_2 - 3\alpha Q m_3 - \alpha m_4 &= 0, \\
2(D/\beta)m_1 - Km_3 - 3\alpha Q m_4 - \alpha m_5 &= 0, \\
3(D/\beta)m_2 - Km_4 - 3\alpha Q m_5 - \alpha m_6 &= 0, \ldots
\end{aligned} \tag{21}
$$

where $m_n = \langle x_1^n \rangle$. Clearly, Eq. (21) is indeterminate because there are two more unknown moments than the number of equations. Hence, the intent of cumulant analysis is to bring about a closure approximation with the aid of Eq. (20). To do this simply, we introduce the nondimensional $\overline{m}_n = m_n/(2\sigma_o^2 k_o/\alpha)^{n/4}$, $X = \sigma^2/(2\sigma_o^2 k_o/\alpha)^{1/2}$, and $\lambda_n = \kappa_n/(2\sigma_o^2 k_o/\alpha)^{n/4}$ into Eqs. (20–21) and obtain

$$-z\overline{m}_1 - 3\sqrt{z/2}\,\overline{m}_2 - \overline{m}_3 = 0 \tag{22a}$$

$$\frac{1}{2} - z\overline{m}_2 - 3\sqrt{z/2}\,\overline{m}_3 - \overline{m}_4 = 0 \tag{22b}$$

$$\overline{m}_1 - z\overline{m}_3 - 3\sqrt{z/2}\,\overline{m}_4 - \overline{m}_5 = 0 \tag{22c}$$

$$\frac{3}{2}\overline{m}_2 - z\overline{m}_4 - 3\sqrt{z/2}\,\overline{m}_5 - \overline{m}_6 = 0, \ldots \tag{22d}$$

together with

$$
\begin{aligned}
&\overline{m}_1 = \overline{m}, \quad \overline{m}_2 = X + \overline{m}^2, \quad \overline{m}_3 = 3X\overline{m} + \overline{m}^3 \\
&\overline{m}_4 = \lambda_4 + 3X^2 + 6X\overline{m}^2 + \overline{m}^4 \\
&\overline{m}_5 = 5\lambda_4\overline{m} + 15X^2\overline{m} + 6X\overline{m}^3 + \overline{m}^5 \\
&\overline{m}_6 = \lambda_6 + 15X^3 + 15\lambda_4(X + \overline{m}^2) + 45X^2\overline{m}^2 + 15X\overline{m}^4 + \overline{m}^6, \ldots
\end{aligned} \tag{23}
$$

B. Sequential Cumulant Solutions

First, inserting Eq. (23) into Eq. (22a) yields

$$z\overline{m} + 3\sqrt{z/2}(X\overline{m} + \overline{m}^2) + 3X\overline{m} + \overline{m}^3 = 0 \tag{24}$$

from which we find

$$\overline{m} = -\sqrt{\frac{z}{2}} \tag{25}$$

in agreement with Eq. (11). Next, Eq. (22b) together with Eq. (25) give

$$3X^2 - \frac{z}{2}X - \frac{1}{2} + \lambda_4 = 0 \tag{26}$$

However, it is easily checked that Eq. (22c) is redundant to Eq. (26), but Eq. (22d) yields a cubic equation

$$15X^3 + 3zX^2 - \left(\frac{3}{2} + \frac{3z^2}{4} - 15\lambda_4\right)X - \frac{3z}{4} + z\lambda_4 + \lambda_6 = 0 \qquad (27)$$

Similarly, from the moment equation $5\overline{m}_4/2 - z\overline{m}_6 - 3\sqrt{z/2}\,\overline{m}_7 - \overline{m}_8 = 0$ which is two orders beyond that presented in Eq. (22), we obtain the following quartic equation

$$105X^4 + \frac{135z}{2}X^3 - \left(\frac{15}{2} + \frac{15z^2}{4} - 210\lambda_4\right)X^2 - \left(\frac{15z}{2} + \frac{5z^3}{8} - \frac{135z}{2}\lambda_4 - 28\lambda_6\right)X$$
$$-\frac{5z^2}{8} - \frac{5}{2}\left(1 + \frac{z^2}{2}\right)\lambda_4 + \frac{9z}{2}\lambda_6 + \lambda_8 = 0 \qquad (28)$$

with the use of $\overline{m}_7$ and $\overline{m}_8$ (Appendix C). Note that the effect of the abridged Edgeworth series will first show up in Eq. (28), whereas Eqs. (26) and (27) are not at all affected by the abridgment.

There are in all three equations (26–28) but for four unknowns X, λ_4, λ_6, and λ_8 -- a closure problem of the stochastic nonlinear dynamics.[21] We shall, therefore, suppress the highest order cumulant at each level of the closure approximation. For the lowest order closure, Eq. (26) under $\lambda_4 = 0$ yields $X = [z + (z^2 + 24)^{1/2}]/12$ which reduces to $X = z/6$ for large z. As shown in Fig. 8, this falls well below the Fokker–Planck variance $X = z/2$ for large z. Since Eq. (26) represents the second-order cumulant analysis, it should be considered as a natural equivalent linearization as opposed to the brute-forced one to be presented in Sec. IV, when the mean is nonzero. Now, for the fourth-order cumulant analysis we consider Eqs. (26–27) under $\lambda_6 = 0$, and Eqs. (26-28) are solved under $\lambda_8 = 0$ for the sixth-order cumulant analysis. Summarized in Fig. 8 are variance estimates by the second- through eighth-order cumulant analyses, approaching the Fokker–Planck variance from below as we include more and more of the higher order cumulants in the closure truncation. It is seen from Fig. 8 that the sixth-order cumulant analysis can give an accurate variance estimate for $z < 1$. But, for $z > 4$ even the eighth-order cumulant analysis cannot approximate the Fokker–Planck variance within 10%. Figure 8 clearly indicates that the moment formulation converges very slowly at best for small z but is totally inadequate at worst for a high-temperature buckled plate at large z. This difficulty at high temperature is not unexpected in that many terms are needed in the Edgeworth series to approximate a bimodal distribution. To be specific, we have displayed in Fig. 9 the successive development of the abridged Edgeworth distributions for $z = 2$. We note that the second- through sixth-order cumulant analyses simply lower the central peak of the Gaussian basis form, and it is the eighth-order cumulant analysis that first begins to generate bimodality. As shown in Table 1, the eighth-order cumulant

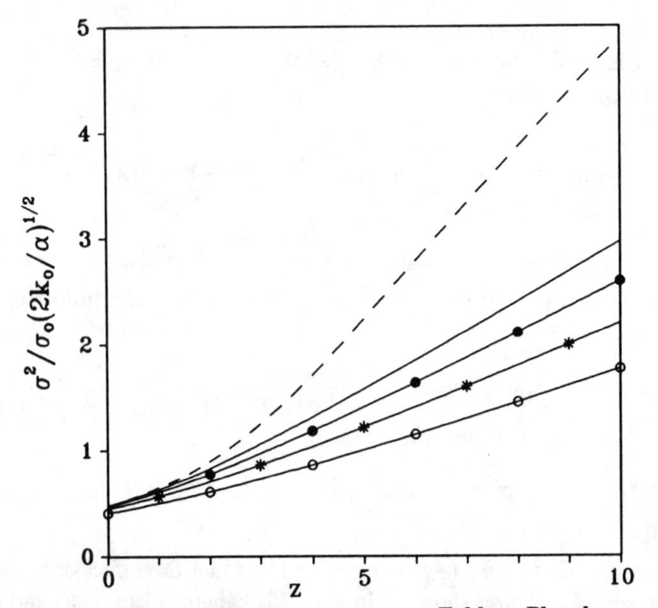

Fig. 8 Successive variance estimations: – – – – Fokker–Planck, ——— eighth-order cumulant analysis, —o— sixth-order cumulant analysis, —*— fourth-order cumulant analysis, and —•— second-order cumulant analysis.

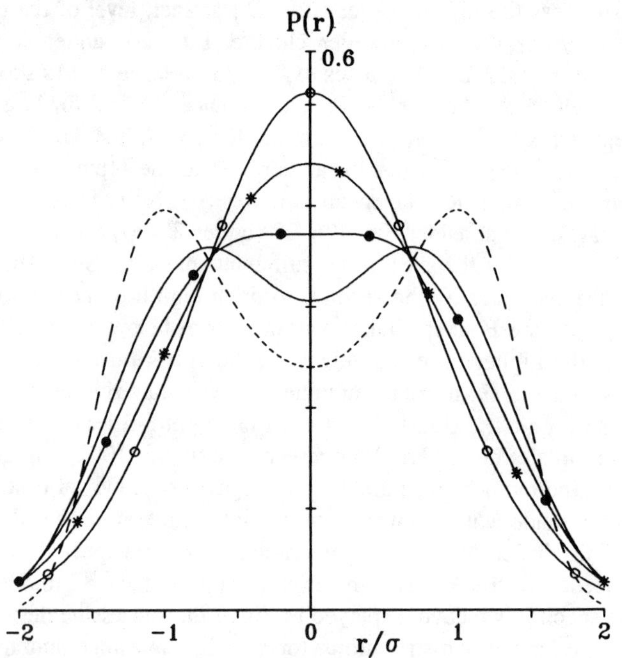

Fig. 9 Abridged Edgeworth series distributions at z=2: – – – – Fokker–Planck, ——— eighth-order cumulant analysis, —o— sixth-order cumulant analysis, —*— fourth-order cumulant analysis, and —•— second-order cumulant analysis.

Table 1 Variance and cumulant ratios at z=2

System	X	κ_4/σ^4	κ_6/σ^6	κ_8/σ^8
Fokker–Planck distribution	0.893	–1.254	7.649	–101.06
Eighth-order cumulant analysis	0.827	–1.060	5.433	–22.04
Sixth-order cumulant analysis	0.774	–0.875	3.372	- - -
Fourth-order cumulant analysis	0.703	–0.567	- - -	- - -
Second-order cumulant analysis	0.608	- - -	- - -	- - -

analysis does not give κ_6 and κ_8 which are large enough in absolute magnitude to adequately capture the bimodal Fokker–Planck distribution.

In spite of the inherent difficulty in moment formulation, the unfolding of the Fokker–Planck distributions depicted in Fig. 2 offers simple moment expressions which are asymptotically valid in the limit as $z \to \infty$. We notice that the two peaks of the Fokker–Planck distribution move away from each other and become sharply peaked as z increases. Eventually, the Fokker–Planck distribution degenerates into two delta functions of 1/2 magnitude located at $r = \pm\overline{m}$; hence, $\overline{m}_n = \int_{-\infty}^{\infty} \delta(r \mp \overline{m}) r^n \mathrm{d}r = \overline{m}^n$ for even n as $z \to \infty$. In words, moments are given by the nth powers of static displacement alone. For instance, $X = z/2$ as we have already seen, and the fourth-order moment is $\overline{m}^4 = (z/2)^2$ in the high-temperature limit. This will be used in Sec. V for the asymptotic estimation of the rms strain and stress.

IV. Brute-Forced Application of the Equivalent Linearization Procedure

The so-called equivalent linearization[22] begins by replacing Eq. (6) with a surrogate linear oscillator equation

$$\ddot{x} + \beta \dot{x} + K_e x = f(t) \tag{29}$$

where the equivalent stiffness K_e is yet to be determined. Implicit in Eq. (29) is the assumption that $<x> = 0$ under the zero-mean forcing. The deviation of Eq. (29) from Eq. (6) is denoted by the error term $e = (K - K_e)x + 3\alpha Q x^2 + \alpha x^3$ which we wish to minimize. By the usual argument, we then arrive at

$$K_e = K + \frac{3\alpha Q <x^3>}{<x^2>} + \frac{\alpha <x^4>}{<x^2>} \tag{30}$$

which is still useless due to the third- and fourth-order moments. To make a further progress, we assume that the third-order moment vanishes $<x^3> = 0$ and the fourth- and second-order moments are related by $<x^4> = 3<x^2>^2$. Then Eq. (30) reduces to the familiar result $K_e = K + 3\alpha <x^2>$ of equivalent linearization. Now, inserting K_e into the input–output variance relationship[23]

$<x^2> = D/\beta K_e$ for Eq. (29), we obtain the quadratic equation

$$3(\alpha/k_o)<x^2>^2 + 2(s-1)<x^2> - \sigma_o^2 = 0 \tag{31}$$

Although Eq. (31) was derived in a roundabout fashion, one finds that it is just the second equation of Eq. (21), if $m_3 = 0$ and $m_4 = 3m_2^2$ are invoked and then $m_2 = <x^2>$ is substituted. In terms of X and z, Eq. (31) becomes $3X^2 + zX - 1/2 = 0$ with the positive root $X_{el} = [(z^2+6)^{1/2} - z]/6$, where the subscript el denotes equivalent linearization. This should be compared with the root of Eq. (26) under $\kappa_4 = 0$, which was earlier termed natural equivalent linearization. That is, we have here $X_{el} \to 0$ as $z \to \infty$, which is at odds with $X \to z/6$ of Eq. (26) under the same limit. This is attributed to the assumption $<x> = <x^3> = 0$

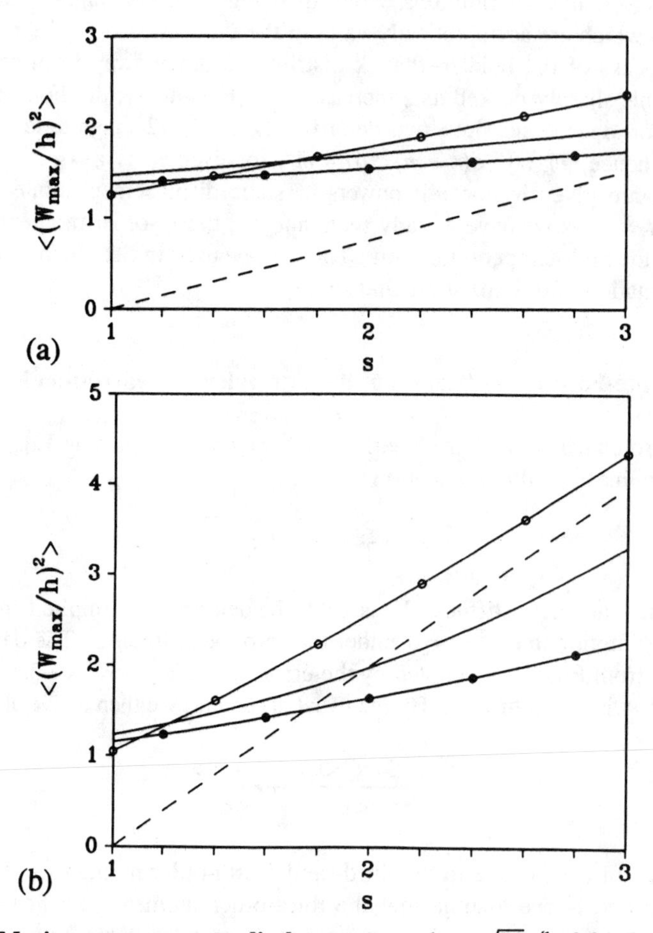

Fig. 10 Maximum mean square displacement, $\gamma = 1$, $\mu = \sqrt{0.1}$, $\xi = 0.04$, $D = 1/2$; ——— Fokker–Planck, —o— equivalent linearization, and —•— fourth-order cumulant analysis: a) simply supported plate, and b) clamped plate.

invoked by the zero-mean Gaussian statistics. Hence, the equivalent linearization procedure presented here is a brute-forced attempt.

Now, for the total second-order moment, averaging both sides of Eq. (5) yields $<q^2>_{el} = Q^2 + <x^2>_{el}$ under the assumption $<x>_{el}=0$, hence

$$\frac{<q^2>_{el}}{\sqrt{2\sigma_o^2 k_o/\alpha}} = \frac{z}{2} + X_{el} \tag{32}$$

In Fig. 3 we compare $<q^2>_{el}$ with the Fokker–Planck variance. Because of the factor $z/2$, $<q^2>_{el}$ lies above the Fokker–Planck for all but the very small z in the figure. The maximum error in the total second-order moment by equivalent linearization is about 33.6% overestimation at $z \approx 2$ and then decreases gradually to about 3.2% at $z = 10$. Since $X_{el} \to 0$ as $z \to \infty$, Eq. (32) observes the correct high-temperature moment relation $<q^2>_{el} \approx Q^2$. This is indeed a fortuitous accident, although the assumption $<x> = <x^3> = \kappa_4 = 0$ cannot be justified other than by the end.

V. Some Plate Examples

We have heretofore expressed the dimensionless moments of both simply supported and clamped plates by a single parameter $z = (s-1)\sqrt{2k_o/\alpha\sigma_o^2}$. This unified representation was possible because of the restriction $f_o = 0$. For the parameter values $\gamma = 1$, $\mu = \sqrt{0.1}$, $\xi = 0.04$, and $D = 1/2$ used here, we have $\sqrt{2k_o/\alpha\sigma_o^2} \approx 0.49$ and 1.55 for the simply supported and clamped plates, respectively. Hence, for a given s a clamped plate will have z about three times larger than the simply supported plate. We shall first compare the maximum mean square displacements estimated by the equivalent linearization technique (ELT), Fokker–Planck distribution, and fourth-order cumulant analysis (FCA). Further, a similar comparison will be made for the rms strain and stress distributions.

Maximum Mean Square Displacements

Upon multiplying the variance by the factors 4 and 64/9 for simply supported and clamped plates [Eq. (59) of Ref. 1], we present in Fig. 10 the maximum mean square transverse displacement $<(W_{max}/h)^2>$, where h is the plate thickness. Note that the mean square displacements of the ELT are those already presented in Ref. 1. For the simply supported plate, Fig. 10a shows that FCA gives a better displacement estimate (−12.5%) than the ELT (+19.6%) over the range of $s = 1$–3. However, the Fokker–Planck displacement for a clamped plate is underestimated (−31.8%) by FCA and overestimated just as much (+31.4%) by the ELT (Fig. 10b). This clearly shows that the FCA is also inadequate for a clamped plate for $s = 1$–3. However, in view of Fig. 8, a higher order cumulant analysis could improve the variance estimate when the thermal load is large. However, such will not be attempted here.

Root Mean Square Strain and Stress Distributions

For a square plate ($\gamma = 1$), the normal strain ($\varepsilon_x, \varepsilon_y$) and stress ($\sigma_x, \sigma_y$) are symmetric in the x and y coordinates, so that it suffices to consider only the ε_x and σ_x in (x/a) at $(y/b) = 1/2$. (Note that x and y have been used previously in a different context.) The normal strain and stress can be put in a symbolic form

$$S = C_0 + C_1 q + C_2 q^2 \tag{33}$$

where $S = \varepsilon_x b^2/\pi^2 h^2$ or $\sigma_x b^2/E\pi^2 h^2$ (E is the modulus of elasticity) and C_i are given in Appendix D for the simply supported and clamped plates. By squaring and averaging, we obtain

$$< S^2 > = C_0^2 + 2C_0 C_1 < q > + (C_1^2 + 2C_0 C_2) < q^2 > + 2C_1 C_2 < q^3 > + C_2^2 < q^4 > \tag{34}$$

Upon suppressing the odd moments by using Eq. (15), Eq. (34) simplifies to

$$< S^2 > = C_0^2 + (C_1^2 + 2\,C_0 C_2) < q^2 > + C_2^2 < q^4 > \tag{35}$$

Strictly speaking, Eq. (35) should not be used under the ELT, for Eq. (15) is not valid. However, to proceed formally we shall assume that Eq. (35) is applicable in equivalent linearization and, hence, further relate the second- and fourth-order moments by the Gaussian statistics. Hence, a parallel expression

$$< S^2 > = C_0^2 + (C_1^2 + 2\,C_0 C_2) < q^2 >_{\mathrm{el}} + 3\,C_2^2 < q^2 >_{\mathrm{el}}^2 \tag{36}$$

will be adopted when only the variance is computed from the equivalent linearization procedure.

For the thermal parameters $T_o = 2$, $\delta_v = 0.2$, and $\delta_g = 0$ (defined in Appendix A) which give rise to $s \approx 2$ for both simply supported and clamped plates, we compute the mean square strain and stress by using either Eq. (35) or Eq. (36), depending on whether moments are evaluated by the Fokker–Planck distribution and FCA or the ELT. The rms strain and stress distributions are pre-

Table 2 Maximum over (+) and under (–) estimations

Plate	s	z	Strain, stress, or variance	ELT, %	FCA, %
Simply supported	2.03	0.51	$<\varepsilon_x>$	+27.4	–3.3
			$<\sigma_x>$	+42.2	–2.4
			$<q^2>$	+6.3	–9.9
Clamped	2.06	1.64	$<\varepsilon_x>$	+64.5	–8.8
			$<\sigma_x>$	+58.4	–6.3
			$<q^2>$	+30.5	–16.8

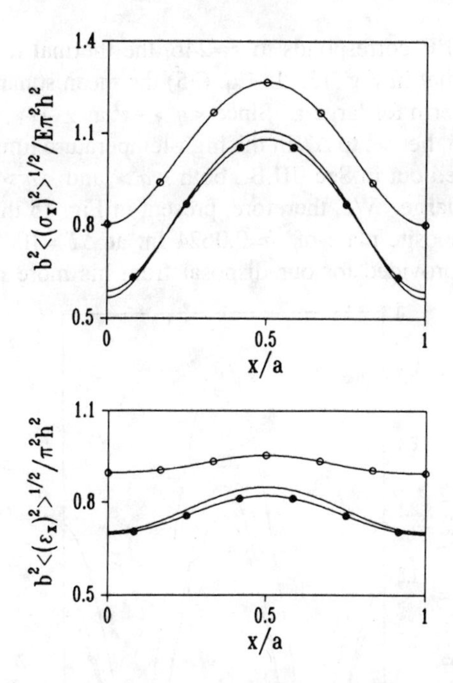

Fig. 11 Root mean square extreme-fiber strain and stress distributions on a simply supported plate, γ =1, $\mu=\sqrt{0.1}$, ξ =0.04, D=1/2, T_o =2, δ_v =0.2, δ_g =0: ——— Fokker–Planck, —o— equivalent linearization, and —•— fourth-order cumulant analysis.

sented in Figs. 11 and 12 for the simply supported and clamped plates, respectively, and the maximum deviations of ELT and FCA from the Fokker–Planck in the rms strain and stress are summarized in Table 2. In contrast to the variance estimates for which neither the ELT nor FCA has an edge (Fig. 10), Table 2 clearly indicates that the rms strains and stresses are overestimated by the ELT more so than underestimated by FCA. This is partially due to that both the Fokker–Planck approach and FCA yield $<q^4>$ which is smaller than $3<q^2>^2$. In fact, κ_4 is typically nonzero, and has the limiting value of −2 as $z\to\infty$. Hence, $<q^4>_{el}$ can be as large as three times the actual $<q^4>$.

High-Temperature Normal Stress

Figure 13 displays the maximum rms stress at the midplate location x/a = y/b =1/2 for both the simply supported and clamped plates. It appears that the maximum stress increases linearly in the figure, although the clamped plate has a much larger stress value than simply supported plate under the identical acoustic and thermal loading. This linear increase in the rms stress was also observed in the numerical simulation of Vaicaitis.[16] Shown in Fig. 14 for the convenience of readers is reproduction of Fig. 11 of Vaicaitis,[16] which depicts the maximum rms stress vs ΔT for the composite plate of h = 0.0416 in. Since the upper ΔT= 2000°F corresponds to $s \approx 246$, the thermal loading in Fig. 14 extends far beyond that in Fig. 13. In Eq. (35) the mean square stress is dominated by the $<q^4>$ term for large z. Since $<q^4> \sim z^2$ as $z \to \infty$, the rms stress is proportional to z and, hence, to ΔT in the high-temperature limit.

Finally, as pointed out in Sec. III.B., both $<q^4>$ and $<q^2>$ are independent of σ_o^2 as z becomes large. We, therefore, present in Fig. 15 the maximum rms stresses for the composite plate of h=0.0624 in. at ΔT =1000°F, which Vaicaitis[24] has kindly provided for our disposal from his more recent numerical

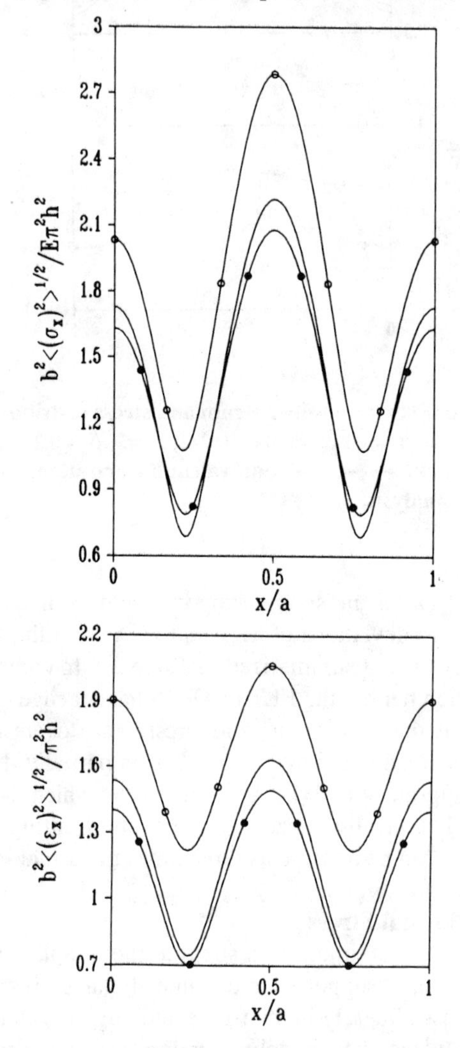

Fig. 12 Root mean square extreme-fiber strain and stress distributions on a clamped plate, γ =1, $\mu=\sqrt{0.1}$,ξ =0.04, D=1/2, T_o =2, δ_v =0.2, δ_g =0: ——— Fokker–Planck, ——o—— equivalent linearization, and ——•—— fourth-order cumulant analysis.

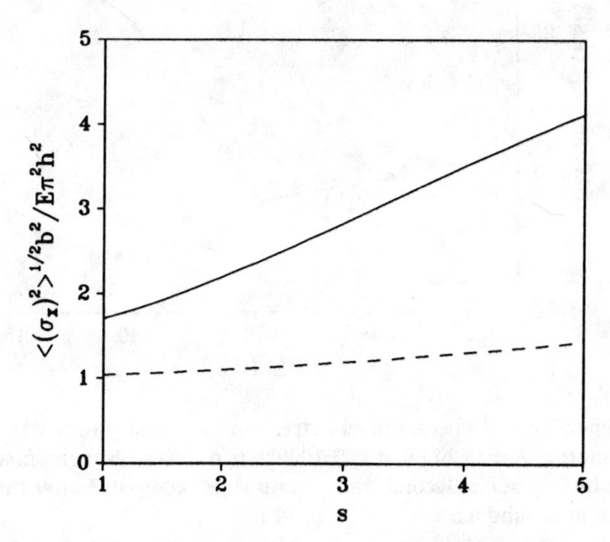

Fig. 13 Maximum rms extreme-fiber stress at the midplate location, $\gamma=1$, $\mu=\sqrt{0.1}$, $\xi=0.04$, $D=1/2$, $T_o=2$, $\delta_v=0.2$, $\delta_g=0$: – – – – simply supported plate, and ——— clamped plate.

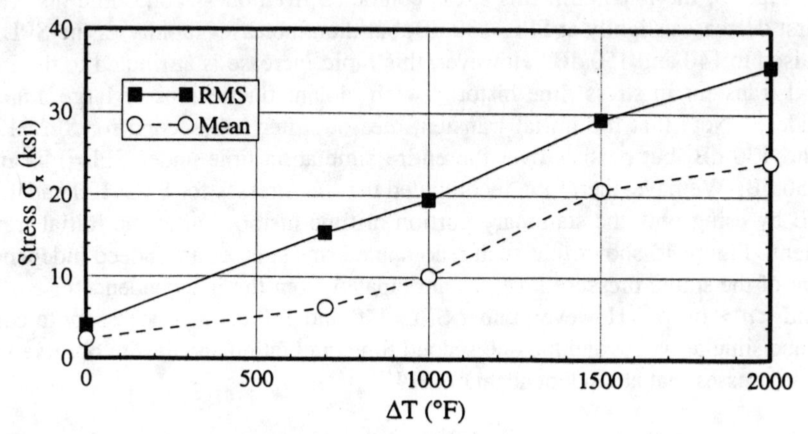

Fig. 14 Maximum rms stress of composite plate of thickness h=0.0416 in.; reproduction of Fig. 11 for SPL=130 dB in Ref. 16.

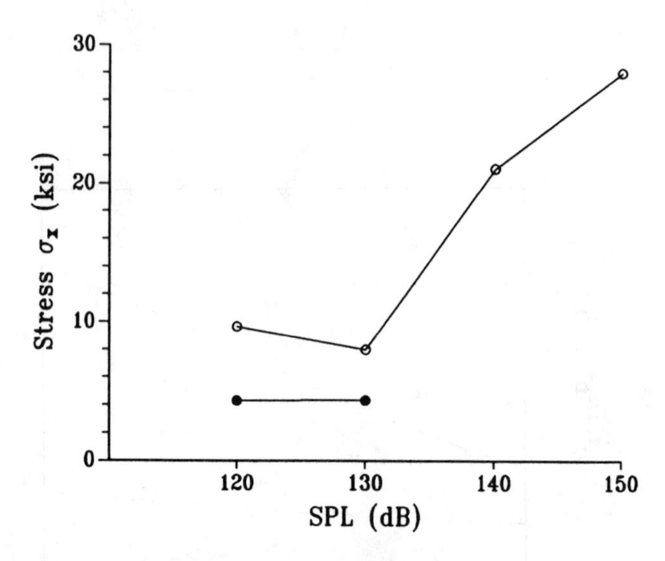

Fig. 15. Dependence of maximum rms stress on the sound pressure level for the composite plate of h=0.0624 in. at DT=1000°F: o rms stress computed from the entire time history of one second, and • rms stress computed after the transient has died out at around 0.5 s.

simulation carried out under the sound pressure level (SPL) over 120 to 150 dB. It must be pointed out that the rms stress is computed directly from a stress time history of one-second record length in Vaicaitis' simulation.[16] As shown in Fig. 15, the maximum rms stress computed from one-second time histories first decreases slightly at SPL=130 dB, but then increases rapidly as the SPL is raised to 140 and 150 dB. However, this rapid increase is attributed to the initial transient in stress time histories with violent fluctuations of large amplitudes. Note that the initial transient dies out after half second for SPL=120 and 130 dB, but persists over the entire simulation time under SPL=140 and 150 dB. We have, therefore, recomputed the rms stresses for SPL=120 and 130 dB by using only the stationary portion of time histories after the initial transient. Figure 15 shows that such recomputed rms stresses are indeed independent of the sound pressure level, as anticipated from the independence of $<q^2>$ and $<q^4>$ on σ_o^2. However, under SPL=140 and 150 dB it is necessary to continue simulation beyond the one-second time integration and thus to observe the rms stresses that are independent of SPL.

VI. Conclusions

The present investigation is restricted to f_o=0 which represents zero temperature gradient across the plate thickness. Owing to the small thermal conductivity of composites,[25] a large temperature gradient is more likely to be sustained in a composite plate than in a metallic plate. Hence, the general case of

$f_o \neq 0$ is under investigation. Under $f_o = 0$ the Fokker–Planck distribution is symmetric and, more important, the structural and thermal parameters become lumped into a single parameter z. The Fokker–Planck distribution is unimodal at $z = 0$ and becomes bimodal for $z > 0$. That is, it splits into two symmetric peaks as the plate temperature rises above the critical buckling temperature. The successive cumulant analysis is presented in Sec. III based on the abridged Edgeworth series proposed in Ref. 4. Although a lower order (e.g., the fourth-order) cumulant analysis can provide an adequate variance estimate for very small $z \ll 1$, the cumulant analysis converges very slowly as z becomes large. This is because a bimodal distribution requires many Hermite basis functions in the Edgeworth series expansion.

Surprisingly, the equivalent linearization gives a remarkably good estimate of the total second-order moment for large $z > 8$. However, this is not a feat of the equivalent linearization. In fact, it is a fortuitous accident that total moments are determined by the static snap-through displacement alone for large z, and thereby obscuring the fault of equivalent linearization. In any event, the equivalent linearization cannot generate the fourth-order moment, other than by the Gaussian approximation, which is needed for the computation of the mean square stress and strain. Hence, the rms stress and strain are greatly overestimated by the equivalent linearization for a simply supported and clamped plate.

Notwithstanding the closure difficulty in moment approximations, a simplifying picture emerges from the Fokker–Planck distribution which peaks out sharply at the static snap-through positions as z becomes very large. Hence, the asymptotic form of the moments involves the static displacement alone, independent of the forcing. That is, at high temperature the rms displacement and stress/strain become independent of the acoustic load. In particular, the rms displacement increases proportional to the square root of the plate temperature, and the rms stress and strain are linearly proportional to plate temperature above the critical buckling temperature. Although the asymptotic high-temperature moment behavior was inferred from a single-mode analysis, it has been confirmed by the multimode numerical simulation of composite plates over the range of sound pressure level of 120 -150 dB.[16]

Appendix A: Plate Parameters

For the simply supported plate we have

$$k_o = (\gamma^2+1)^2, \quad s = T_o[1+(1-\mu)\delta_v/8]$$
$$f_o = (\gamma^2+1)^2 \delta_g T_o/24, \quad \alpha = 3[(1-\mu^2)(\gamma^4+1)+2(\gamma^4+1+2\mu\gamma^2)]$$

and

$$k_o = 16(\gamma^4+2\gamma^2/3+1)/3, \quad s = T_o[1+\delta_v(1-\mu)(1+\gamma^2(\gamma^2+1)^{-2})/6]$$
$$f_o = (\gamma^4+2\gamma^2/3+1)\delta_g T_o/6, \quad \alpha = (32\gamma^2/3)\{(\gamma^2+\gamma^{-2}+2\mu)$$
$$+(4/9)(1-\mu^2)[17(\gamma^2+\gamma^{-2})/8+4(\gamma+\gamma^{-1})^{-2}+(\gamma+4\gamma^{-1})^{-2}+(4\gamma+\gamma^{-1})^{-2}]\}$$

for the clamped plate. Here, $\gamma = b/a$ is the aspect ratio of plate sides a and b, and μ is Poisson's ratio. The uniform plate temperature T_o and the scale magnitudes of temperature distribution δ_v and temperature gradient δ_g are all measured in units of the critical buckling temperature.[1] Finally, the usual viscous damping constant $\beta = \xi\sqrt{k_0}$ has a small coefficient $\xi << 1$.

Appendix B: Moment Equations

One can derive the moment equations from the Fokker–Planck equation by partial integrations. It is, however, more direct to begin with the evolution equation for a scalar function $g(x_1, x_2)$ along the trajectory of Eq. (7) obtained by the Ito calculus (Lemma 4.2 of Ref. 10; Chap. 7.4.1 of Ref. 11)

$$<\dot{g}> = <x_2 \frac{\partial g}{\partial x_1}> - <\left(Kx_1 + 3\alpha Q x_1^2 + \alpha x_1^3 + \beta x_2\right)\frac{\partial g}{\partial x_2}> + D<\frac{\partial^2 g}{\partial x_2^2}> \tag{B1}$$

Let us consider $g = x_1^r x_2^s$ and denote $m_{rs} = <x_1^r x_2^s>$. The first-order moment equations for r+s=1 are

$$\dot{m}_{10} = m_{01} \tag{B2a}$$

$$\dot{m}_{01} = -Km_{10} - \beta m_{01} - 3\alpha Q m_{20} - \alpha m_{30} \tag{B2b}$$

For stationary solution the right-hand side of Eq. (B2a) gives $m_{01} = 0$, hence

$$-Km_{10} - 3\alpha Q m_{20} - \alpha m_{30} = 0 \tag{B3}$$

follows from Eq. (B2b). Next, the second-order moment equations for r+s=2 are

$$\dot{m}_{20} = 2m_{11} \tag{B4a}$$

$$\dot{m}_{11} = m_{02} - Km_{20} - \beta m_{11} - 3\alpha Q m_{30} - \alpha m_{40} \tag{B4b}$$

$$\dot{m}_{02} = -2Km_{11} - 2\beta m_{02} - 6\alpha Q m_{21} - 2\alpha m_{31} + 2D \tag{B4c}$$

First, $m_{11} = 0$ from Eq. (B4a). In view of $m_{21} = m_{31} = 0$, the right-hand side of Eq. (B4c) reduces to $m_{02} = D/\beta$. Hence, Eq. (B4b) gives rise to

$$D/\beta - Km_{20} - 3\alpha Q m_{30} - \alpha m_{40} = 0 \tag{B5}$$

which is the stationary contribution from Eq. (B4).

Similarly, among the third-order moment equations for r+s=3

$$\dot{m}_{30} = 3m_{21} \tag{B6a}$$

$$\dot{m}_{21} = 2m_{12} - Km_{30} - \beta m_{21} - 3\alpha Q m_{40} - \alpha m_{50} \tag{B6b}$$

$$\dot{m}_{12} = m_{03} - 2Km_{21} - 2\beta m_{12} - 6\alpha Q m_{31} - 2\alpha m_{41} + 2Dm_{10} \tag{B6c}$$

$$\dot{m}_{03} = -3Km_{12} - 3\beta m_{03} - 9\alpha Q m_{22} - 3\alpha m_{32} + 6Dm_{01} \tag{B6d}$$

we find an independent moment relation

$$2(D/\beta)m_{10} - Km_{30} - 3\alpha Q m_{40} - \alpha m_{50} = 0 \qquad \text{(B7)}$$

from Eqs. (B6b) and (B6c). Note that Eq. (B6d) is redundant to Eq. (B3). Finally, the independent moment relation from the fourth-order moment equations for r+s=4 is

$$3(D/\beta)m_{20} - Km_{40} - 3\alpha Q m_{50} - \alpha m_{60} = 0 \qquad \text{(B8)}$$

Eqs. (B3), (B5), (B7), and (B8) constitute the hierarchical moment equations of order up to 4.

Appendix C: Some Higher Order Moments

We present here some moments of order higher than those given in Eq. (23)

$$\overline{m}_7 = 7\lambda_6\overline{m} + 35\lambda_4\overline{m}(3X + \overline{m}^2) + 105X^3\overline{m} + 105X^2\overline{m}^3 + 21X\overline{m}^5 + \overline{m}^7$$

$$\overline{m}_8 = \lambda_8 + 28\lambda_6(X + \overline{m}^2) + 70\lambda_4(3X^2 + 6X\overline{m}^2 + \overline{m}^4) + 105X^4 + 420X^3\overline{m}^2 + 420X^3\overline{m}^2 + 210X^2\overline{m}^4 + 28X\overline{m}^6 + \overline{m}^8$$

Appendix D: Normal Strain and Stress Distributions

For the single-mode representation we compute strain ε_x and stress σ_x in the normalized x coordinate while y=1/2 is fixed. (Note that x and y are normalized by the plate sides a and b.) We present Eqs. (66–69) of Ref. 1 with a change in notation $\beta = \gamma$.

Simply supported plate:

$$\frac{\sigma_x b^2}{E\pi^2 h^2} = -\frac{(\gamma^2+1)T_o}{12(1-\mu^2)}\left\{1 + \frac{(1-\mu)\delta_v}{4}\left[1 - \frac{\cos 2\pi x}{\gamma^2+1}\right] + \bar{z}\delta_g \sin\pi x\right\} + \frac{2\bar{z}(\gamma^2+\mu)\sin\pi x}{1-\mu^2}q + \left[\frac{\gamma^2+\mu}{1-\mu^2} + \gamma^2\right]\frac{q^2}{2}$$

$$\frac{\varepsilon_x b^2}{\pi^2 h^2} = \frac{(\gamma^2+1)T_o\delta_v}{12(1+\mu)}\left\{\sin^2\pi x + \frac{1}{4}\left[(1+\mu\cos 2\pi x) - \frac{1-\mu\gamma^2}{\gamma^2+1}\cos 2\pi x\right]\right\} + 2\bar{z}\gamma^2(\sin\pi x)q + (2\gamma^2 + \mu\cos 2\pi x)\frac{q^2}{2}$$

Clamped plate:

$$\frac{\sigma_x b^2}{E\pi^2 h^2} = -\frac{(\gamma^4 + 2\gamma^2/3 + 1)T_o}{3(1-\mu^2)(\gamma^2+1)}\left\{1 + \frac{(1-\mu)\delta_v}{4}\left[1 - \frac{\cos 2\pi x}{\gamma^2+1}\right] + \bar{z}\delta_g \sin^2\pi x\right\} - \frac{16\bar{z}(\gamma^2\cos 2\pi x - \mu\sin^2\pi x)}{3(1-\mu^2)}q + \frac{32}{9}\left\{\frac{5\gamma^2}{16} + \frac{3(\gamma^2+\mu)}{16(1-\mu^2)} - \frac{\cos 2\pi x}{2(\gamma+\gamma^{-1})^2} - \frac{\cos 2\pi x}{(\gamma+4\gamma^{-1})^2} + \frac{\cos 4\pi x}{4(4\gamma+\gamma^{-1})^2}\right\}q^2$$

$$\frac{\varepsilon_x b^2}{\pi^2 h^2} = \frac{(\gamma^4+2\gamma^2/3+1)T_o\delta_v}{3(1+\mu)(\gamma^2+1)}\left\{\sin^2\pi x+\frac{1}{4}\left[(1+\mu\cos2\pi x)-\frac{1-\mu\gamma^2}{\gamma^2+1}\cos2\pi x\right]\right\}$$
$$-\frac{16\bar{z}\gamma^2\cos2\pi x}{3}q+\frac{32}{9}\left\{\frac{3\gamma^2}{16}+\frac{\gamma^2+\mu\cos2\pi x}{4}+\frac{\gamma^2-\mu\cos4\pi x}{16}\right\}q^2$$
$$+\frac{32}{9}\left\{-\frac{\cos2\pi x(1-\mu\gamma^2)}{2(\gamma+\gamma^{-1})^2}-\frac{\cos2\pi x(1-\frac{\mu\gamma^2}{4})}{(\gamma+4\gamma^{-1})^2}+\frac{\cos4\pi x(\frac{1}{4}-\mu\gamma^2)}{(4\gamma+\gamma^{-1})^2}\right\}q^2$$

Here, E is the modulus of elasticity, h is the plate thickness, and μ denotes Poisson's ratio. The uniform plate temperature T_o and the scale magnitudes of temperature variation δ_v and temperature gradient δ_g are measured in units of the critical buckling temperature. Finally, $\bar{z}$ is the normalized thickness coordinate z over $\pm h/2$; hence, $\bar{z}=1/2$ for the extreme fiber strain and stress in the main text.

Acknowledgments

This work is supported in part by the Air Force Office of Scientific Research task (2304N1). I wish to thank Chris Clay and Ken Wentz for their providing funds (Extreme Environment Structures Core) for the numerical simulation work of Rimas Vaicaitis who has kindly carried out some additional simulation runs for the root mean square stress data presented in Figure 15.

References

[1]Lee, J., "Large-Amplitude Plate Vibration in an Elevated Thermal Environment," *Applied Mechanics Review*, Vol. 46, P. 2, Nov., 1993, pp. s242–s254.

[2]Crandall, S. H., and Zhu, W. Q., "Random Vibrations: A Survey of Recent Developments," *Journal of Applied Mechanics*, Vol. 50, Dec., 1983, pp. 953–962.

[3]Socha, L., and Soong, T. T., "Linearization in Analysis of Nonlinear Stochastic Systems," *Applied Mechanics Review*, Vol. 44, Oct., 1991, pp. 399–422.

[4]Lee, J., "Improving the Equivalent Linearization Technique for Stochastic Duffing Oscillator," to appear in the Journal of Sound and Vibration, 1995.

[5]Holmes, P. J., "A Nonlinear Oscillator with a Strange Attractor," *Philosophical Transactions of the Royal Society of London*, Series A, Mathematical and Physical Sciences, Vol. 292, Oct., 1979, pp. 419–448.

[6]Raty, R., von Voehm, J., and Isomaki, H. M., "Absence of Inversion-Symmetric Limit Cycles for Even Periods and the Chaotic Motion of Duffing's Oscillator," *Physics Letters* A, Vol. 103, July, 1984, pp. 289–292.

[7]Nayfeh, M. A., Hamdan, A. M. A., and Nayfeh, A. H., "Chaos and Instability in a Power System – Primary Resonant Case," *Nonlinear Dynamics*, Vol. 1, 1990, pp. 313–339.

[8]Souza, M. A., and Mook, D. T., "Post-Buckling Nonlinear Vibrations of Initially Imperfect Structural Systems," *32nd Structures, Structural Dynamics, and Materials Conference* (Baltimore, MD), AIAA, Washington, DC, 1991, pp. 2772–2775, AIAA paper 91–1228–CP.

[9]Morton, J. B., and Corrsin, S., "Consolidated Expansions for Estimating the Response of a Randomly Driven Nonlinear Oscillator," *Journal of Statistical Physics*, Vol. 2, No. 2, 1970, pp. 153–194.

[10]Jazwinski, A .H., *Stochastic Processes and Filtering Theory*, Academic, New York, 1970.

[11]Soong, T. T., *Random Differential Equations in Science and Engineering*, Academic, New York, 1973.

[12] Caughey, T. K., "Derivation and Application of the Fokker–Planck Equation to Discrete Nonlinear Dynamical Systems Subjected to White Random Excitation," *Journal of the Acoustical Society of America*, Vol. 35, No. 11, 1963, pp. 1683–1992.

[13]Andronov, A. A., Vitt, A. A., and Pontryagin, L. S., "On the Statistical Investigation of Dynamical Systems," *Journal of Experimental and Theoretical Physics* (Russian ZETP), Vol. 3, No. 3, 1933, p. 165; *A. A. Andronov Selected Works*, Academy of Science, USSR, 1956, pp. 142–160.

[14]Haken, H., *Synergetics, An Introduction*, Springer Series in Synergetics, Vol. 1, 3rd Ed., Springer-Verlag Berlin, Germany, 1983.

[15]Cramer, H., *Mathematical Methods of Statistics*, Princeton Univ. Press, Princeton, NJ, 1946.

[16]Vaicaitis, R., "Response of Composite Panels Under Severe Thermo-Acoustic Loads," Aerospace Structures Information and Analysis Center, Report TR–94–05, Wright-Patterson AFB, OH, Feb., 1994.

[17]Assaf, S. A., and Zirkle, L. D., "Approximate Analysis of Non-Linear Stochastic Systems," *International Journal of Control.*, Vol. 23, No., 1976, pp. 477–492.

[18]Beaman, J. J., and Hedrick, J. K., "Improved Statistical Linearization for Analysis and Control of Nonlinear Stochastic Systems," *Journal of Dynamic Systems, Measurement, and Control*, Vol. 103, Mar., 1981, pp. 14–21.

[19]Stratonovich, R. L., *Topics in the Theory of Random Noise*, Vol. 1, Gordon and Breach, New York, 1963.

[20]Kendall, M. G., and Stuart, A., *The Advanced Theory of Statistics, Distribution Theory*, Vol. 1, Charles Griffin, London, England, UK, 1963.

[21]Kraichnan, R. H., "The Closure Problem of Turbulence Theory," *Proceedings of Symposia in Applied Mathematics*, Vol. XIII, 1962, pp. 199–225, American Mathematical Society, Providence, RI.

[22]Caughey, T. K., "Equivalent Linearization Techniques," *Journal of Acoustical Society of America*, Vol. 35, No. 11, 1963, pp. 1706–1711.

[23]Lin, Y. K., *Probabilistic Theory of Structural Dynamics*, Robert E. Krieger Publishing, Huntington, NY, 1976.

[24]Vaicaitis, R., Department of Civil Engineering and Engineering Mechanics, Columbia University, New York, NY, Private communication, Nov., 1994.

[25]Thornton, E. A., "Thermal Buckling of Plates and Shells", *Applied Mechanics Review*, Vol. 46, Oct., 1993, pp. 485–506.

Thermally Induced Vibrations of Structures in Space

Ramesh B. Malla* and Anindya Ghoshal†
Ut.:versity of Connecticut, Storrs, Connecticut 06269-2037

Abstract

Structures in space environments, whether operating in orbit or on the extraplanetary surfaces are exposed to a wide range of temperature variations. The severity of the temperature effect has been demonstrated and underlined by the failure of several space structures in fulfilling their intended missions in the past. It has been realized that the extreme temperature gradient possible in a structure subjected to space environment gives a host of structural problems that are not normally found in the terrestrial structures. These include thermal buckling, deformation, stresses, vibration and oscillatory motion, nutational body motion, and material degradation. This paper begins with the survey of relevant published literature within the last four decades on temperature effects in space structures. It discusses salient features of the thermal effects, the problems that were identified, and the remedies adopted. Furthermore, the paper attempts to identify new potential basic problems that may arise in space structures due to extreme temperature variations and their possible solution in light of the technological advancement of the modern era. Emphasis is given to the potential thermally induced vibration effects in large and lightweight structures in orbit and on the moon. Coupled thermal- mechanical structural vibration is examined. Work in progress to develop a refined analysis of an orbital truss structure with thermal effects is outlined.

I. Introduction

As space structures orbit around the Earth in space they undergo cycles of entering into and exiting from the Earth's shadow thus exposing themselves to a wide range of temperature changes that cause thermally induced vibrations,

---

*Assistant Professor, Department of Civil and Environmental Engineering. Member AIAA.

†Graduate Research Assistant, Department of Civil and Environmental Engineering, Student Member AIAA.

thermal buckling, surface material degradation, thermally induced nutational body motion, and thermal stresses. The temperature variation may give rise to thermally induced deformation, possible skin buckling, and overall distortion of the shape of the structure that may change the geometry sufficiently so as to alter the attitude and, thus, compromise the capability of a system to execute its assigned mission. Space researchers had been dealing with the relative importance of the three known sources of solar radiation in space environment, direct solar, Earth albedo, and direct Earth. Their effects vary depending on the orbital radius [low Earth Orbit (LEO) or geosynchronous Earth Orbit (GEO)], area to mass ratio of the structure, property of the space structure material, etc. For example, when a large antenna got distorted due to thermal effects it started reflecting signals inaccurately.[1] Spacecraft booms exposed to solar radiation have been found to exhibit thermally induced oscillatory instabilities (thermal flutter).[2-4] Space structures are generally very flexible, and flexibility effects may compound in greater magnitude in large space structure[es-9] especially when they consists of structural components with open sections. Even as late as the early 1990s space scientists encountered thermally induced oscillations when Hubble Space Telescope (HST) was launched in low Earth orbit. Ultimately astronauts replaced the solar arrays early 1994.

In the present paper the authors have attempted to trace investigations of the phenomenon of thermally induced vibration on space structures by examining the background of thermal effects on structures in space- and lunar-based environments. The paper begins with a review of the published literature during the last four decades dealing with temperature effects in space structures. It discusses the salient features of the thermal effects, and the problems that may arise in space structures. The problems that were identified, and the remedies adopted are also discussed. Especially the paper attempts to identify new potential basic problems that may arise in space structures due to extreme temperature variations and their possible solution in light of the technological advancements of the modern era. Emphasis is given to the potential thermally induced vibration effects in large and lightweight structures in orbit and on the moon. Coupled thermal-mechanical structural vibration is examined. The paper is intended to give state-of-the-art information in the related subject matter. A brief description of the authors' work in progress on the development of a refined analysis technique of an orbital truss structural system with thermal effects is outlined.

II. Historical Perspectives of Space Structures

A. Echo

The first successful launch and deployment of the Echo-I communication satellite on August 12, 1960 heralded the onset of a medium size, inflatable thin walled structure into orbit.[10,11] It was placed near the 1668-km circular orbit. It was a balloon-shaped satellite with a diameter of 30.48 m (100 ft) and was constructed from a 0.00127-cm (0.0005-in.) thick polyethylene terepthalate (PET) plastic film with aluminum vapor deposited on the outside surface in order to provide reflectance to radio frequency transmission and to protect the plastic film from damaging ultra-violet radiation. Inside the balloon sublimating powders were added so as to develop sufficient vapor pressure for inflation in the vacuum of space. The magnitude of equilibrium pressure developed inside the satellite was dependent on the magnitude of the membrane skin temperature. The variation of temperature around the orbit was from 132° to − 107° C or a change of 239° C. The final design of Echo-I indicated that its rigidity was sufficient to retain its shape in a 1668-km-altitude orbit, but it would require at a slight increase in rigidity to prevent buckling at a lower altitude orbit. In Echo II, this additional rigidity to prevent buckling at lower altitude orbit was provided by two layers of very thin aluminum foil with an inner layer of nylon. The Echo II was 41 m in diameter.

B. Passive Geodetic Earth Orbiting Satellites

The ECHO missions were followed by the passive geodetic Earth orbiting satellite (PAGEOS) missions. PAGEOS I was identical to ECHO I in shape, size, and basic principles of function. It was also fabricated of 0.00127-cm-PET film coated on the outside surface with vapor deposited aluminum to make it highly reflective of sunlight. It was placed in a near-polar and circular orbit of 4237-km altitude on June 24, 1966.[12] PAGEOS I was a technologically advanced version of ECHO I. The objective of this mission was to provide a luminous target in near-Earth space that fulfills the requirement of worldwide satellite triangulation. The temperature variation in this orbit was predicted to be from 119° to -130°C. The predicted life of this satellite was about 5 years whereas that of the ECHO I was considerably less. Another difference in Echo I and PAGEOS I was that PAGEOS I spent a large percentage of its orbital lifetime in the inner Van Allen, a region of high intensity, magnetically trapped, particulate radiation which extends from 2000 to 4800 km above the Earth.

C. Other Proposed Orbital and Interplanetary Travel Structures

At about the time of development of the ECHO and PAGEOS vehicles, consideration was given to design of outer space structures in the form of reflectors, solar collectors, scientific equipment packages, and an orbiting astronomical observatory. These systems were relatively smaller sizes orbiting in the altitude range of 300−800 km. For this range of altitudes, a number of pertinent investigations were carried out to design against the hazards of meteoroid particles, high-energy radiation, thermal effects and environmental factors. The radio astronomy Explorer (RAE) satellite which was launched by

NASA in July 1968 was of relatively larger size. The main purpose of this satellite was to carry out measurements of long wave radio signals from outer space[7,9]. It had four 750-ft storable tubular extendable member (STEM) Be-Cu antennas in the orbital plane. Major design requirement of larger lightweight structures have been summarized by Hedgepeth[6] and Bush et al[8]. Variations in the orbits of ECHO I and PAGEOS were identified as Earth's albedo radiation force perturbations.[13]

The solar sail concept discussed by Garvin[14] and by Tsu[15] was studied in detail and designed by the NASA Jet Propulsion Laboratory as a square kite-like spacecraft for a practical interplanetary voyage.[16,17] Heligyro, which was another proposed solar vehicle,[7,9] consisted of 12-km twisted blades which rotated in a solar radiation field generating the required propulsive forces. The tethered orbiting interferometer (TOI) was yet another proposed RAE satellite consisting of two end bodies joined by a flexible lightweight cable 5-km long.[7,9]

D. Storable Tubular Extendable Member

Nearly all of the satellites in orbit that have extendable booms (appendages) use the basic STEM principle. In the past, these booms have been used on orbiting geophysical observatories (OGO) IV, V, and VI as electric field probes on the Department of Defense gravity experiment (DODGE), applications technological satellite (ATS), and gravity-gradient stabilization experiment (GGSE III) as well as on the radio astronomy Explorer to form the large V-shaped antenna for radio-frequency reception. The completely deployed boom was modeled as a clamped-free, thin-walled cylinder with an open section. Its length is arbitrary but, in practice, is usually between 30 and 100 ft. Problems associated with such booms have been studied in great extent (for example, see refs. 2-4) and will be described in more detail later in the section that deals with thermally induced vibrations of orbital structures.

E. Hubble Space Telescope

On April 24, 1990 the Hubble space telescope (HST), see Fig. 1, was launched into LEO from onboard the Space Shuttle Discovery and was deployed successfully from the Discovery payload bay. The problem of spherical aberration was found after the initial imaging data were examined. At this point another problem was discovered: a pointing jitter induced by thermally driven bending of the solar arrays.[18-20] The vibrations were induced when the solar array structure reacted to the temperature change which occurred as the spacecraft passed from shadow to sunlight. The arrays consisted of thin solar panels that were originally furled within a drum and later unfurled by 20-ft BISTEMs. The BISTEM consisted of two concentric thin-walled tubes stored on a drum and later unfurled in a flat configuration. The ends of the solar arrays were reported to move about 10 in. The arrays vibrated upto 6 min. This vibration, in turn, set

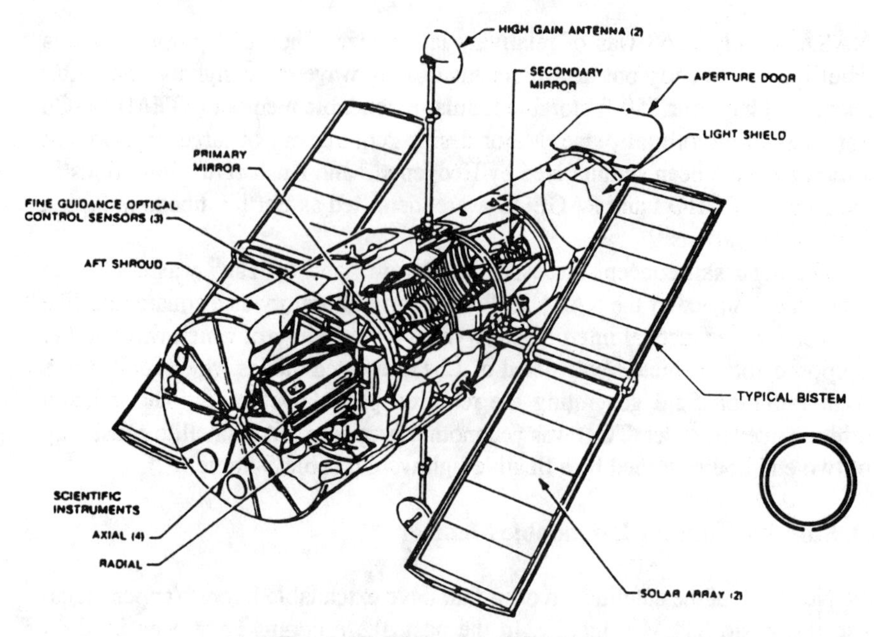

Fig. 1. Hubble space telescope

up secondary random oscillations that occurred for a minute or two at a time over periods of up-to 20 min in sunlight. Although efforts were made to damp out the oscillations[20] ultimately replacement of the solar arrays were done in early 1994.[21]

F. Proposed Space Station

The international community, under the leadership of NASA, has continued to bring the proposed space station (Fig. 2) to reality. The configuration of the main baseline structure is a truss-like structure. The support structures for the solar panels are also truss type. Because of its relatively large size, high flexibility, and structural complexity (needed to satisfy many mission objectives) as well as the intended use of new structural material, it is envisioned that new problems related to thermal-vibration will surface. A thorough investigation of thermal-structure interaction will be of utmost importance for this structure.

III. General Thermal-Structure Interaction Problems

The phenomenon of coupling between thermodynamic and mechanical processes in a solid media was first predicted by Duhamel in 1837 (see Allen[22]). The significance of this coupling depends on the type of material behavior encountered during the process as well as the time required to perform the

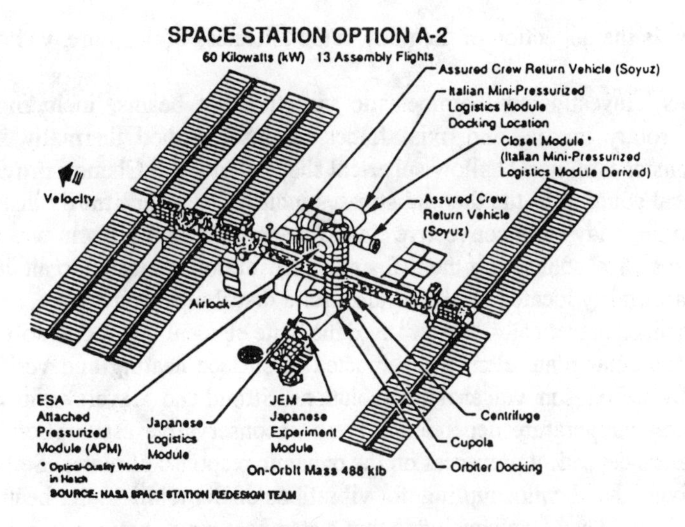

FIg. 2. Space Station Freedom option A-2

process. In case of elastic solids, this coupling is insignificant except under conditions when inertial effects are not negligible. However, inelastic media circumstances may arise wherein thermodynamic/mechanical coupling occurs even under quasi-static conditions. Allen[22] presented a review on this topic.

A. Thermally Induced Vibrations on Beams, Plates, and Shells

In 1956, Boley[23] first provided the analytical modeling of thermally induced response of beams including inertia (see also Refs. 24-27). Boley established fundamental concepts of thermally induced vibration behavior. He studied thin beam or thin plate subjected to uniform step heating on one surface. The transient distribution through the thickness caused a time-dependent thermal moment which acted as the dynamic forcing function. For a beam, the governing equation of motion under the simultaneous action of heat and applied loads can be written as

$$\frac{\partial^2}{\partial x^2}\left(EI\frac{\partial^2 v}{\partial x^2}\right) + \rho A\frac{\partial^2 v}{\partial t^2} = \frac{\partial^2 M_T}{\partial x^2} \tag{1}$$

where A is the area of cross section, EI the flexural stiffness, t time, x the coordinate along the beam, v transverse displacement, ρ density of the beam material, and M_T the thermal moment.

For a simply supported rectangular plate the governing equation is

$$D\nabla^4 w + \rho h\frac{\partial^2 w}{\partial t^2} = -\frac{1}{1-\nu}\nabla^2 M_T \tag{2}$$

where w is the deflection of the plate, h the thickness of the plate, ν Poisson's ratio.

Jones[28] investigated thermoelastic vibrations of beams, including shear effects, rotary inertia, and axial forces. Kraus[29] studied thermally induced vibrations of thin, non-shallow spherical shells. Mow and Cheng[30] provided an analytical solution to the thermal stresses induced by an arbitrarily distributed fast-moving body in an convective elastic space. Fourier transform was used to get the analytical solution for thermal stress distribution subjected to an unit heat source arbitrarily located on the surface; for a distributed heat source, solutions were obtained numerically. Mason[31] used the finite element method to solve beam and rectangular plate elements subjected to surface heating and verified his results by comparison with analytical solutions. Stroud and Mayers[32] showed that neglecting temperature dependence is an unconservative assumption for the temperature-dependent properties on the dynamic response of plates. Seibert and Rice[33] found that thermocoupling for vibrations of Bernoulli-Euler beams were insignificant. They also concluded that a step heat input may cause significant vibrations in short beams as well as long beams, which were treated as Timoshenko's beams. Bargmann[34] surveyed the developments for thermally induced waves and vibrations and included an extensive bibliography. Jadeja and Loo[35] investigated the vibrations of a rectangular plate subjected to sinusoidal surface heating and the role of such vibrations in terms of fatigue failure. Thornton and Foster[36] provided an extensive review of the dynamic response of rapidly heated space structures.

B. Computational Methods

Several computational methods for analysis of thermally induced vibrations of beam structures were presented by Manolis and Beskos[37] and Namburu and Tamma.[38] These investigations arre done on the assumption that the applied heating and temperature field are independent of the structure's deformation resulting in stable thermally induced oscillations. However, when applied heating and the temperature field depend on the structure deformation, which means coupling of deformations with heating and temperature, it may result in unstable thermally induced oscillations, i.e., oscillations of increasing magnitude (thermal flutter). Namburu and Tamma[38] described an effective unified methodology for the modeling/analysis of thermally induced structural dynamics of flexural configurations with emphasis on Euler-Bernoulli type beam models. Glass and Tamma[39] numerically investigated the effect of temperature dependent thermal conductivity, specific heat and density on coupled, non-Fourier dynamic thermo-elastic models involving relaxation times. Beam[40] of NASA Ames demonstrated unstable thermally induced oscillations in the laboratory in 1968. Augusti[41] studied the radiantly heated rigid links with elastic springs and demonstrated that dynamic instability was possible. Thornton[42] provides a review of progress in aerospace thermal structures during the last three decades.

1. *Finite Element Methods*

Emery et al.[43] described a special one-dimensional element which can be applied to sparse structures, such as sparse frames or fibrous insulation, to compute the radiation view factors for curved surfaces. Normally, the radiation exchange in large space structures and frames is done by representing these surfaces by an assemblage of flat planes. For cylinders whose length to diameter ratio is a large, this leads to inaccurate results because of the unfavorable conditions. Thornton et al.[44] presented a thermoviscoplastic computational method for hypersonic structures. It employs a unified viscoplastic constitutive model implemented in finite element analysis for a quasistatic thermal structures analysis. It also provides an approach to analyze convectively cooled hypersonic structures and transient inelastic structural behavior at elevated temperatures. Ignaczak[45] reviewed results of finite wave speeds of thermoelastic disturbances based on different models of a thermoelastic solid in order to study the effect of thermoelastic coupling on the speed of thermoelastic waves in a non-homogenous and anisotropic medium. The study also includes the modeling of waves of more complex nature such as electromagneto-thermoelastic waves and others.

Kasza[46] proposed an alternative finite element formulation for spatial beams and truss structures using Betti's law of reciprocal work under various loadings, including thermal loads. Thornton and Kolenski[47] developed a thermo-viscoplastic finite element method employing the Bodner–Partom constitutive model to investigate the response of simplified thermal structural models subjected to intense local heating. They found that with rapid rises in temperature, nickel alloy structures display initially higher yield stresses due to strain rate effects. As the temperature approaches elevated values, yield stress and stiffness degrade rapidly and pronounced plastic deformation occurs. This is a particularly useful study for vehicles which accelerate at high speeds in the atmosphere, where shocks sweep across the hypersonic vehicles interacting with local shocks and boundary layers exposing the structural surfaces to severe local pressures and heat fluxes. A spacecraft re-entry vehicle may possibly face the same situation.

2. *Continuum Model*

Noor et al.[48] developed continuum models for large repetitive beams and plate-like lattices with arbitrary configurations subjected to static, thermal, and dynamic loadings. Nayfeh and Hefzy[49] introduced a rather straightforward construction procedure in order to derive continuum equivalence of discrete truss-like repetitive structures. The construction procedure, after the overall structure is specified, includes identification of parallel members, derivation of unidirectional "effective continuum" properties for each of those sets, and orthogonal transformations used to determine the contribution of each set to the overall effective continuum properties of the structure. Breakwell and Andeen[50] analyzed the dynamics of a passive space communicator in synchronous orbit.

3. *Finite Difference Methods*

Jianping and Harik[51,52] developed a finite difference technique for the analysis of axisymmetric conical shells with variable thickness and spherical shells. Variation of temperature loading due to supersonic flow is assumed to vary along the meridian and across the thickness of the shell, where material properties are considered to be temperature dependent. Classical thin-shell theory is assumed in the analysis. Boyce[53] developed a methodology that provides for quantification of uncertainty in the lifetime material strength degradation of structural components of aerospace propulsion systems subjected to a number of diverse random effects, including high temperature.

IV. Thermally Induced Vibration on Orbital Structures

For Earthbound structures quasistatic thermal-structural analysis is normally good enough as thermal response time are typically much larger than the characteristic structural response time. Structures designed for the Earth environment must be stiff enough to withstand gravitational loads and, hence, have relatively high frequencies and short natural periods. In space with very low gravitational loads, however, structures can be designed to be large in size with slender components and, therefore, are generally more flexible with low natural frequencies. Under such conditions thermally induced structural oscillations may be more pronounced.

A. History of Theoretical Development

1. *Observations*

The phenomenon of thermally induced vibrations in orbiting space structures became widely known during the flight of the OGO-IV spacecraft in the late 1960s. On that flight the 60-ft experiment boom broke into a sustained solar-induced large-amplitude oscillation which severely compromised the performance of the spacecraft. When the 60-ft electron probe was in solar eclipse, all noticeable satellite motion was damped out. Whenever the satellite left eclipse or entered into full sunlight, however, the control system experienced a strong periodic disturbance which was large enough to cause the error signal of the attitude control jets to exceed the deadband limits which resulted in a large consumption of energy. The disturbance torque had the same frequency as the boom-spacecraft system and a magnitude which corresponded to that which would be produced by vibrations with an amplitude of about ± 20 ft. NASA Goddard Space Flight Center had theorized that the observed vibrations were thermally induced. The anomalous motion were observed on OGO IV and several other three-axis stabilized satellites deploying booms.[2,3] A less dramatic but still serious thermally induced instability occurred during the flight of Explorer 45 (SSS-A). On this flight, minute oscillations of the four experiment booms caused by the modulations of the thermal input were phased in such a manner that a nutational destabilizing effect was produced. Flately[54] gives detailed analysis of this problem.

2. Analytical Development and Mathematical Modeling

Ellerton[55] in 1946 and Sternberg[56] in 1947 initiated some of the preliminary thoughts on thermal effects from the solar radiation on space rockets. Sandroff and Prigge[57] used the "thermo bottle principle" in convection with a whipple meteor shield to isolate the vehicle from the heating and cooling effects of thermal radiation. They concluded that thermal control in a space vehicle was a relatively easy engineering problem but not without its penalties in the form of design time, cost, and weight. Hanel[58] discussed the thermoelastic temperature control of satellites and space vehicles. He determined that the temperature of the satellite's instruments could be kept constant within a few degrees of the design value, regardless of orbital conditions, internally dissipated power, and some erosion of skin coatings. Schmidt and Hanawalt[59] investigated the temperatures that existed on a particular satellite configuration and studied various pertinent parameters in order to determine their relative importance. The analysis was largely dependent on geometry involving spacewise as well as timewise variations in skin temperature. Charles and Raynor[60] studied a nonlinear problem of solar heating in a space vehicle idealized as a thin-walled circular cylinder rotating with uniform velocity about its geometric axis for a situation in which heat transfer by convection and heat exchange within the cylinder was negligible. Florio and Hobbs[61] presented a closed-form analytical solution for steady-state thermal balance analysis of temperature distribution in a discontinuous cylindrical tube. The tube was considered to model a gravity gradient passive altitude control rod, such as was isolated in space at a significant distance from the spacecraft structure. Yu[62] did an analytical study of thermally induced vibrations and their stability including flutter of a flexible appendage on the basis of Hamilton's principle and also including damping.

Motivated by the thermally induced phenomenon in OGO-IV spacecraft, Donahue and Frisch[2] and Frisch[3] have presented detail analytical investigation of thermally induced response of long thin cylinders of open section. According to Frisch[3] the theory behind the observed thermally induced vibration can be summarized as follows. When an open cylinder is deployed in direct sunlight, thermal gradients are built up around and along it, at a rate determined by the thermal characteristics of the boom material. The boom would experience thermal stresses if it were rigidly held; as it is free, it relieves thermal stresses by bending and twisting toward its instantaneous thermal equilibrium shape. Because of its mass inertia, its response takes time. As it twists in the directional thermal field established by the sun, a change in the heat flow around the perimeter occurs. This causes a change in the overall temperature distribution and, hence, a change in the position of thermal equilibrium toward which the boom is moving. It is shown that under a broad range of conditions the position of thermal equilibrium can change with a strong component at the first natural bending frequency of the boom. The system is then, in effect, driven at resonance, and large- amplitude thermally induced bending motion can occur. Graham[63] also studied stability of in-plane bending oscillations of long flexible members (STEMs) when subjected to solar heating. Frisch[4] later presented an in-depth analysis method for thermally induced response of flexible structures.

Chambers et al.[64] developed a streamlined thermal analysis method to predict accurately the temperature data needed to perform the thermoelastic deformation analysis of large space structure in a cost-effective manner. Thornton and Tamma[65] proposed a finite element analysis for elevated structures with thermal protection systems (TPS). Then performance of the TPS/structural element was demonstrated for a Shuttle wing truss subjected to re-entry heating, reducing model size and computational time. Thornton and Paul[66] reviewed advances in modeling, analysis, and understanding of the thermal-structural response in the mid 1980s. They expressed a definite need for more thermal-structural experiments to validate thermal-structural design and analysis. Mikulas, Jr.[67] showed that thermal buckling decreases with increase in truss structure size.

Malla et al.[68] studied the radiation thermal effects simultaneously on orbital motion, attitude motion, and axial deformation of a very large, axially flexible space structure describing a planar pitch motion around the Earth. Relative contributions from the direct solar, Earth albedo, and direct Earth radiation were assessed. Bainum et al.[69] studied the dynamics and control of a large space structures under radiation thermal shocks.

Thornton and Kim[70] presented an analytical method for determining the thermal-structural response of a flexible rolled up solar array due to a sudden increase in external heating. Calculations were presented for the solar arrays on the Hubble space telescope. The flexural equation of motion for a beam representing the central truss of the solar array was noted as

$$EI\frac{\partial^4 v}{\partial x^4} + P\frac{\partial^2 v}{\partial x^2} + \frac{\partial^2 M_T}{\partial x^2} + m_x\frac{\partial^2 v}{\partial t^2} = 0 \tag{3}$$

and

$$M_T(x,t)=\int_A E\alpha_T \Delta Tz \; dA \tag{4}$$

where v is the beam deflection, EI the flexural stiffness, m_x the mass per unit length, P the axial load, M_T the thermal moment, t time, x the coordinate along the length of the beam, α the coefficient of thermal expansion or contraction, ΔT the temperature difference, and A the area of cross section.

Gulick[71] conducted an analytical study of thermally induced motion for a beam model of an axial boom of a spin-stabilized spacecraft. The study included an uncoupled thermal-structural analysis which assumed that the temperature distribution is independent of the deflections of the boom and a coupled thermal-structural analysis which included the dependence of the incident heat flux and temperature distribution on the boom's deflected position. The results showed that unstable thermally induced vibrations were possible for parameter characteristics of a typical spacecraft axial boom. Thornton et al.[72] developed an analytical approach for investigation of thermally induced vibrations of a split-blanket solar array due to self-shadowing of the central truss.

3. *Nutational Body Motion*

Tshuchiya[73,74] presented techniques necessary for the preflight analysis of the potentiality of a thermally induced instability. Tsuchiya[73] examined the spinning spacecraft with flexible appendages exposed to solar radiation which caused thermally induced vibrations of the appendages. The vibration occurred at a spin frequency causing a periodic variation of the moments of inertia of the spacecraft at that frequency and this, in turn, produced a self-excitation of nutational body motion through the instrumentality of parametric excitation. The amplitude of nutational body motion was determined by means of the method of averaging. The stability of motion was examined. Tsuchiya also studied thermally induced vibrations of a flexible appendage attached to a spacecraft containing a rotor, with particular emphasis on the behavior near resonance.[74] Resonance occurred when the natural frequency of the appendage was approximately equal to twice/half the nutational frequency. The investigator established the criteria for onset of unstable motion of the appendage for the first case and, in the second case, showed that the vibration of the appendage builded up to the finite value through the resonant excitation.

4. *Earth Albedo and Emitted Radiation Effects*

In 1962 Levin[75] concluded from his investigation that the principal component of the Earth's reflected radiation on a spherical satellite is in the radial direction and that this component diminishes to approximate 10% of the direct radiation at an altitude about one Earth radius. Other components of the reflected radiation maybe neglected. Dennison[76] proposed an efficient numerical procedure for computing the magnitude of the illumination of a space vehicle surface due to Earth-reflected sunlight. Snoddy[77] assumed Rayleigh scattering in the atmosphere and diffuse reflection from the Earth's surface and, thereby, computed and presented the intensity distribution over the visible portion of the Earth's surface as seen by a small element of area at various altitudes above the atmosphere. Linton[78] developed a method for calculating the Earth's albedo using thermal data from the PEGASUS I environmental effects sensors.

Mills et al.[79] in 1971 gave the annual averages of Earth albedo and radiation for use in estimating long term effects on space vehicle equipment and surfaces. The study also treats spectral distributions of albedo and Earth radiation. Furukawa[80] developed the dimensionless radiosity function and an analytical solution based on generalized functions and integro-differential equations in quasisteady state to conduct a thermal analysis of large space telescope in orbit.

Some numerical results showed the longitudinal temperature distribution of the telescope heated cyclically. Modest[81] dealt with the derivation of the algorithm describing solar radiation impinging on an infinitesimal surface after reflection from a gray and diffuse planet.

B. Experimental Investigation

Space structures have a tendency to become more flexible when designed to satisfy the requirement to be large and light. They are generally poor in passive damping because of the material's poor structural damping and the lack of other types of damping in space. Once a vibration of such a structure occurs, it has low natural frequencies and a long decay time that may lead to fatigue or to instability of the structure. By theoretical analysis and experiments in a vacuum under laboratory conditions it has been shown that the thermally induced unstable bending vibration (thermal flutter) occurs when the one side of the thin-walled boom is subjected to the radiation heating (Murozono and Sumi[82]; Sumi et al.[83,84]). This phenomenon was explained as caused by the phase difference between the response of the flexural vibration of the boom section. Conversely, it is expected that the damping of the system is increased if the thermal bending moment, which is in the direction suppressing the bending vibration, is applied to a cantilever beam by applying alternately appropriate temperature difference between both the surfaces of the beam. Murozono and Sumi[82] studied the digital vibration control of a flexible cantilever using a thermal bending moment caused by the temperature gradient across the section of the beam both by experiments and simulations. Foil strain gauges are bonded on the surfaces on the beam as actuators which are capable of producing a thermal gradient. Thermal bending moment are applied in the proper sense to a flexible cantilever beam having very low natural frequency, so that active control of the first mode bending vibration is realized. Experimental results show that the damping ratio is increased by 10 times.

Clayton and Tinker[85] described experimental and analytical characterization of a flexible thermal protection material known as tailorable advanced blanket insulation (TABI). This material utilizes a three-dimensional ceramic fabric-core structure and an insulation filler. Schonberg[86] presented the results of an experimental study in which aluminum dual-wall structures were tested under a variety of high-speed impact conditions to study the effect of multi layer thermal insulation (MLI) thickness and location on perforation resistance.

Foster and Thornton[87] performed laboratory tests on a model spacecraft boom to investigate the stable and unstable thermally induced oscillation of a representative flexible spacecraft. The authors were the first to report the phenomenon of simultaneous excitation of both fundamental and second natural modes of vibration for a fixed-free beam which they termed Thermal strum.

C. Materials

Several composite materials that perform better in extreme temperature variation have found greater use for space structures. Varieties of insulating materials and coatings have also been developed for thermal protection of space structures. Wright[1] gives the physical properties of several metals and composites. For the microwave radiometer spacecraft (MRS), graphite fibers and epoxy are selected as the primary materials for basic structural elements. Aluminum was chosen for joints and fittings. MRS is basically a skeletal framework built from long slender columns. In addition to buckling, the structure may also undergo relatively small deflections due to thermal loadings. Although thermal distortions maybe controlled by thermal coatings or insulations, a structural material with a very low-thermal expansion coefficient, such as graphite-epoxy, would greatly ease the thermal problems. Henninger[88] supplied the data on solar absorptance and thermal emittance of many common types of thermal control coatings together with some sample spectral data curves of absorptance.

Experimental research on advanced composite materials indicates that they may undergo up to a 15% stiffness loss due to thermomechanical fatigue, which causes a variety of damage modes (due to microcracking) in the structure (Kalyansundaram et al.[89]). Additional loss of stiffness may be attributed to elevated temperature and chemical changes due to solar radiation and other environmental effects. This reduction in stiffness affects the dynamic response, which in turn, is critical in the development of control systems for large space systems (LSS). Kalyansundaram et al.[89] presented a sensitivity analysis that investigated the effect of stiffness loss on structural frequencies and mode shapes. They basically investigated the effect of degradation of material properties on the dynamic response of LSS.

Ekvall et al.[90] studied the elevated temperature aluminum alloys that are being developed for advanced fighter aircraft structural applications in the 300–600 °F temperature range. Noda[91] reviewed the papers, mostly after 1980, on thermal stresses in materials with temperature-dependent properties. The thermal and mechanical properties in materials subjected to thermal loads due to high temperature, high gradient temperature, and cyclical changes of temperature are dependent on temperature magnitudes.

Noor and Burton[92] reviewed in depth the computational modeling for high-temperature multilayered composite plates and shells. They focused their review on the hierarchy of composite models, predictor-corrector procedures, the effect of temperature dependence of material properties on the response, and the sensitivity of the thermo-mechanical response of variations in material properties.

Rockoff and David[93] described the approach McDonnell Douglas Aerospace (MDA) has employed to enhance blanket durability for astronaut handling, identifying, and replacing materials that are adversely affected by the low-Earth-orbit environment, and protecting non-replaceable susceptible materials from degradation factors, while maintaining blanket performance and low weight. They also described MDA's use of Unigraphics, a computer aided design (CAD) tool to improve modeling and to speed the design process of multilayer insulation blankets, that have been traditionally difficult to model.

D. Thermal Design Requirements for Large Orbiting Structures

Hedgepeth et al.[94,95] describes the conceptual mechanical design of an Earth-orbiting solar-reflecting spacecraft. Large orbiting structures experience substantial temperature changes as their orbit passes from the sun side of Earth into the Earth's shadow and out again. The design of the reflector spacecraft must take into account the thermal influence on structural shape and loads resulting from thermal strain. In addition, the materials used in the design must be suitable for use through their entire expected temperature range in orbit and angle to withstand prolonged exposure to the vacuum and radiation environment. Low-thermal-expansion materials, such as graphite/epoxy composites, are preferred for structural members because this property tends to minimize differential thermal expansions that create dimensional imperfections.

The combination of low expansion materials and the open nature of the trusses used in the baseline results in acceptable small distortions in the trusses themselves. Certain situations will result in substantial differential changes in length between the truss structures and the stay tapes used, even though both are made of graphite/epoxy (Hedgepeth[94,95]). For example, when the spacecraft first enters the Earth's shadow, the tapes will cool much faster than the rods in the masts. For this reason, the stay tape reels at the tip of each central mast are locked together after deployment to act as a single reel with a constant torque device to provide pretension. In this manner, the tapes can adjust in length at the same time, but not individually, and avoid any unreasonable loads due to thermal strain. The tapes will each have at least one twist along their length so that their view factor with respect to the sun is less position-dependent. It is interesting that the reflector surface tends to ensure an almost even as the relative sun angle changes. In a nonreflector spacecraft, there are certain sun angles where some tapes see nearly full illumination and others see almost none on the same side of the spacecraft.

V. Orbiting Large Truss System

Since truss-like structures have high stiffness to mass ratio they are preferred for large space structure constructions. A detail analytical methodology specialized for truss type structures in orbit is required for obtaining accurate mechanical and thermal response of these structures. This section gives a brief outline of the authors' effort in this area.

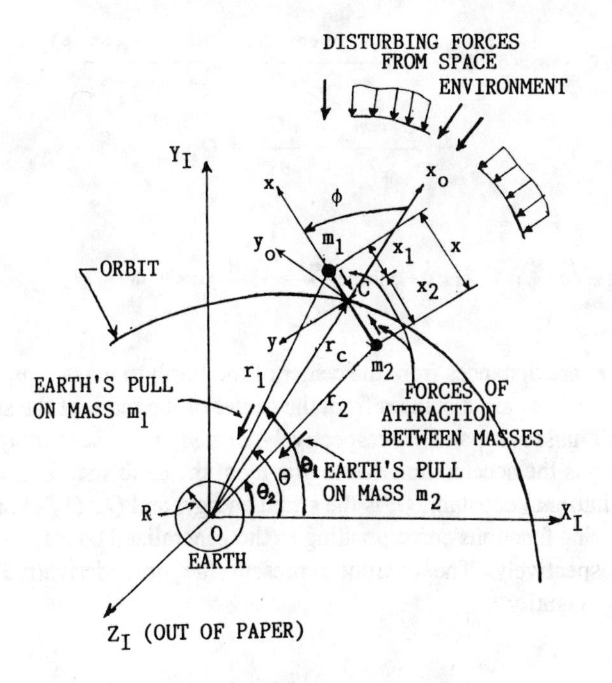

Fig. 3. Dumbbell orbital structure

A. Formulation for a Building Block Space Structure

Representing a space structure by a dumbbell-shaped geometry consisting of two end masses m_1 and m_2 connected by an axially flexible link structure which moves in a general non circular orbit (Fig. 3), Malla et al.[68,96] have derived a set of four equations of motion using Lagrange's principles. This dumbbell structure is assumed to move in the orbital plane (coplanar motion) around a spherical Earth of radius R and center at O. It is also assumed that the orbit of the structure is defined by the motion of its center of mass C.

The equations of motion are derived in terms of the four generalized coordinates of the problem: 1) the radial distance r_c between the center of the Earth and the center of mass of the space structure, 2) the angle θ describing the sweep of the radial line r_c in the plane of motion, 3) the distance x between the two dumbbell masses m_1 and m_2, and 4) the angle ϕ (pitch angle) between the local vertical line OC and the line joining the two masses.

$$m\ddot{r}_c - mr_c\dot{\theta}^2 + \mu\left[\frac{m_1(r_c+x_1\cos\phi)}{r_1^3} + \frac{m_2(r_c-x_2\cos\phi)}{r_2^3}\right] = Q_r \quad (5a)$$

$$mr_c(r_c\ddot{\theta}+2\dot{\theta}\dot{r}_c) - \overline{m}\mu\left[\frac{(-1)}{r_1^3} + \frac{(1)}{r_2^3}\right]xr_c\sin\phi = (Q_\theta - Q_\phi) \quad (5b)$$

$$\bar{m}\ddot{x} - \bar{m}x(\dot{\theta}+\dot{\phi})^2+\bar{m}\mu\left[\frac{(x_1+r_c\cos\phi)}{r_1^3} + \frac{x_2 - r_c\cos\phi)}{r_2^3}\right] + \frac{Gm_1m_2}{x^2} + \frac{\partial U_e}{\partial x} = Q_x \tag{5c}$$

$$\bar{m}[x^2(\ddot{\theta}+\ddot{\phi}) + 2x\dot{x}(\dot{\theta}+\dot{\phi})m\mu[\frac{(-1)}{r_1^3} + \frac{1}{r_2^3}]xr\sin\phi = Q_\phi \tag{5d}$$

Where r_1 and r_2 are distances from the center of the Earth to masses m_1 and m_2, respectively; x_1 and x_2 are distances from the center of the mass of the structure to the dumbbell masses m_1 and m_2, respectively, $m = m_1 + m_2$; $\bar{m} = m_1m_2/m$; $\mu = gR^2$, in which g is the acceleration due to gravity at the earth's surface; G is the universal gravitational constant; U_e is the strain energy; and Q_r, Q_θ, Q_x, and Q_ϕ are external forcing functions corresponding to the generalized coordinates r_c, θ, x, *and* ϕ, respectively. The overdot represent the time derivative of the corresponding quantity.

B. Thermal Analysis and Temperature Distribution in Orbit

Although there are several studies reported in literature on this subject, thermal problems in space structures persist, and studies leading to new methods are sought after. The temperature gradient that is present in the space structure should be considered for accurate modeling of member behavior to assess the system response. Furthermore, the thermal-structural interaction needs to be considered. Malla et al.[68] used the mathematical model [Eq.(1)] with a thermodynamic equation of heat radiation to study thermal effects on a large space structure's orbital, attitude, and axial deformation simultaneously in low Earth and geosynchronous orbits.

For any thermal analysis, an accurate determination of temperature variation around an orbit of a structure in space is most essential. It can be deduced from the fundamental hypothesis of the mathematical theory of heat flow in solids (Boley and Weiner[26]) that if the structure radiates heat $\dot{q}_{out}$ as a black body and that the heat input can be defined by the function $\dot{q}_{inp}$, then the thermodynamic equation of heat conduction and radiation for a one-dimensional structure is

$$K_{th}V\frac{\partial^2 T(x,t)}{\partial x^2} - Mc\frac{\partial T(x,t)}{\partial t} = \dot{q}_{out}(x,t)-\dot{q}_{inp}(x,t) \tag{6}$$

where K_{th} is the thermal conductivity, V the volume of the structure, M the mass of the structure effective to radiation, c the specific heat of material, x the longitudinal length measured from the center of mass of the structure, t the coordinate used to measure time, and, T the absolute temperature at the position x at time t. K_{th} depends on materials, and, in general, c depends on pressure and temperature.

An accounting of energy to and from, or stored in, a space structure serves to determine the temperature of the structure. Assuming the space structure is a black body, the time rate of radiative heat output $\dot{q}_{out}$ was determined. The time rate of radiation heat effects received by a space structure $\dot{q}_{inp}$ are of mainly three types: 1) direct solar, 2) reflected from Earth and atmosphere, 3) emitted from Earth and atmosphere. These heat quantities (input and output) are complicated functions of material properties, structural dimension, and structure's orientation and distance with respect to the Earth and the sun. Modest[81] has shown that Earth reflected heat can be determined with good accuracy using the view factor given by Cunningham[97,98].

If only the axial deformation is present in the structural member, the force Q_{th} in this direction, due to temperature variation in the structure, is given by

$$Q_{th} = E[-A\alpha_{th}(T-T_s)+\int_A \alpha_{th}(T-T_s)dA] \tag{7}$$

Where α_{th} is the coefficient of thermal expansion or contraction, T_s the absolute reference temperature (°K), and A the cross-sectional area.

C. Formulation for Total of Structural System

In the studies conducted by Malla et al.[68,96] the space structure was modeled by a single continuum one-dimension elastic link with the structure's mass lumped at the two ends. It was assumed that the temperature remained constant throughout the structure at a given time (isothermal condition). The structure was considered to have uniform temperature throughout its length (i.e., T was considered a function of time only; therefore, Eq.(6) without the first term on the left side was used).

For an accurate and refined analysis these assumptions and constraints have to be reconsidered. It is believed that a substantial improvement and extension can be achieved over this model. An effort is underway to develop a refined analysis technique which is applicable for the total structural system and has potential for providing a very useful contribution to the study of a space structure's dynamic and thermal response of in an intact state. The following steps (based on the finite element approach) are being adopted to achieve the needed refinement (Malla and Ghoshal[99,100]). 1) Derive a sets of coupled nonlinear equations of motion for the dumbbell element with the generalized coordinates in Eqs. [(5a-5d)] being the degrees of freedom of the end masses. 2) Use the dumbbell formulation (Fig. 3) to represent a member of the total truss-like space structure (Figs. 4). 3) Assemble the equations for all members to

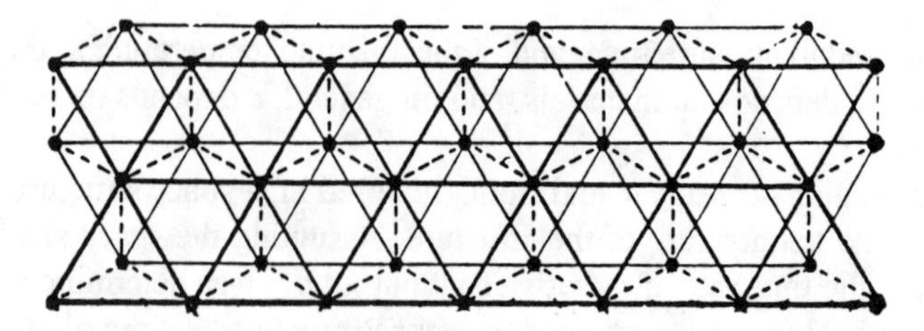

Fig. 4. Large tetrahedral truss-type space structure

obtain the mathematical model for the total structure. 4) Solve these equations of motion using the central difference numerical technique with Newton-Raphson iteration scheme. For the thermal effect analysis, which is planned for the future, the mathematical formulation will be modified to include the temperature variation across and along the member.

VI. Thermal Effects in Lunar Structures

A prime consideration in the design of lunar structures would be that the structures should be able shield against the types of hazards found on the lunar surface: continuous solar/cosmic radiation, meteorite impacts, extreme variations in temperature and radiation (Task Committee[101]). Chow and Lin[102] envision the lunar structure to be constructed of double-skin composite fabric membrane filled with structural foam for pressurized or unpressurized applications. Environmental constraints are going to significantly effect the lunar structure. Extreme temperatures on the lunar surface ranging from -310 °F (-190 °C) to +270 °F (137 °C) exist in the transition between the day and night, which occurs in roughly two-week cycles. This requires the use of construction materials that are capable of operating in the full range of lunar temperatures. Thermal problems can alleviated to a large extent by covering the structure with a thick layer of lunar soil. Solar radiation and meteoroid bombardment presents another serious constraint on construction and for that matter, on life itself. Lunar structures must be designed for rapid construction to minimize the exposure time to these hazards. A design criterion is, therefore, to provide the lunar structure with maximum construction tolerances and minimum requirements for work precision.

The gradients of temperature are rapid during the transition between day and night caused by the lack of an atmosphere. Some individual surface rocks reach 130 °C during daytime. Any structural design must thereby account for such predictable changes, both regarding thermal expansion/contraction and regarding fatigue aspects (Benaroya and Ettouney[103]). A change of temperature (ΔT) will result in a change of length (ΔL) equal to $\alpha(\Delta T)L$ and, if the ends are fixed for a change in the total stress ($\Delta \sigma$) equal to $EA\alpha(\Delta T)$, where A is cross-sectional area and E the modulus of elasticity. The large range of temperature fluctuations will have a direct effect on the relative displacements of the roof of the structure. This, in turn, may require different design detailing for the structural joints and connections. In this context it may be mentioned that fatigue must be considered in design since lunar surface environment experiences periodic and rapid temperature gradients between days and nights. Although the internal structure will be at human-habitable temperatures, the extremities of the structure, as well as structures not intended for habitation, need to be designed to survive such temperature differences and to sill function as needed. Large expansions or contractions will generally not be tolerable and will need to be considered in structural designs. Estimated design life and, thus, the expected number of loading cycles will determine the seriousness of the fatigue problem. In the likely situation that a layer of regolith (lunar soil) is placed atop the structure for shielding, the added weight would partially balance the forces on the structure caused by the internal pressurization. Johnson et al.[104] states that some of the challenges for the design engineers of the Lunar Transit Telescope (LTT) are 1) Passive thermal cooling, 2) Passive thermal self-compensation of the structure for stability, and 3) Tailoring of thermal expansion coefficients (feasible with some composite materials).

VII. Areas of Further Research

Even though several studies are reported on thermal problems of space structures, the problem persists.

A. Research to Study Basic Concepts

Some important areas for thermal structural research have been identified.[66] They include 1) spacecraft self-shadowing effects on structural response, 2) effects of large prime-power systems on spacecraft thermal-structural behavior, 3) better knowledge of material properties and their effects on long-term structural response, and 4) better understanding of thermally induced structural vibrations.

B. New Computational Capability

Additional computations with large structures are needed to further delineate problems because computations with preliminary structural designs have only partially identified problems in analysis capabilities. Improved computer program capability to model and analyze nonlinear response of structural components is desirable.

C. Active Control

Accuracy requirements envisioned for future large space antenna structures may require active control or static adjustment of the surface. One means is to accomplish such a control is by changing lengths of individual members of a supporting truss structure. This is easily accomplished if the supporting structure is statically determinate so that essentially no force is required to change the shape of the surface. A statically determinate structure also has the advantage of being free of thermal stress for all of the temperature distributions. This maybe of special importance for configurations having very flexible slender members that can support only small compressive loads without buckling. Haftka and Adelman[105] proposed a concept that utilizes controlled temperature changes in selected members to effect changes in the member lengths in order to minimize surface errors. Anderson and Nimmo[106] developed a general configuration that results in a statically determinate rectangular platform structure. The vibration modes and frequencies of such structures are determined and compared to those of a redundant platform having the same size and joint arrangement. Thus, a relative stiffness of the statically determinate platform is obtained. Temperature change may be used to control such structures.

D. Technology Transfer

It is believed that the research and technology on thermal-structure interaction developed for space structures can be used, with modification, for terrestrial structures built in extreme temperature environments. This is especially true in light of the widespread use of composite and lighter materials for Earthbound structures. Of particular interest are effects of thermal loading on large-scale civil infrastructure, such as bridges and buildings. Ho Duen and Lin[107] studied thermal loadings acting on highway bridges as random variables. Values of such loadings for a 50-year return period are retained based on an analysis of the statistics of extremes.

VIII. Conclusions

Performance and behavior of structures in space are greatly affected by the extreme temperature variations that exist in the space environment. Research on thermally induced vibrations on structures has been done from 1950s beginning with the analytical studies of Boley. With the deployment of spacecraft booms, thermally induced unstable bending-torsional oscillations were traced to be one of the reasons for anomalous behavior for Earth satellites as far back as in the 1960s (OGO-IV). There have been considerable analytical and computational studies on thermally induced oscillations and since the 1980s researchers have turned toward experimental analysis of the phenomenon as well. Continued recurrence of undesirable thermally induced oscillations on spacecraft structures, however, suggests that a number unresolved problems remains unanswered. Further analytical, computational, and experimental investigations on the phenomenon of thermally induced oscillations are necessary for better understanding of thermally induced vibration on space structures and, thus, providing future designers with better design criteria for space structures that would eliminate such undesirable oscillations. Control algorithms for active and/or passive suppression of such oscillations are potential areas for further research work. The authors are currently incorporating the thermal effects in their model of refined analysis of response of orbital truss type structures. Such a refined modeling is necessary for more accurate on-orbit determination of the response of the orbital truss type structures under thermal loadings and other space environmental disturbances.

Acknowledgment

This material is based upon work supported by the State of Connecticut under grant number 93K007 (Apollos Kinsley Collaborative Grants). The support is gratefully acknowledged. Any opinions, findings, and conclusions or recommendations expressed in the publication are those of the authors and do not necessarily reflect the views of the State of Connecticut.

References

[1]Wright, R. L., "Evolution and Design Characteristics of Microwave Radiometer Spacecraft," NASA Reference Publication No.1079, December 1981, pp. 51-65.

[2]Donahue, J., and Frisch, H., "Thermoelastic Instability of Open Section Booms," NASA TND-5310, Dec. 1969.

[3]Frisch, H. P., "Thermally Induced Response of Long Thin-Walled Cylinders of Open Section," Journal of Spacecraft and Rockets, Vol. 7, No. 8, 1970, pp. 897-905.

[4]Frisch, H. P., "Thermally Induced Response of Flexible Structures: A Method for Analysis," AIAA Journal, Vol. 3, No. 1, 1980, pp. 92-94.

[5]Noll, R., et al., "Effects of Structural Flexibility on Spacecraft Control Systems," NASA-SP-8016, N69-37030, April 1969.

[6]Hedgepeth, J. M., "Survey of Future Requirements for Large Space Structures, "NASA CR-2621, January 1976.

[7]Mishra, A. K. and Modi, V. J., "The Influence of Satellite Flexibility on Orbital Motion," Proceedings of the AIAA Symposium, AIAA, New York, June 1977, pp. 59-74.

[8]Bush, H., Mikulas, M, and Heard, W., "Some Design Considerations of Large Space Structures," AIAA Journal, Vol. 16, No. 6, 1978, 352-359.

[9]Meirovitch, L. (Ed.), Dynamics and Control of Large Flexible Spacecraft, AIAA Symposium Proceedings, June 1977, June 1979, and June 1981, VPI & SU, Blacksburg, VA.

[10]Bryant, R., "A Comparison of Theory and Observation of the ECHO-I Satellite," Journal of Geophysical Research, Vol. 66, No. 9, 1961, pp. 3066-3069.

[11]Clemmons, D., Jr., "The ECHO I Inflation System," NASA TND-2194, June 1964.

[12]Teichman, L., "The Fabrication and Testing of PAGEOS I," NASA TND-4596, June 1968.

[13]Prior, E. J., "Earth Albedo Effects on the Orbital Variations of ECHO I and PAGEOS I," Dynamics of Satellites, edited by B. Morando, Springer-Verlag, New York, 1970, pp. 303-312.

[14]Garvin, R. L., "Solar Sailing - A Practical Method of Propulsion within the Solar System," Jet Propulsion, Vol. 28, No. 3, 1958, pp. 188-190.

[15]Tsu, T. C., "Interplanetary Travel by Solar Sail," ARS Journal., Vol. 29, June 1959, pp. 422-427.

[16]Rodriguez, G., Marsh, E., and Gunter, S., "Solar Sail Attitude Dynamics and Control," Dynamics and Control of Large Flexible Spacecraft edited L. Meirovitch, AIAA, New York, June 1977, pp. 187-302.

[17]Trubert, M. and Blount, M., "Thrust and Moment on the Square Sail," Solar Sail Study 662-34, Jet Propulsion Lab., California Inst. of Technology, Pasadena, CA, Oct. 1977.

[18]Asker, J.R., "Hubble Beset by Radiation Effects, Solar Array Vibrations, but Controllers Expect Early Resolution of Both Problems," Aviation Week and Space Technology, Vol. 26, June 18, 1990.

[19]Sawyer, K., "Hubble Space Telescope's Flutter Will Require Solar-Panel Fix," The Washington Post, Nov. 11, 1990, pp. A6.

[20]Polidam, R. S., "Hubble Space Telescope Overview," 29th Aerospace Sciences Meeting, AIAA Paper 91-0402, Reno, NV, January 1991.

[21]Cuviello, M.J., "Maintaining Hubble Space Telescope Performance Through In-Orbit Servicing," 29 Aerospace Sciences Meeting, AIAA Paper 91-0405, Reno, NV, Jan 1991.

[22]Allen, D. H., "Thermomechanical Coupling in Inelastic Solids," edited by R. B. Hetnarski, "Review of Thermal Stresses," ASME Applied Mechanics Review, Vol. 44, No. 8, 1991, pp. 361-373.

[23]Boley, B. A., "Thermally Induced Vibrations in Beams," Journal of the Aeronautical Sciences, Vol. 23, No. 2, 1956, pp. 385-393.

[24]Boley, B. A., and Barker, A. D., "Dynamic Response of Beams and Plates to Rapid Heating," Journal of Applied Mechanics, Vol. 24, No. 3, 1957, pp. 413-416.

[25]Boley, B. A., and Barker, A. D., "Discussion," Journal of Applied Mechanics, Vol. 25, No. 2, 1958, pp. 309-310.

[26]Boley, B. A., and Weiner, J. H., Theory of Thermal Stresses, Wiley, New York, 1960.

[27]Boley, B. A., "Approximate Analyses of Thermally Induced Vibrations in Beams and Plates," Journal of Applied Mechanics, Vol. 39, No. 1, 1972, pp. 212-216.

[28]Jones, J. P., "Thermoelastic Vibrations of a Beam," Journal of the Acoustical Society of America, Vol. 39, No. 3, 1966, pp. 500-505.

[29]Kraus, H., "Thermally Induced Vibrations of Thin Non-Shallow Spherical Shells," AIAA Journal, Vol. 4, No. 3, 1966, pp. 542-548.

[30]Mow, V. C., and Cheng, H. S., "Thermal Stresses in An Elastic Half-Space Associated With An Arbitrarily Distributed Moving Heat Source," Journal of Applied Mathematics and Physics, Vol. 18, Fasc. 4, 1967, pp. 500-507.

[31]Mason, J. B., "Analysis of Thermally Induced Structural Vibrations by Finite Element Techniques," NASA TM X-63448, August 1968.

[32]Stroud, R. C., and Mayers, J., "Dynamic Response of Rapidly Heated Plate Elements," AIAA Journal, Vol. 9, No. 1, 1971, pp. 76-83.

[33]Seibert, A. G. and Rice, J. S., "Coupled Thermally Induced Vibrations by Finite Elements," AIAA Journal, Vol. 7, No. 7, 1973, pp. 1033-1035.

[34]Bargmann, H., "Recent Developments in the Field of Thermally Induced Waves and Vibrations," Nuclear Engineering and Design, Vol. 27, No. 3, 1974, pp. 372-385.

[35]Jadeja, N. D., and Loo, T. C., "Heat Induced Vibration of a Rectangular Plate," Journal of Engineering for Industry, Vol. 96, No. 3, 1974, pp. 1015-1021.

[36]Thornton, E. A., and Foster, R. S., "Dynamic Response of Rapidly Heated Space Structures," Computational Nonlinear Mechanics in Aerospace Engineering, edited by S. N. Atluri, Progress in Astronautics and Aeronautics, Vol. 146, AIAA, Washington, DC, 1992, pp. 451-477.

[37]Manolis, G. D., and Beskos, D. E., "Thermally Induced Vibrations of Beam Structures," Computer Methods in Applied Mechanics and Engineering., Vol. 21, No. 3, 1980, pp. 337-355.

[38]Namburu, R. R., and Tamma, K. K., "Thermally-Induced Structural Dynamic Response of Flexural Configurations Influenced by Linear/NonLinear Thermal Effects," Proceedings. of AIAA/ASME/ASCE/AHS 32nd Structures, Structural Dynamics and Materials Conference (Baltimore, MD), Pt. IV, AIAA, Washington, DC, Apr. 8-10, 1991, pp. 2667-2679.

[39]Glass, D., and Tamma, K., "Dynamic Thermoelasiticity with Temperature Dependent Thermal Properties," Proceedings of the AIAA/ASME/ASCE/AHS 32nd Structures, Structural Dynamics and Materials Conference, (Baltimore, MD), AIAA, Washington, DC, P. IV, April 8-10, 1991, pp. 2654-2665.

[40]Beam, R. M., "On the Phenomenon of Thermoelastic Instability (Thermal Flutter) of Booms with Open Cross Sections," NASA TN D-5222, June 1969.

[41]Augusti, G., "Instability of Struts Subject to Radiant Heat," Mechanica, Vol. 3, No. 2, 1968, pp. 167-176.

[42]Thornton, E. A., "Thermal Structures: Three Decades of Progress," Journal of Aircraft, AIAA, Vol. 29, No. 3, 1992, pp. 485-498.

[43]Emery, A. F., Mortazavi, H. R., and Nguyen, M. N., "Radiation Exchange in Large Space Structures and Frames," AIAA Journal, Vol. 23, No. 6, 1985, pp. 947-953.

[44]Thornton, E., Oden, J., Tworzydlo, W., and Youn, S., "Thermoviscoplastic Analysis of Hypersonic Structures Subjected to Severe Aerodynamic Heating," Journal of Aircraft, Vol. 27, No. 9, 1990, pp. 826-835.

[45]Ignaczak, J., "Domain of Influence Results in Generalized Thermoelasticity- A Survey," ASME Applied Mechanics Review, Vol. 44, No. 9, 1991, pp. 375-382.

[46]Kasza, F., "An Alternative Finite-Element Formulation for Spatial Beams and Truss Structures," Space Structures 4 edited by G. Parke and C. Howard, Elsevier, London, UK, 1993, pp. 1354-1364.

[47]Thornton, E. A., and Kolenski, D. J., "Viscoplastic Response of Structures for Intense Local Heating," Journal of Aerospace Engineering, Vol. 7, No. 1, 1994, pp. 50-71.

[48]Noor, A. K., Anderson, M. S., and Greene, W. H., "Continuum Models for Beam and Platelike Lattice Structures, AIAA Journal, Vol. 16, No. 12, 1978, pp. 1219-1228.

[49]Nayfeh, A., and Hefzy, M., "Continuum Modeling of the Mechanical and Thermal Behavior of Discrete Large Structures," AIAA Journal, Vol. 19, No. 6, 1981, pp. 766-773.

[50]Breakwell, J. V., and Andeen, G. B., "Dynamics of a Flexible Passive Space Array," Journal of Spacecraft and Rockets, Vol. 14, No. 5, 1977, pp. 556-561.

[51]Jianping, P., and Harik, I. E., "Thermal Stresses in Spherical Shells," Journal of Aerospace Engineering, Vol. 6, No. 1, 1993, pp. 106-110.

[52]Jianping, P., and Harik, I. E., "Axisymmetric Pressures and Thermal Gradients in Conical Missile Tips," Journal of Aerospace Engineering, Vol. 4, No. 3, 1991, pp. 237-255.

[53]Boyce, L., "Probabilistic Material Degradation Under High Temperature, Fatigue and Creep," Journal of Aerospace Engineering, Vol. 6, No. 4, 1993, pp. 347-362.

[54]Flatley, T.W., "Nutationally Destabilizing Thermo-Elastic Effects of Explorer 45 (SSA-A)," NASA/GSFC X-732-77-138, Greenbelt, MD, May 1975.

[55]Ellerton, G. C., "Effect of Sun's Heat on Space Rockets," Journal of American Rocket Society, Dec. 1946, pp. 24-28.

[56]Sternberg, R. L., "Some Remarks on the Temperature Problems of the Interplanetary Rocket," Journal of the American Rocket Society, Vol. 17, No. 70, 1947, pp. 34-35.

[57]Sandroff, P. E., and Prigge, J. S. Jr., " Thermal Control In a Space Vehicle," Journal of the Astronautical Sciences, Vol. 3 No. 1, 1956, pp. 4-8.

[58]Hanel, R. A., "Thermostatic Temperature Control of Satellites And Space Vehicles," ARS Journal, Vol. 29, May 1959, pp. 358-361.

[59]Schmidt, C. M., and Hanawalt, A. J., "Skin temperature of a satellite," ARS Journal, Oct.ober 1957, pp. 1079-1083.

[60]Charles, A., and Raynor, S., "Solar Heating of a Rotating Cylindrical Space Vehicle,' ARS Journal, May 1960, pp. 479-484.

[61]Florio, F. A., and Hobbs, R. B., "An Analytical Representation of Temperature Distribution in Gravity Gradient Rods," AIAA Journal, Vol. 6, No.1, 1968, pp. 99-102.

[62]Yu, Y. Y., "Thermally Induced Vibration and Flutter of a Flexible Boom, " Journal of Spacecraft, Vol. 6, No. 6, 1969, pp. 902-910.

[63]Graham, J. D., "Solar Induced Bending Vibrations of a Flexible Member," AIAA Journal, Vol. 8, No. 11, Nov. 1970, pp. 2031-2036.

[64]Chambers, B. C., Jensen, C. L., and Coyner, J.V., "An Accurate and Efficient Method for Thermal/Thermoelastic Performance Analysis of Large Space Structures," Proceedings. of the AIAA 16th Thermophysics Conference (Palo Alto, CA), AIAA, Washington, DC, June 23-25, 1981, pp. 1-7 (AIAA Paper 81-1178).

[65]Thornton, E. A., and Tamma, K. K., "Finite Element Analysis of Structures With Thermal Protection Systems," AIAA/ASME Joint Fluids, Plasma, Thermophysics and Heat Transfer Conference, St. Louis, MO, June 7-11, 1982, (AIAA Paper 82-0835).

[66]Thornton, E. A., and Paul, D. B., "Thermal Structural Analysis of Large Space Structures: An Assessment of Recent Advances," Journal of Spacecraft and Rockets, Vol. 22, No. 4, 1985, pp. 385-393.

[67]Mikulas, M. M. Jr., "Space Station Truss Selection Considerations," NASA Lyndon Johnson Space Center, September 1986, pp. 1-21.

[68]Malla, R. B., Nash, W. A., and Lardner, T. J., "Thermal Effects on Very Large Space Structures," Journal of Aerospace Engineering, Vol. 1, No. 3, 1988, pp. 171-189.

[69]Bainum, P., Hamsath, N., and Krishna, R., "The Dynamics and Control of Large Space Structures After the Onset of Thermal Shock," Acta Astronautica, 1989, Vol. 19, No. 1, pp.1-8.

[70]Thornton, E. A., and Kim, Y. A., "Thermally Induced Bending Vibrations of A Flexible Rolled-Up Solar Array," Journal of Spacecraft and Rockets, Vol. 30, No. 4, 1993, pp. 438-448.

[71]Gulick, D. W., "Thermally Induced Vibrations of An Axial Boom on a Spin-Stabilized Spacecraft," Proceedings of the AIAA/ASME/ASCE/AHS 35th Structures, Structural Dynamics and Materials Conference (Hilton Head, SC), AIAA, Washington, DC, April. 18-21, 1994 (AIAA Paper 94-1556).

[72]Thornton, E. A., Chini, G. P., and Gulick, D. W., "Thermally Induced Vibrations of A Self-Shadowed Split-Blanket Solar Array," Proceedings of AIAA/ASME/ASCE/AHS 35th Structures, Structural Dynamics and Materials Conference, (Hilton Head, SC), AIAA, Washington, DC, April 18-21, 1994 (AIAA Paper 94-1379).

[73]Tsuchiya, K., "Thermally Induced Nutational Body Motion of a Spinning Spacecraft with Flexible Appendages," AIAA Journal, Vol. 13, No. 4, 1975, pp. 448-453.

[74]Tsuchiya, K., "Thermally Induced Vibrations of a Flexible Appendage Attached to a Spacecraft," AIAA Journal, Vol. 15, No. 4, 1977, pp. 505-510.

[75]Levin, E., "Reflected Radiation Received by An Earth Satellite," ARS Journal, Vol. 32, 1962, pp. 1328-1331.

[76]Dennison, A. J. Jr., "Illumination of a Space Vehicle Surface Due to Sunlight Reflected from Earth," ARS Journal, Vol. 32, April 1962, pp. 635-637.

[77]Snoddy, W. C., "Irradiation Above Atmosphere due to Rayleigh Scattering and Diffuse Terrestrial Reflections," Proceedings of the AIAA Thermophysics Specialist Conference (Monteray, CA), AIAA, New York, Sept. 13-15, 1965, pp. 239-279.

[78]Linton, R. C., "Earth Albedo Studies Using Pegasus Thermal Data," AIAA Thermophysics Specialist Conference, AIAA Paper 67-332, New Orleans, LA, Apr. 17-20, 1967, pp. 475-490.

[79]Mills,S. (chair), "NASA Space Vehicle Design Criteria (Environment): Earth Albedo and Emitted Radiation," NASA SP-8067, July 1971, pp. 1-44.

[80]Furukawa, M., "Thermal Analysis of Large Space Telescope on Orbit," Proceedings. of the 12th International Symposium on Space Technology and Science, Tokyo, Japan, 1977, pp. 733-738.

[81]Modest, M. F., "Solar Flux Incident on An Orbiting Surface After Reflection from a Planet," AIAA Journal, Vol. 18, No. 6, 1980, pp. 727-730.

[82]Murozono, M., and Sumi, S., "Active Vibration Control of a Flexible Cantilever Beam by Applying Thermal Bending Moment, "Proceedings of the 2nd Joint Japan/U.S. Conference on Adaptive Structures, (Nagoya, Japan), edited by Y. Matsuzaki and B. Wada, Nov. 12-14, 1991, pp. 315-331.

[83]Sumi, S., Murozono, M., and Imoto, T., "Thermally Induced Bending Vibration of Thin-Walled Boom Caused by Radiant Heating," Transaction of Japanese Society of Mechanical Engineers, Vol. 56, 1990, pp. 300-307 (In Japanese with English Abstract).

[84]Sumi, S., Murozono, M., Imoto, T., and Nakazato, S., "Thermally Induced Bending Vibration of Thin-Walled Boom with Closed Section Caused by Radiant Heating," (In English), Memoirs of the Faculty of Engineering, Kyushu Univ., Vol. 49, No. 4, 1989, pp. 273-290.

[85]Clayton, J., and Tinker, M. L., "Characterization and Modeling of An Advanced Flexible Thermal Protection Material for Space Application," Proceedings of AIAA/ASME/ ASCE/AHS 32nd Structures, Structural Dynamics and Materials Conference (Baltimore, MD), AIAA, Washington, DC, P. 1, April 8-10, 1991, pp. 139-147.

[86]Schonberg, W., "Sensitivity of Dual-Wall Structures Under Hypervelocity Impact to Multi-Layer Thermal Insulation Thickness and Placement," Proceedings of AIAA/ASME/ASCE/AHS 34th Structures, Structural Dynamics and Materials Conference, (La Jolla, CA), Pt. 4, AIAA, Washington, DC, April 19-22, 1993, pp. 1844-1853.

[87]Foster, R. S., and Thornton, E. A., "An Experimental Investigation of Thermally Induced Vibrations of Spacecraft Structures," Proceedings of the AIAA/ASME/ASCE/AHS 35th Structures, Structural Dynamics and Materials

Conference (Hilton Head, SC), AIAA, Washington, DC, April 18-21, 1994, (AIAA Paper 94-1380).

[88]Henninger, J. H., "Solar Acceptance and Thermal Emittance of Some Common Spacecraft Thermal-Control Coatings," NASA Reference Publication No. 1121, 1984, pp. 1-50.

[89]Kalyansundaram, S., Lutz, J. D., Haisler, W. E., and Allen, D. H., "Effect of Degradation of Material Properties on the Dynamic Response of Large Space Structures," Journal of Spacecraft, Vol. 23, No. 3, 1986, pp. 297-302.

[90]Ekvall, J. C., Ranen, R. A., Chellman, D. J., Flores, R. R., Gerbach, M. J., "Elevated Temperature Aluminum Alloys For Advanced Fighter Aircraft," Journal of Aircraft, Vol. 7, No. 9, 1990, pp. 836-843.

[91]Noda, N., "Thermal Stresses in Materials with Temperature-Dependent Properties," ASME Applied Mechanics Review, Vol. 44, No. 9, 1991, pp. 383-397.

[92]Noor, A. K., and Burton, W. S., "Computational Models for High Temperature Multi-Layered Composite Plates And Shells," Applied Mechanics Review, Vol. 45, No. 10, 1992, pp. 419-446.

[93]Rockoff, L. M., and David, K. E., "Multilayer Insulation Blanket Design for the International Space Station Program," Proceedings of AIAA/ASME/ASCE/AHS 35th Structures, Structural Dynamics and Materials Conference, (Hilton Head, SC), AIAA, Washington, DC, April 18-21, 1994, pp. 1-7.

[94]Hedgepeth, J. M., Miller, R.K., and Knapp, K., "Conceptual Design Studies for Large Free-Flying Solar-Reflector Spacecraft," Ames Research Center R-1015, NASA Contract NAS 1-15347, March 1981.

[95]Hedgepeth, J. M., "Support Structures for Large Infrared Telescopes," NASA CR-3800, Ames Research Center, Carpinteria, CA, July 1984.

[96]Malla, R.B., Nash, W. A., and Lardner, T. J., "Motion and Deformation of Very Large Space Structures," AIAA Journal, Vol. 27, No. 3, 1989, pp. 374-376.

[97]Cunningham, F.G., "Power Input to a Small Flat Plate from a Diffusely Radiating Sphere, with Application to Earth Satellites," NASA TND-710, 1961.

[98]Cunningham, F.G., "Earth Reflected Solar Radiation Input to Spherical Satellite," NASA TND-1099, 1961.

[99]Malla, R. and Ghoshal, A., "Dynamic Response of Truss-Type Orbital Structures," Proceedings of 12th U.S. National Congress of Applied Mechanics, University of Washington, Seattle, WA, June 26-July 01, 1994, pp 6.

[100]Malla, R., and Ghoshal, A., "An Analytical Procedure to Determine On-Orbit Dynamic Response of Planar Truss-Type Space Structures," Proceedings of the AIAA/ASME/ASCE/AHS 36th Structures, Structural Dynamics and Materials Conference (New Orleans, LA), AIAA, Washington, DC, April. 10-12, 1995 (AIAA Paper 95-1439).

[101]Task Committee on Lunar Base Structures, ASCE, "Overview of Existing Lunar Base Concepts," Journal of Aerospace Engineering, Vol. 5, No. 2, 1992, pp. 159-174.

[102]Chow, P. Y., and Lin, T. Y.,"Structural Engineer's Concept of Lunar Structures," Journal of Aerospace Engineering, Vol. 2, No. 1, 1989, pp. 1-9.

[103]Benaroya, H., and Ettouney, M., "Design and Construction considerations for Lunar Outpost," Journal of Aerospace Engineering, Vol. 5, No. 3, 1992, pp. 261-273.

[104]Johnson, S., Chua, K., and Wetzel, J., "Engineering Issues for Early Lunar-Based Telescopes," Journal of Aerospace Engineering., Vol. 4, No. 3, 1992, pp. 323-336.

[105]Haftka, R. T., and Adelman, H. M., "An Analytical Investigation of Shape Control of Large Space Structures by Applied Temperatures," AIAA Journal, Vol. 23, No. 3, 1985, pp. 450-457.

[106]Anderson, M. S., and Nimmo, N. A., "Dynamic Characteristics of Statically Determinate Space-Truss Platforms," Journal of Spacecraft, Vol. 23, No. 3, 1986, 303-307.

[107]Ho, D., and Chi-Ho, L., "Extreme Thermal Loadings in Highway Bridges, "Journal of Structural Engineering, Vol. 115, No. 7, 1989, pp. 1681-1695.

Transient Thermal–Structural Response of a Space Structure with Thermal Control Materials

Yool A. Kim* and Hugh L. McManus†
Massachusetts Institute of Technology, Cambridge, Massachusetts 02139

Abstract

This paper presents an analytical approach to obtain the transient thermal–structural response of an insulated space structure operating in the Space Shuttle payload bay. The structure is exposed to the space thermal environment, which is complex due to shadowing effects. The payload bay thermal environment is characterized based on an energy balance for a differential element at an arbitrary location in the bay. Assuming that the thermal resistance of the insulation is very large, the nonlinear radiation environment model can be decoupled from the linear internal conduction model. The insulation is modeled as a linear resistance between the environment and the structure. This simplification allows an efficient thermal analysis for multiple mission scenarios and structural designs. Temperature gradients obtained from a finite element thermal analysis are applied to a structural finite element model to determine the structural response. Results are presented for a conceptual structural design of the stellar interferometer tracking experiment. The temperature variations in the structure are found to depend strongly on the Shuttle orientation. The resulting thermal deformations have a significant effect on the performance of the stellar interferometer tracking experiment.

Introduction

Many Earth orbiting space structures must maintain dimensional stability to allow accurate and precise pointing. Such structures are subject to a continuously changing space thermal environment and may, if improperly

*Graduate Research Assistant, Space Engineering Research Center, Department of Aeronautics and Astronautics.

†Assistant Professor of Aeronautics and Astronautics, Space Engineering Research Center , Senior Member AIAA.

designed, experience unacceptable deformations. Thermal control materials such as multilayer insulation (MLI) thermal blankets are commonly used to reduce the temperature variations. Nevertheless, spacecraft design usually entails complex thermal analyses.

Thermal analyses of space structures can be difficult and time consuming for two reasons. First, the transient heating loads, the shadowing effects, and differential heating due to spacecraft geometry create a complex thermal environment. The nonlinear radiation coupling between the environment and the structure further increases the complexity of the analysis. Second, the spacecraft is subject to numerous time-varying orbital attitudes, and for spacecraft design purposes, multiple structural configurations must be analyzed. Repeated runs of analyses can demand a large amount of time and computing power. As an example, the stellar interferometer tracking experiment (SITE) is a flight experiment of an optical interferometer operating in the Space Shuttle payload bay. The SITE requires the Shuttle to point at many target stars; thus, SITE is exposed to many different aspects of the space environment. To function, the SITE structure must remain both thermally and dynamically quiet.[1] Hence, the design of SITE entails the thermal analysis of multiple structural configurations in the complex cargo bay thermal environment under multiple mission scenarios.

Background

Current methods for determining the thermal response of space structures usually require two models, a geometry mathematical model and a thermal mathematical model.[2] In the first model, the orbital radiation heat loads, which include direct solar, Earth emitted, and Earth reflected solar or albedo, are calculated. The shadowing effects and the view factors are also determined. The environmental loads and view factors are input to the thermal mathematical model and the heat transfer problem, which includes conduction, convection, radiation, and internal heat generation, is solved, usually via a finite difference technique. The structural response, on the other hand, is usually obtained using a finite element method. Two approaches to determine the thermal deformations are as follows: 1) the temperature distribution obtained from the thermal analyzer is input to a structural finite element model to determine the structural response, or 2) the thermal and the structural analyzers share the same finite element model to determine the thermal–structural response. The advantages and the disadvantages of these approaches are discussed in Ref. 3. In general, thermal engineers use commercial thermal analysis tool packages which provide the computing capabilities. However, these tools require a large computing time to perform multiple transient analyses, which involve many iterations to solve the nonlinear heat transfer problem.

Some analytical methods have been developed to simplify the computing process and to get good approximate results. For steady-state analyses of simple shapes, and many transient one-dimensional analyses, analytical methods can be used. Usually, assumptions about the heating loads and orbital conditions must be made to reduce the complexity of the problem. For instance, Mahaney and Thornton[4] applied a solar attenuation factor to the direct solar flux impinging on an orbiting truss structure to account for members shadowing other members. Reference 5 shows an investigation of thermally induced vibrations of a flexible rolled-up solar array using analytical methods. An analytical form of the boom's thermal response can be written based on several assumptions about the thermal response of the solar array boom.[5] However, these studies involve many assumptions and simplifications which are specific to the spacecraft being investigated. For spacecraft design purposes, a general but simpler form of current thermal analysis tools is desirable. The objective of this paper is to develop an analytical approach which allows the efficient thermal analysis of space structures for multiple mission scenarios and structural designs.

Approach

In this paper an analytical approach to obtain the transient thermal–structural response of a space structure operating in the Space Shuttle payload bay is developed. The structure is protected by MLI, and the thermal analysis assumes that the thermal resistance of the MLI is very large, which allows a decoupling of the nonlinear radiation environment model from the linear internal conduction model. A transient thermal response of an isothermal differential element at an arbitrary location in the bay is calculated to characterize the environment in the bay. The MLI is mathematically modeled as a linear thermal resistance between the environment and the structure. A thermal finite element analysis is used to obtain the temperature solution. Then the thermal gradients are input to a finite element structural model to obtain a static structural response. This simple thermal analysis technique is applied to a conceptual structural design of SITE for a preliminary evaluation of the thermal–structural response. Thermal analyses for various Shuttle orientations are performed to investigate and to understand the critical parameters such as the orbit parameters, Shuttle orientation, and SITE structural design.

Characterization of the Environment

In this section an analytical approach to determine the thermal response of the MLI outer layer is presented.

Geometry Model

In the analysis, a structure of an arbitrary shape lies in an open payload bay of a Space Shuttle. The structure is insulated with MLI. The bay is modeled as a box with length L, width W, and height H (see Fig. 1). The box has an open top to represent the open cargo door. The origin of the bay coordinate system is located at the geometric center of the box. The X_B axis points out of the box and the Y_B axis points toward the nose of the Shuttle. The analysis considers a small

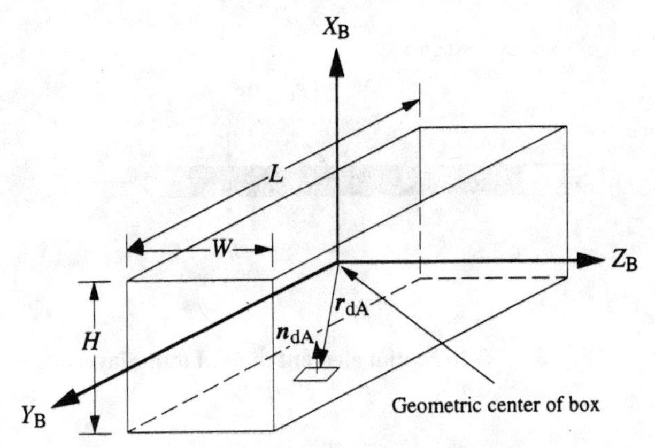

Fig. 1 Geometrical model of Shuttle payload bay.

area of the MLI outer layer, dA, within the bay with arbitrary orientation $\boldsymbol{n}_{dA}$ and position $\boldsymbol{r}_{dA}$ (see Fig. 1).

Determining the view factors and the shadowed regions requires bookkeeping of parameters such as the solar vector $\boldsymbol{S}$, the Earth pointing vector $\boldsymbol{e}_r$, and the surface normal vector $\boldsymbol{n}_{dA}$ (see Fig. 2). These geometric parameters depend on the orbital elements and the Shuttle orientation, and the bookkeeping may become complicated. Therefore, a set of reference frames, in which the vectors can be readily identified, is established. For calculations of the heat loads, all vectors are transformed into the bay coordinate system defined earlier.[6]

Assumptions

For orbital heating load calculations, the model assumes 1) the Shuttle is in a relatively low circular orbit, 2) the orbit is low enough that the Shuttle is in the Earth's shadow for half of the orbit period and the Earth can be considered as an isothermal infinite plate, and 3) the bay walls are isothermal gray surfaces and maintain a constant temperature. Based on the assumption that the thermal resistance of the insulation is very large, the heat exchange between the outer layer of the insulation and the structure q_{MLI} is initially neglected ($q_{MLI} << q_{external}$, see Fig. 3). This assumption will be investigated in a later section. Also, because the outer layer of the MLI is thin, the in-plane

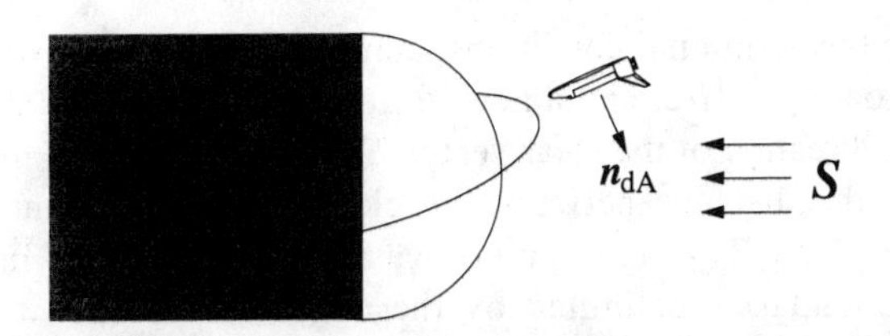

Fig. 2 Sun, Earth, and surface normal orientation vectors.

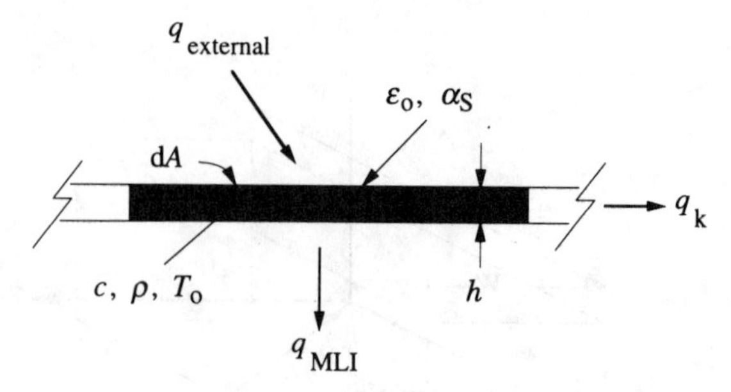

Fig. 3 Differential element of MLI outer layer.

conduction q_k is assumed to be negligible ($q_k << q_{external}$), and the material is assumed to be isothermal through its thickness.

Governing Differential Equation

Consider an isothermal differential element of the MLI outer layer with specific heat c, density ρ, thickness h, and area dA (see Fig. 3). These properties are assumed to be constant. The thermal environment of the element consists of the sun, the Earth, the bay, and deep space. Direct solar radiation and albedo (solar radiation reflected off the Earth) impinge on the element, and the element exchanges heat with the Earth, the bay, and deep space. A simple energy balance equation for the differential plane element is written to calculate its temperature T_o.

$$c\rho h \, dA \frac{dT_o}{dt} = \left(q_{sun} + q_{albedo} + q_{Earth} + q_{bay} - q_{loss} \right) dA \tag{1}$$

The element receives no direct solar radiation when the sun is obscured by the Earth or the bay surfaces. No solar radiation reaches the element while the Shuttle is in the Earth's shadow. While the Shuttle is in the sunlight, two geometric conditions must be met for the direct solar to reach the element. As can be seen in Fig. 4, the first condition is that the solar vector in the bay reference frame S_B must come in from the open top or

$$S_B \cdot X_B < 0 \tag{2}$$

To determine the additional conditions, consider a plane P_{top} which is defined by the equation $X_B = H/2$, and a new vector S_{inbox} (see Fig. 4). The vector S_{inbox} has a direction of the solar vector S_B that impinges on the element. Its magnitude is the distance between the element and the point at which S_B intersects the plane P_{top} (e.g., points A_1 and A_2). If S_{inbox} intersects P_{top} outside the boundaries delimited by the edges of the open top, the sun is obscured by the bay wall and, hence, the element is in shadow (see Fig. 4).

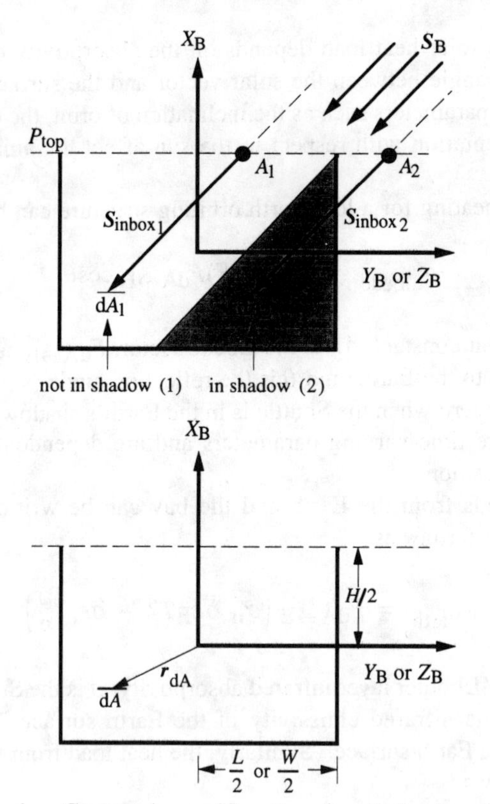

Fig. 4 Geometric considerations for shadowing effect.

Otherwise, the following inequalities are satisfied:

$$-\frac{1}{2}L < \left(\boldsymbol{r}_{dA} \cdot \boldsymbol{Y}_B - \boldsymbol{S}_{inbox} \cdot \boldsymbol{Y}_B\right) < \frac{1}{2}L \tag{3a}$$

$$-\frac{1}{2}W < \left(\boldsymbol{r}_{dA} \cdot \boldsymbol{Z}_B - \boldsymbol{S}_{inbox} \cdot \boldsymbol{Z}_B\right) < \frac{1}{2}W \tag{3b}$$

Given the solar absorptivity of the MLI α_S, when these conditions are met, the amount of direct solar absorbed by the element is

$$q_{sun} = \alpha_S \left| \boldsymbol{n}_{dA} \cdot \boldsymbol{S}_B \right| \tag{4a}$$

otherwise

$$q_{sun} = 0 \tag{4b}$$

Equations (2) and (3) indicate that the shadowing effect is dependent on the geometry of the bay, the location of the element, and the solar incidence angle, as expected. The solar heat load depends on the absorptivity of the insulation material and the angle between the solar vector and the surface normal of the element. Hence, parameters such as the inclination of orbit, the day of orbit, and the structure orientation with respect to the sun affect the amount of incident solar flux.

The albedo heating for a low-Earth orbiting structure can be approximated as

$$q_{\text{albedo}} = \alpha_S\, S(AF)F_{dA\rightarrow E}\cos\theta \tag{5}$$

where S is the solar constant, AF is the albedo factor, $F_{dA\rightarrow E}$ is the view factor from the element to the Earth, and θ is the reflection angle (see Ref. 7 and Fig. 5). The albedo is zero when the Shuttle is in the Earth's shadow. Note that both $F_{dA\rightarrow E}$ and θ are time-varying parameters and are dependent on the Shuttle orientation and position.

The heat loads from the Earth and the bay can be written based on the radiative heat transfer law as

$$q_{\text{Earth}} = F_{dA\rightarrow E}\left(\alpha_{ir}\,\sigma\varepsilon_E T_E^4 - \sigma\varepsilon_o T_o^4\right) \tag{6}$$

where α_{ir} is the MLI outer layer infrared absorptivity, σ is the Stefan-Boltzmann constant, ε_E is the infrared emissivity of the Earth surface, T_E is the mean temperature of the Earth surface. Similarly, the heat load from the bay surfaces is written as follows

$$q_{\text{bay}} = \sum_i F_{dA\rightarrow P_i}\left(\alpha_{ir}\,\sigma\varepsilon_{\text{bay}} T_{\text{bay}}^4 - \sigma\varepsilon_o T_o^4\right) \tag{7}$$

where $F_{dA\rightarrow P_i}$ is the view factor from the element to the ith surface of the bay P_i, ε_{bay} is the infrared emissivity of the bay surface, and T_{bay} is the bay surface

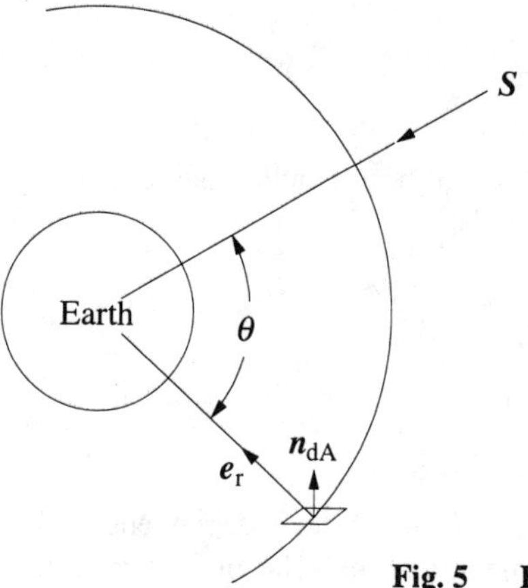

Fig. 5 Reflection angle θ for Earth albedo radiation.

temperature. The heat loss to the deep space can be written as

$$q_{loss} = \left[1 - \left(\sum_i F_{dA \to P_i} + F_{dA \to E}\right)\right] \sigma \varepsilon_o T_o^4 \tag{8}$$

where ε_o is the infrared emissivity of the MLI outer layer. Gathering all the heat load terms yields the final governing differential equation.

$$\frac{dT_o}{dt} = \frac{1}{c\rho h}\Big[\alpha_S \left| n_{dA} \cdot S_B \right| + \alpha_S S\,(AF) F_{dA \to E} \cos\theta + \alpha_{ir}\left(\sum_i F_{dA \to P_i} \sigma \varepsilon_{bay} T_{bay}^4 + F_{dA \to E} \sigma \varepsilon_E T_E^4\right) - \sigma \varepsilon_o T_o^4 \Big] \tag{9}$$

Equation (9) is a first-order nonlinear differential equation, and it applies to one side heating of the differential element when the element is exposed to the sun. This equation is numerically integrated in time using a fourth-order Runge–Kutta finite difference scheme. The view factors, $F_{dA \to E}$ and $F_{dA \to P_i}$, are calculated using the contour integral method, which is tractable because the model assumes the bay surfaces and the Earth to be planar. The entire outer layer of MLI is modeled as a mesh of differential elements, and its temperature distribution is obtained using Eq. (9) in conjunction with Eqs. (2) and (3). A computer program was written to calculate the heating loads and the temperature distribution. User input parameters are orbital parameters, Shuttle orientation, structural geometry, location of the structure in the bay, and thermo-optical and physical properties of the multilayer insulation. The program also allows the Shuttle's orientation to change at any time for a user specified time duration.

Linear Model of Multilayer Insulation

Multilayer insulation acts as a link between the radiation environment and the structure (see Fig. 6). MLI consists of multiple thin layers of material with (in space) vacuum between them. Radiation is the only heat transfer mechanism between the layers. The analysis assumes that the thermal resistance of MLI is large, and the insulated structure and the inner layer of MLI are in perfect contact such that the inner layer temperature T_i and the structural temperature T_s are equal.

An effective emissivity ε_{eff} of the MLI can be derived in terms of the emissivity of the outer layer ε_o, the emissivity of the inner layer ε_i, and the

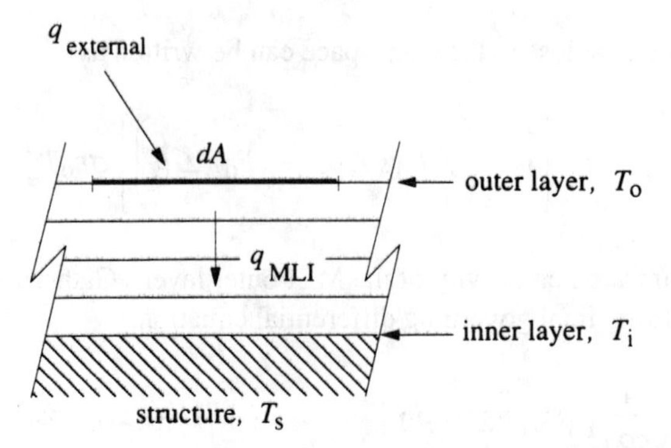

Fig. 6: Analytical model of multilayer insulation.

number of layers n (Ref. 8).

$$\varepsilon_{\text{eff}} = \frac{1}{n}\left(\frac{1}{1/\varepsilon_{\text{o}} + 1/\varepsilon_{\text{i}} - 1}\right) \tag{10}$$

The use of the effective emissivity allows the heat transfer between the outer layer and the inner layer to be expressed as a function of the temperatures of the outer layer and the inner layer, T_{o} and T_{i}, respectively

$$q_{\text{MLI}} = \sigma\varepsilon_{\text{eff}}\left(T_{\text{o}}^4 - T_{\text{i}}^4\right) \tag{11}$$

Recall that the assumption $q_{\text{MLI}} << q_{\text{external}}$ has already been made, allowing the calculation of T_{o} independent of the response of the underlying structure. To compute the structural temperatures, q_{MLI} must be computed as a nonlinear function of both T_{o} and T_{i}. A linear approximation of Eq. (11) can be obtained by introducing a coefficient h_{eff}, or

$$q_{\text{MLI}} = h_{\text{eff}}(T_{\text{o}} - T_{\text{i}}) \tag{12a}$$

where

$$h'_{\text{eff}} = \frac{\sigma\varepsilon_{\text{eff}}\left(T_{\text{o}}^4 - T_{\text{i}}^4\right)}{(T_{\text{o}} - T_{\text{i}})} \tag{12b}$$

However, h'_{eff} is a function of the known but time-varying variable T_{o} and unknown structural temperature T_{i}. Several assumptions are made to calculate a constant h_{eff}. The outer layer temperature T_{o} is written as the sum of a constant average temperature $T_{\text{o}}^{\text{avg}}$, which is the average temperature of all points in consideration, and a temperature perturbation $\Delta T_{\text{o}}(r_{dA}, t)$, which is the temperature deviation from the average temperature

$$T_o(\mathbf{r}_{dA},t) = T_o^{avg} + \Delta T_o(\mathbf{r}_{dA},t) \tag{13a}$$

Similarly, the inner layer temperature is written as

$$T_i(\mathbf{r}_{dA},t) = T_i^{avg} + \Delta T_i(\mathbf{r}_{dA},t) \tag{13b}$$

Upon substituting Eqs. (13a) and (13b) into Eq. (12b), the binomials are expanded. Assuming that the average temperatures are equal ($T_i^{avg} = T_o^{avg}$) and that the temperature perturbations are relatively small compared to the average temperature ($\Delta T_o < T_o^{avg}$ and $\Delta T_i < T_o^{avg}$), the higher order terms of ΔT_i and ΔT_o are neglected. A simple expression for a constant h_{eff} results

$$h_{eff} = 4\sigma\varepsilon_{eff}\left(T_o^{avg}\right)^3 \tag{14}$$

Note that the assumption of a large thermal resistance of MLI allows $\Delta T_i / T_o^{avg}$ to be relatively small. However, $\Delta T_o / T_o^{avg}$ may not be always small, resulting in a significant error in the approximation of q_{MLI}. An example and the limitations of employing a constant h_{eff} are further discussed in the numerical section.

In the case where the constant h_{eff} does not model q_{MLI} sufficiently, an effective outer layer temperature T_o^{eff} is introduced to rewrite Eq. (12a) as

$$q_{MLI} = h_{eff}\left(T_o^{eff} - T_i\right) \tag{15a}$$

where the effective temperature is directly computed from Eqs. (11) and (15a).

$$T_o'^{eff} = \frac{\sigma\varepsilon_{eff}}{h_{eff}}\left(T_o^4 - T_i^4\right) + T_i \tag{15b}$$

Here the assumption that $\Delta T_i < T_o^{avg}$ is small is applied to determine an approximate T_o^{eff} in terms of the known variables.

$$T_o^{eff} = \frac{\sigma\varepsilon_{eff}}{h_{eff}}\left(T_o^4 - T_i^{avg^4}\right) + T_i^{avg} \tag{16}$$

The application of T_o^{eff} is also presented in the numerical section.

Finite Element Analysis

The transient thermal response is obtained using a finite element thermal model of the structure. The structure is a three-dimensional conduction model with a time-dependent convective boundary condition. The ambient temperatures and the convection coefficient are defined by T_0^{eff} and h_{eff}, respectively. The convection elements using h_{eff} and T_0^{eff} replace the radiation elements to model the thermal loading on the structure and, thus, the finite element analysis becomes linear. Then the temperature distribution is input to a compatible structural finite element model to calculate the thermal deformations.

Numerical Results

Numerical results for a conceptual structural design of SITE in a Space Shuttle payload bay are presented to illustrate the methods developed in the paper and to understand the key parameters in the analysis. The external environment of the SITE structure in two orbit cases is considered. The thermal response of the MLI outer layer over the entire surface of the structure was calculated. The methods for a linear approximation of the heat transfer between the outer layer and the structure are compared, and their limitations are discussed. The decoupling of the radiation environment and the insulated structure is investigated. The temperature distribution of SITE from the finite element analysis is presented, and the effect of the thermal deformations on the performance of SITE is assessed.

Geometry

The SITE structure is an aluminum plate box with length L_S, width W_S, height H_S, and plate thickness t_S (see Fig. 7). The box is wrapped in aluminized

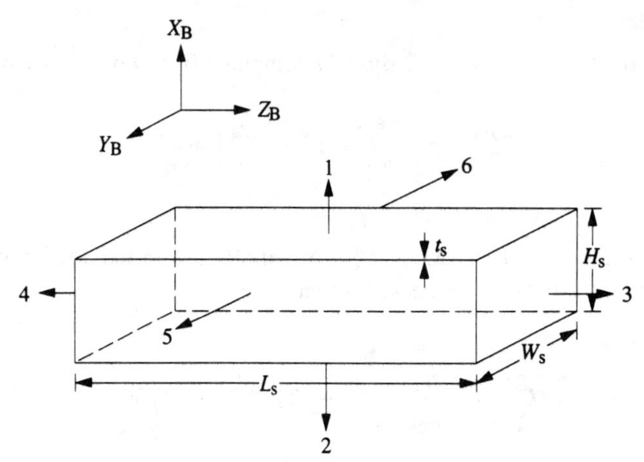

Fig. 7 SITE conceptual structural design.

Kapton MLI. The top, bottom, right, left, front, and back surfaces are numbered as 1, 2, 3, 4, 5, and 6, respectively. The numerical data for the box and the aluminized Kapton are listed in Table 1 (see Ref. 2).

The two orbits chosen for the numerical calculations have an inclination of 28.5 deg and an altitude of 278 km, resulting in a orbital period of 90 min. The Shuttle enters the sunlight at $t = 0$. In orbit 1, referred to as the sun view/cold orbit, the Shuttle bay points directly at the sun while in the sunlight. The orbiter then rotates 180 deg about the Y_B axis and maintains a constant attitude while in the shadow (see Fig. 8a). In orbit 1, SITE only observes while in the Earth's shadow because SITE does not observe while pointing directly at the sun. Figure 8b shows the geometry of orbit 2, referred to as the gravity gradient orbit. The Shuttle bay points at a target and continually rotates about the X_B axis. As a result, the Shuttle's nose continuously points at the Earth. Table 2 lists the bay dimensions and the orbital heating load parameters.

Characterization of the Environment

The finite difference model was used to obtain the thermal response of the MLI outer layer at 174 points on the surfaces of the SITE box for orbits 1 and 2. The sun view/cold attitude induces a wide T_o range, with extreme temperature values of 189 – 384 K, which both occur on surface 1 (top surface) of the SITE box. In the gravity gradient case, T_o reaches 369 K, but the lower extreme is only 238 K. The high extreme appears on surface 1 (top surface) at different times in the orbit. The lower extreme appears on surface 6 (back surface).

Figures 9a and 9b present the thermal responses for selected four nodes on surface 1 (top surface) in orbits 1 and 2, respectively. For the sun view/cold orientation, because the Shuttle points directly at the sun, the surface receives a constant amount of direct solar for half orbital period and, hence, all of the nodes attain their steady-state values. As the Shuttle enters the Earth's shadow the temperatures drop drastically. The sinusoidal shape of the curves are results of the albedo and Earth-emitted radiation, because the position of the Earth relative to the structure varies in sinusoidal nature for orbit 1.

Table 1 Aluminized Kapton MLI data

L_S, m	4.1707
W_S, m	0.9982
H_S, m	0.6477
t_S, m	3×10^{-3}
α_S	0.31
ε_o	0.45
h,m	6.35×10^{-6}
ρ, kg/m^3	1412
n	10

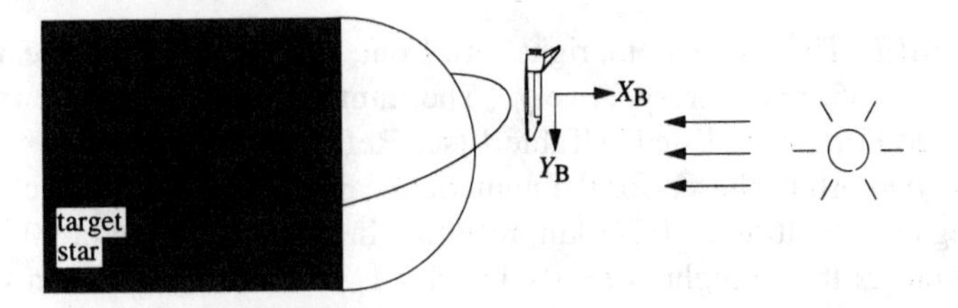

Fig. 8a Orbit 1: sun view/cold.

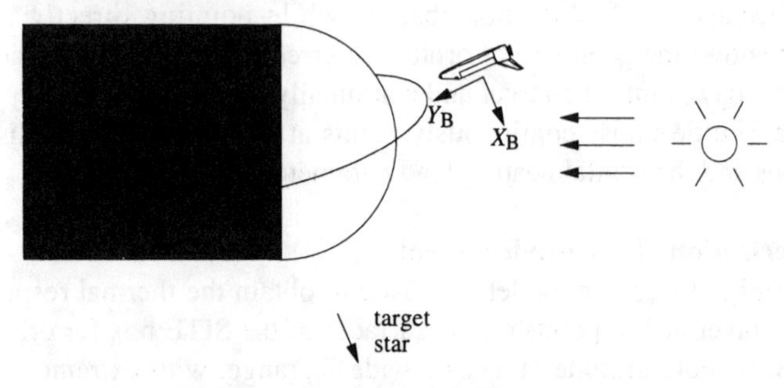

Fig. 8b Orbit 2: gravity gradient.

The gravity gradient orbit shows an interesting result because the Shuttle orientation is more complex than the direct sun viewing attitude. Figure 9b illustrates the local shadowing effect which occurs due to the time varying relative position of the sun from surface 1 (top surface) of the box. Nodes enter and leave the local shadow at different times in the orbit and, thus, the nodes attain the same peak temperature occur at different times in the orbit. The sinusoidal behavior occurs due to the time-varying relative position of the sun. As expected, temperatures drop drastically as soon as nodes enter shadow. During the orbital night the sinusoids do not appear in the temperature histories because the relative position of the Earth is constant. These figures indicate that the type of the orbit and the Shuttle orientation influence the magnitude of the thermal response as well as the time and space variation of the response.

Table 2 Bay thermal environment model data

H, m	4.572
L, m	18.29
W, m	4.572
AF	0.25
S, W/m^2	1390
σ, W/m^2-K^4	5.67×10^{-8}

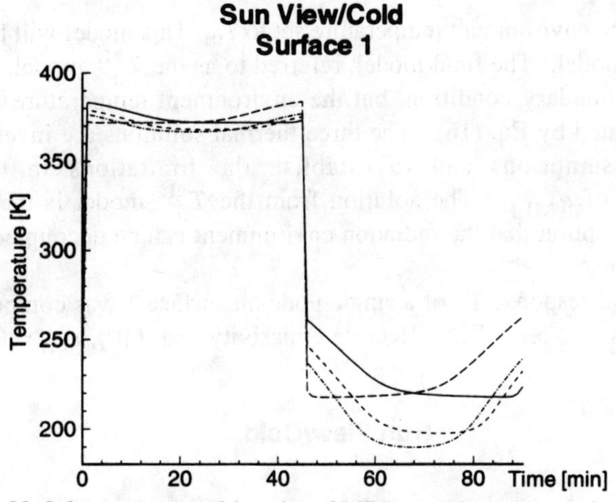

Fig. 9a Nodal temperature histories of MLI outer layer on surface 1 in sun view/cold.

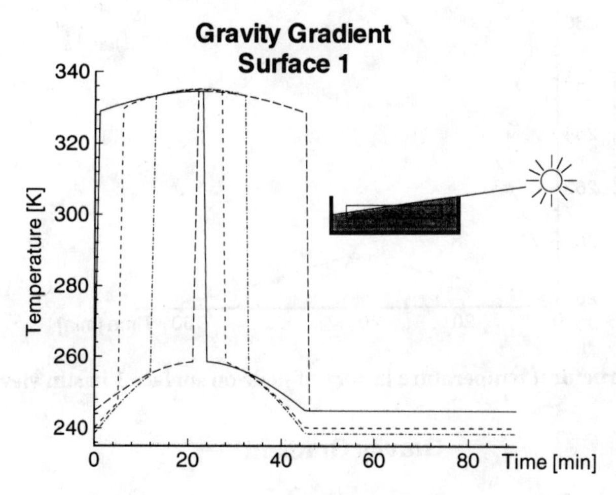

Fig. 9b Nodal temperature histories of MLI outer layer on surface 1 in gravity gradient.

Linear Model of Multilayer Insulation

The finite element thermal analysis package ADINA-T was used to develop a thermal model of the SITE structure. Three methods to model the heat transfer between the MLI outer and inner layer, q_{MLI}, are applied to the finite element thermal model to obtain the structure's thermal response. The first thermal model used a radiative boundary condition with the environment temperature set to T_o and $\varepsilon = \varepsilon_{eff}$. This boundary condition results in thermal loads q_{MLI} as calculated by Eq. (11). This model will be referred to as the radiative model. The second model used a convective boundary condition with a constant h_{eff},

Eq. (14), and the environment temperature set to T_0. This model will be referred to as the h_{eff} model. The final model, referred to as the T_0^{eff} model, also used a convective boundary condition, but the environment temperature was set to T_0^{eff} as calculated by Eq. (16). The three thermal solutions are investigated to assess the assumptions and to establish the limitations in the linear approximation of q_{MLI}. The solution from the T_0^{eff} model is then used to assess the assumption that the radiation environment can be decoupled from the internal structure.

The thermal response T_s of a single node on surface 1 was computed using the three q_{MLI} models. The effective emissivity, Eq. (10), $\varepsilon_{eff} = 0.029$ was

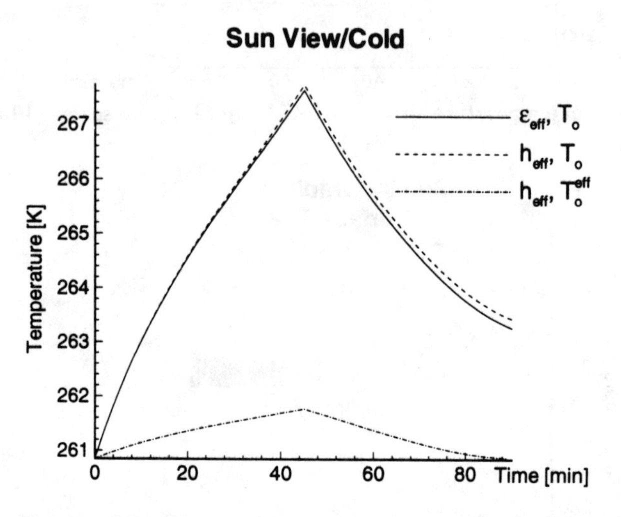

Fig. 10a Structural temperature history of node on surface 1 in sun view/cold.

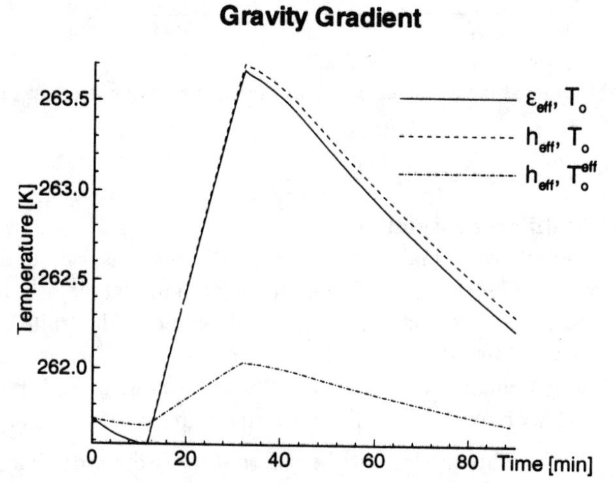

Fig. 10b Structural temperature history of node on surface 1 in gravity gradient.

used in the radiative model. The effective convection coefficients h_{eff}, Eq. (14), were computed to be 0.0292 W/m^2-K and 0.0295 W/m^2-K using T_i^{avg} of 260.8 and 261.7 K for orbits 1 and 2, respectively. The effective outer layer temperature T_o^{eff}, Eq. (16), was computed using the computed h_{eff} values. Figure 10 illustrates the comparison of the models in determining the structure temperature T_s. First of all, the structural temperature variation is much less than the environment temperature variation. Secondly, for both orbits, the linear T_o^{eff} model agrees well with the nonlinear radiation model. On the other hand, although the h_{eff} model follows the general trend of the radiation model, the errors of the h_{eff} model calculations are significant. The maximum errors for orbits 1 and 2 are approximately 6 and 1.7 K, respectively. The T_o^{eff} model is effective because the structural temperature deviation is small compared to the average temperature due to the effective insulation and the large thermal mass of the SITE structure. The maximum values of the ratio $\Delta T_i / T_o^{avg}$ for orbits 1 and 2 are 0.02 and 0.01, respectively. The error of the h_{eff} model is significant due to a large ΔT_o. The maximum values of $\Delta T_o / T_o^{avg}$ are 0.47 and 0.41, respectively. The disagreements of the results can arise because 1) for large ΔT_o, a constant h_{eff} does not sufficiently model q_{MLI}, and 2) calculation of T_o^{eff} requires relatively small ΔT_i. Thus, the ratios $\Delta T_o / T_o^{avg}$ and $\Delta T_i / T_o^{avg}$ establish the limitations on the use of the h_{eff} and T_o^{eff} models, respectively, in the linear approximation of q_{MLI}.

The heat transfer q_{MLI}, Eq. (15a), was calculated for all surface nodes every 15 min. The T_o^{eff} model was used to compute q_{MLI}. This heat transfer was compared to the total external heat load for the orbit cases. The external heat load is defined here as the sum of the absolute values of the heat load terms in Eq. (1). The maximum absolute value of the ratio $q_{MLI} / q_{external}$ for both orbits is approximately 0.025. The heat transferred through the insulation is only about 2.5 % of the heat exchange between the surface and the environment and, thus, the environment calculation can be decoupled from the internal structural temperature calculations, as the thermal analysis initially assumed.

Finite Element Analysis

Thermal Analysis

The temperature distributions of the SITE structure for both example orbits were calculated based on a linear thermal model using the h_{eff} and T_o^{eff} values (T_o^{eff} model) computed earlier. Figure 11 shows the temperature distribution at the time at which the maximum temperature differential appears, i.e., the time when the difference between the maximum temperature and the minimum temperature anywhere in the structure is at a maximum. Only the SITE observation period was considered. Hence, the thermal response while surface 1 (top surface) is in direct sunlight was neglected.

The structure attained its maximum temperature T_i at the time when the maximum temperature differential appeared. The maximum structural temperature is 268 and 264 K for orbits 1 and 2, respectively. The maximum temperature differential for orbit 1 (see Fig. 11a) is 7 K, occurring between

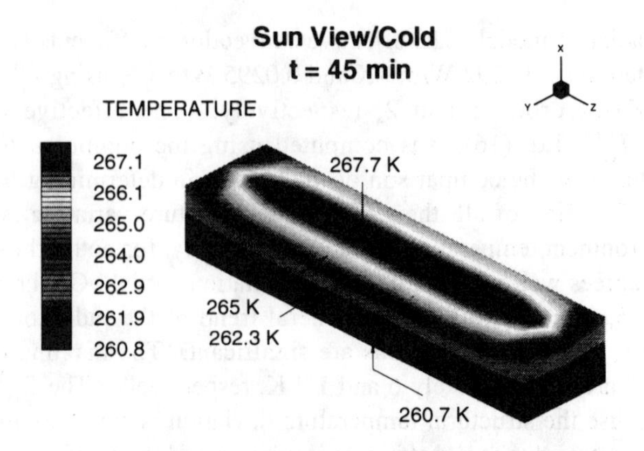

Fig. 11a Thermal response of SITE in sun view/cold at *t* = 45 min.

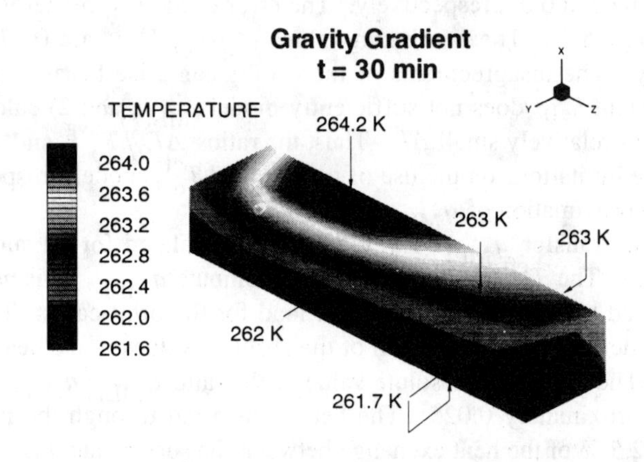

Fig. 11b Thermal response of SITE in gravity gradient at *t* = 30 min.

surface 1 (top surface) and surface 2 (bottom surface). The Shuttle orientation induces a differential heating between the two surfaces because surface 1 (top surface) receives the maximum amount of direct solar whereas surface 2 (bottom surface) receives no solar heating (see Fig. 8a). For the gravity gradient attitude, a maximum temperature differential of 3 K appears across the top left corner of surface 1 (top surface) and bottom right corner of surface 5 (front surface) at t = 30 min (see Fig. 11b). At t = 30 min, parts of surfaces 1 (top surface) and 6 (back surface) are exposed to direct solar radiation while the other surfaces are in shadow (see Fig. 8b).

Structural Analysis

To perform the structural analysis, the thermal response would be input to the same finite element mesh, which is one of the advantages of using the finite element model in a thermal–structural analysis. Results shown here, however, were calculated from a different finite element structural mesh due to the temporary unavailability of ADINA-T and an existing structural model on another finite element package. The nodal temperatures and the temperature gradients were transferred to the structural mesh to obtain approximate static thermal deformations. Given the thermal deformations, the effect on the performance of SITE was assessed based on a conceptual optical layout.

One of the measures of SITE performance is represented in terms of the differential path length (DPL), or the difference between the paths of two light beams which enter an interferometer and combine at a detector (see Ref. 1). The SITE requires the DPL to be controlled on the order of wavelengths λ of the target light ($\lambda = 550$ nm). The differential path lengths were computed using the thermal deformations obtained in the structural analysis. The computed DPL for orbits 1 and 2 were approximately 19 λ and 16 λ, respectively.

The preliminary results indicate that for the first-cut SITE design the thermally induced path length errors are too large for a successful mission without active control. Thus, the thermal deformation of the structure has been identified as a critical issue in the design of SITE. Selection of low-thermal-load orbits, use of low-thermal expansion materials, and/or addition of more thermal control levels will be required. The internal optical control system of SITE can also be modified to reduce the DPL.

Summary

An analytical approach to obtain the transient thermal–structural response of an insulated space structure operating in the Space Shuttle payload bay is presented. The payload bay thermal environment is characterized based on an energy balance for a differential element at an arbitrary location in the bay. Assuming that the thermal resistance of the insulation is very large, the nonlinear radiation environment model is decoupled from the linear internal conduction model. Two methods are developed to approximate the insulation as a linear resistance between the environment and the structure. The transient thermal response of the structure is obtained from a finite element thermal analysis. The finite element model reduces to a linear conduction model with a linear boundary condition. The temperature distribution is applied to a compatible structural mesh to obtain the structural response.

Numerical results are presented for a conceptual structural design of the stellar interferometer tracking experiment. Two Shuttle orientations were considered to obtain the thermal–structural response of the SITE structure: sun view/cold and gravity gradient. As expected the Shuttle orientation has a significant effect on the temperature distribution of the structure. The maximum outer layer temperature perturbations for sun view/cold and gravity gradient attitudes were 123 K and 107 K, respectively. The maximum structure temperature perturbations for these orbits were 7 K and 2.5 K. The insulation filters the external heat load significantly and, thus, allows the radiation environment to be decoupled from the linear internal model. The structure attained a maximum temperature differential of 7 and 3 K in orbits 1 and 2, respectively. The differential appeared between surfaces 1 and 2 of the structure for the sun view/cold attitude. The gravity gradient attitude induced a maximum temperature differential across the top left corner of surface 1 (top surface) and bottom right corner of surface 5 (front surface). The locations of the maximum temperature differentials depend on the position of the sun with respect to the surfaces and, hence, the orbital parameters and the Shuttle orientation.

Despite the relatively low differentials in the structure, the performance of SITE was unacceptable without active control. The maximum differential path lengths induced by the thermal deformations of the structure were predicted to be 16 to 19 wavelengths of the target light. These errors can be controlled by 1) a better thermal–structural design, or 2) an appropriate internal optical control system, or 3) both approaches. Thus the thermal–structural analysis is essential in the design process, and a successful design will require iterative thermal–structural analyses to meet the functional requirements of SITE.

The analytical approach presented in the paper allows an efficient thermal analysis for multiple mission scenarios and structural designs. The computer program written to calculate the temperatures of the MLI outer layer is a simple tool, and it allows an understanding of the key parameters which characterize the payload bay thermal environment for various orbital attitudes. The nonlinear heat transfer between the outer layer and the structure can be modeled linearly using an effective constant convection coefficient under the condition that the outer layer temperature perturbation is not too large. To achieve a better approximation, an effective outer layer temperature can be used when the structure temperature perturbation is small compared to the outer layer temperature perturbation. The decoupling of the radiation environment and the internal conduction model simplifies the computation involved in the finite element thermal analysis of the structure.

Acknowledgments

This research was funded by NASA grant NAGW-1335 through the Massachusetts Institute of Technology Space Engineering Research Center. The numerical analysis was done in part with the use of the computing facility at the Massachusetts Institute of Technology Vortical Flow Research Laboratory. The structural finite element model was developed by Brett Masters at the Massachusetts Institute of Technology Space Engineering Research Center.

References

[1]Blackwood, G., Hyde, T., Miller, D., Crawley, E., Shao, M., and Laskin, R., "Stellar Interferometer Tracking Experiment: A Proposed Technology Demonstration Experiment," 44th Congress of the International Astronautical Federation, Paper No. IAF-93-I.5-247, Oct. 1993.

[2]Gilmore, D. G., and Collins, R. L., "Thermal Design Analysis," *Satellite Thermal Control Handbook*, Aerospace Corp. Press, El Segundo, CA, 1994, Chap. 2.

[3]Warren, A. H., Arelt, J. E., Eskew, W. F., and Rogers, K. M., "Finite Element-Finite Difference Thermal/Structural Analysis of Large Space Truss Structures," AIAA Paper 92-4763, 1992.

[4]Mahaney, J., and Thornton, E. A., "Self-Shadowing Effects on the Thermal–Structural Response of Orbiting Trusses," *Journal of Spacecraft and Rockets,* Vol. 24, No. 4, 1987, pp. 342–348.

[5]Thornton, E. A., and Kim, Y. A., "Thermally Induced Bending Vibrations of a Flexible Rolled-Up Solar Array," *Journal of Spacecraft and Rockets*, Vol. 30, No. 4, 1993, pp. 438–448.

[6]Kim, Y. A., "Transient Thermo–Structural Analysis of an Insulated Space Structure," Master's Thesis, Massachusetts Institute of Technology, Cambridge, MA, June 1994.

[7]Mahoney, J., and Strode, K., "Fundamental Studies of Thermal–Structural Effects on Orbiting Trusses, " AIAA Paper 82-650, May 1982.

[8]Haviland, R. P., and Housem C. M., *Handbook of Satellites and Space Vehicles,* D. Van Nostrand, Princeton, NJ, 1965, Chap. 9.

Chapter 2. Experimental Studies of Thermal Structures

Boundary Conditions for Aerospace Thermal-Structural Tests

Max L. Blosser[*]

NASA Langley Research Center, Hampton, Virginia 23681-0001

Abstract

One of the most challenging aspects of thermal-structural testing of aerospace structures is providing the desired boundary conditions for the test specimen. The requirements for thermal boundary conditions often conflict or interact with the requirements for structural boundary conditions. In the current paper, some of the theoretical boundary conditions commonly used in thermal and structural analyses, as well as some experimental approaches to applying thermal and structural boundary conditions, are reviewed. Conflicts and interactions between the thermal and structural boundary conditions are identified and discussed. Several thermal-structural tests of aerospace structures are discussed to illustrate practical methods which have been used to address the challenges described in this paper.

Nomenclature

A	= area
B	= geometric parameter related to surface curvature, see Ref. 11
d	= thickness of contacting body in Eq. (6d)
h	= heat transfer coefficient
k	= thermal conductivity
N	= force resultant along the edge of a plate or shell
M	= moment resultant along the edge of a plate or shell
Q	= heat flow
Q_α	= transverse shear force resultant
q	= heat flux
P	= location of a point P on the boundary
R	= radius of curvature
R_c	= thermal contact resistance

---

[*] Aerospace Engineer, Structures Division, Thermal Structures Branch

T	= temperature
t	= time
u, v, w	= displacements in the α, β, and z directions, respectively
$\overline{\alpha}$	= absorbtivity
α, β, z	= surface coordinates for plate or shell
ε	= emissivity
σ	= Stefan—Boltzman constant

Subscripts

f	= fluid
i	= incident on surface
n	= normal to the boundary
s	= on the boundary surface
1	= referring to the primary contacting body
2	= referring to the secondary contacting body
α, β	= in the directions of the surface coordinates for plate or shell

Subsubscripts

a referring to the contacting surface
b referring to the surface on the secondary body away from the point of contact

Introduction

Advanced supersonic and hypersonic vehicles, such as the high-speed civil transport (HSCT), single-stage-to-orbit (SSTO) vehicle and the National Aero-Space Plane (NASP), will encounter significant aerodynamic heating. Thermal-structural tests are required to design and validate the structures for these vehicles. Providing the desired thermal and structural loads and boundary conditions to the structural test specimens is a challenge. Requirements for thermal loads and boundary conditions often conflict or interact with those for mechanical loads and boundary conditions. Considerable effort and ingenuity is needed to develop experiments which provide an acceptable compromise between the thermal and structural requirements.

Aircraft skins which are subjected to aerodynamic heating are typically lightweight, thin-gauge structures. Therefore, a considerable amount of work has been done to understand the behavior of plate and shell structures subjected to thermal and mechanical loading. Some of the early efforts in thermal-structural testing at three of the major thermal-structures laboratories in the United States are described in Refs. 1—4. More recent experiences with thermal-structural tests are described in Refs. 5 and 6. A review of analytical and experimental research on thermal buckling of plates and shells is given in Ref. 7. Research on convectively cooled structures, including many thermal-structural tests, is reviewed in Ref. 8.

The terms load and boundary condition are often referred to separately. However, it is difficult to draw a clear distinction between the two. In the current paper the term boundary condition is defined to include loading.

Thermal-structural tests are usually performed to validate a particular design or a design method. Some type of analysis is used in the design process to predict the behavior of the test component. Analytical methods often employ simple, well-defined boundary conditions to reduce the complexity and cost of the analysis. When attempting to correlate analytical and experimental results, it is important to consider both the analytical and experimental boundary conditions. Examples of correlating thermal-structural analyses and experiments are given in Ref. 9. In the current paper, some of the theoretical boundary conditions commonly used in thermal and structural analyses, as well as some experimental approaches to applying thermal and structural boundary conditions, are reviewed. Conflicts and interactions between the thermal and structural boundary conditions are identified and discussed. Several thermal-structural tests of aerospace structures are discussed to illustrate practical methods which have been used to address the challenges described in this paper.

Analytical Boundary Conditions

To predict the thermal-structural performance of an aerospace structure, an analyst must calculate the temperature distribution as well as the deformation and internal stress state of the structure. In the most general case there is a coupling between the deformation and temperature distribution.[10] This coupling is associated with an absorption or release of heat due to a mechanically induced dilatation, or volume change, of a material. In most tests of aerospace structures, however, the temperature changes from this thermoelastic coupling are negligible. If this thermoelastic coupling is negligible, the temperatures can be calculated without considering the deformations of the structures. The resulting temperature distribution is used, along with any additional mechanical loading, to calculate the deformations and stresses in the structure. The variation in temperature affects the behavior of the structure primarily through two mechanisms: thermal expansion and variation of material properties with temperature. If there are significant temperature variations within the structure during a test, an accurate temperature distribution over the entire specimen is required to calculate accurate displacements and stresses. The thermal and structural analyses each have their own set of loads and boundary conditions.

Thermal

Within the material of the structure, heat is transferred by conduction. Other modes of heat transfer may occur on the surfaces of the structure. The thermal analysis consists of solving the classical heat conduction equation subject to an initial temperature distribution and appropriate boundary

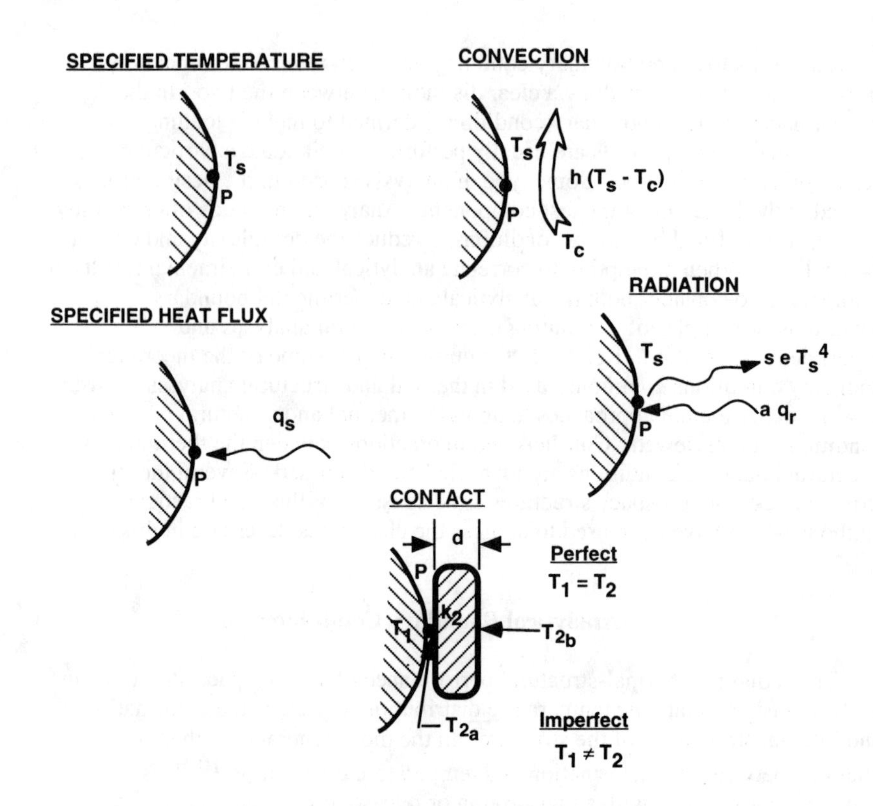

Fig. 1 Thermal boundary conditions.

conditions. Thermal boundary conditions[10] include prescribed temperature, prescribed heat flow, convection, radiation, or contact with another body as illustrated in Fig. 1. Equations for the boundary conditions are given in both differential form, which applies to an infinitesimally small point on the surface, and discrete form, which is an approximation over a finite-size region of the boundary. The discrete form is often useful for determining how to modify experimental thermal boundary conditions.

The simplest thermal boundary condition, from an analyst's point of view, is prescribed temperatures. Temperatures are the fundamental unknowns of the analysis, and so each prescribed temperature directly reduces the size of the problem. This boundary condition may be used to approximate the behavior of a surface of the structure attached to a large heat sink or maintained near a temperature by a very effective coolant flow or an actively controlled heater. To simplify the analysis, temperature histories measured during an experiment are sometimes prescribed on the heated surface of the thermal model to impose thermal loading,

$$T(P,t) = T_S \tag{1}$$

Another simple boundary condition is a prescribed heat flux. A prescribed heat flux is often used for two commonly encountered situations. A prescribed heat flux of zero is used to represent perfectly insulated boundaries and symmetry boundaries. Approximating a well-insulated boundary as perfectly insulated can greatly simplify the analysis. Using symmetry boundary conditions to model a small representative piece of a much larger specimen can also greatly reduce the analytical effort. When the net heat flux entering or leaving a surface is known, it may be applied directly as a prescribed heat flux rather than modeling the actual heat transfer mode at the surface.

$$q_n = - q_s(P,t) \tag{2}$$

Sometimes the simple specified temperature or heat flux boundary conditions are not sufficient, and it is necessary to model the mode of heat transfer at the surface of the model. The remaining three boundary conditions are modes of heat transfer: convection, radiation, and contact.

Convection occurs when a fluid flows across a surface at a different temperature. Heat is transferred between the fluid and the surface, and then the fluid physically moves away from the surface. The heat transfer from the surface to the fluid is approximated by Eqs. (3a) and (3b). The convection coefficient is a function of material properties and flow parameters. The convection boundary condition is used to model situations such as natural convection of air surrounding a test panel, coolant passages attached to the boundary of a panel, or air blown over the surface of a panel.

$$q_n = h(P,t)\,(T_s - T_f) \tag{3a}$$

$$Q_n = h(P,t)\,A\,(T_s - T_f) \tag{3b}$$

All bodies radiate heat from their exposed surfaces. When a body and its surroundings are at the same temperature, it absorbs as much heat as it emits. However, when the temperature of the body is different than its surroundings, radiation heat transfer may become important. Simple expressions for the radiant heat flux leaving a surface are given in Eqs. (4a) and (4b). In addition, when modeling radiation exchange with surrounding objects, geometric view factors must be considered, thereby complicating the analysis. Radiation heat transfer may be used for modeling situations such as a hot panel with a surface exposed to cooler surrounding, a nonuniformly heated panel with an uninsulated internal cavity, or a panel heated by radiant quartz lamps,

$$q_n = \sigma\varepsilon T_s^4 - \bar{\alpha} q_i \tag{4a}$$

$$Q_n = \sigma\varepsilon A T_s^4 - \bar{\alpha} q_i A \tag{4b}$$

Heat transfer between contacting bodies is another commonly occurring boundary condition . If the bodies are different materials, there is an abrupt change of conductivity. There is additional resistance to heat flow between the bodies if the bodies are not in perfect contact. Imperfect contact may be important for situations such as bolted joints with temperature gradients or fixturing which contacts a panel to introduce mechanical loading or restrict motion.

SPECIFIED DISPLACEMENTS

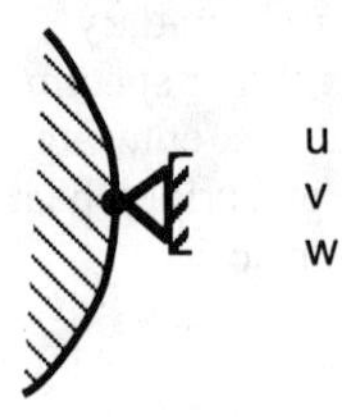

SPECIFIED TRACTIONS (FORCES)

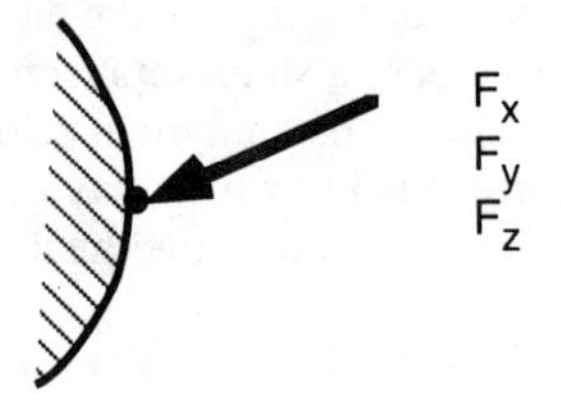

Fig. 2 Three-dimensional structural boundary conditions.

Perfect contact:

$$T_1(P,t) = T_2(P,t) \tag{5a}$$

$$q_n(P,t) = k_2 \frac{\partial T_2}{\partial n}(P,t) \tag{5b}$$

$$Q_n(P,t) = k_2 A / d\left[T_{2_a}(P,t) - T_{2_b}(P,t)\right] \tag{5c}$$

Imperfect contact:

$$q_n(P,t) = 1 / R_c\left[T_2(P,t) - T_1(P,t)\right] \tag{6a}$$

$$q_n(P,t) = k_2 \frac{\partial T_2}{\partial n}(P,t) \tag{6b}$$

$$Q_n(P,t) = A / R_c\left[T_2(P,t) - T_1(P,t)\right] \tag{6c}$$

$$Q_n(P,t) = k_2 A / d\left[T_{2_a}(P,t) - T_{2_b}(P,t)\right] \tag{6d}$$

Structural

The behavior of the structure is governed by equilibrium equations, which assure that internal and external forces remain in equilibrium throughout the structure; strain displacement equations, which relate strains to deformations; and constitutive equations, which relate the strains to stresses in the material. The structural analysis is usually formulated to solve for deformations and/or stresses within the structure. Boundary conditions on the surface may consist of prescribed displacements or tractions. For a three-dimensional body, either displacements or tractions may be specified in any or all three principal directions at any point on the boundary as illustrated in Fig. 2.

Table 1 General plate and shell boundary conditions

Specified plate boundary conditions			Specified shell boundary conditions		
Along edges					
u	or	N_α	u	or	N_α
v	or	$N_{\alpha\beta}$	v	or	$N_{\alpha\beta} + \dfrac{M_{\alpha\beta}}{R_\beta}$
w	or	$Q_\alpha + \dfrac{\partial M_{\alpha\beta}}{\partial \beta}$	w	or	$Q_\alpha + \dfrac{1}{B}\dfrac{\partial M_{\alpha\beta}}{\partial \beta}$
$\dfrac{\partial w}{\partial \alpha}$	or	M_α	$\dfrac{\partial w}{\partial \alpha}$	or	M_α
At corners					
w	or	$M_{\alpha\beta}\Big\vert_{\beta_2}^{\beta_1}$			$M_{\alpha\beta} w\Big\vert_{\beta_2}^{\beta_1}$

Many aerospace structures are thin plates and shells as a result of the need to minimize structural weight. Various plate and shell theories have been developed to reduce the general three-dimensional analysis to a two-dimensional problem. By making simplifying assumptions about the through-the-thickness behavior of the plate or shell, the deformation of the plate or shell is characterized by displacements and rotations at the neutral surface. The boundary conditions along the edge of a plate or shell are specified displacements, rotations, force resultants, and moment resultants about the neutral surface.[10-13] The expressions in Table 1 describe pairs of displacement and force or moment resultants which may be used to completely specify boundary conditions for a plate or shell. The notation and force directions for the force resultants from Ref. 11 are illustrated in Fig. 3.

For plates, the in-plane analysis is usually uncoupled from the bending analysis. For the bending analysis, the general boundary conditions reduce to several commonly encountered special cases: simply supported, clamped, and free.

Simple support:

$$w = 0, \quad M_\alpha = 0 \tag{7a}$$

Clamped:

$$w = 0, \qquad \frac{\partial w}{\partial \alpha} = 0 \tag{7b}$$

Free:

$$Q_\alpha + \frac{\partial M_{\alpha\beta}}{\partial \beta} = 0, \qquad M_\alpha = 0 \tag{7c}$$

Experimental Boundary Conditions

Thermal-structural tests are often performed to assess the performance of aerospace vehicle structures during high-speed flight. Important load cases may

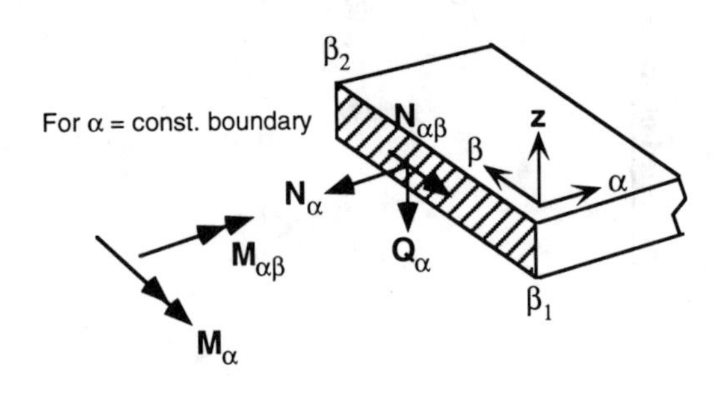

Fig. 3 Notation and positive directions for force resultants.

include: mechanical loading at a uniform elevated temperature with minimal thermal stresses, thermal stresses resulting from nonuniform temperature distributions or other restraints to thermal expansion, or a combination of mechanical loading and thermal stresses. Severe thermal stresses are often the result of transient temperature differences between parts of the structure that increase in temperature at different rates. Transient thermal tests may be used to simulate vehicle heating history and transient thermal response, but less complicated, steady-state tests are often devised to simulate a particular temperature distribution of interest. Tests may involve various combinations of thermal and mechanical load cycles. For airframe structures, test specimens vary from simple tensile coupons to panels to structural components to full-scale vehicles with increasingly complicated boundary conditions. For thermal-structural tests, the desired temperature distributions must be generated to achieve the desired deformations and stresses. Therefore, accurate thermal boundary conditions are a critical prerequisite to obtaining the anticipated structural response.

Where possible, fixturing used to provide experimental thermal and structural boundary conditions should be designed to be easily analyzed. For example, insulating an exposed surface can eliminate natural convection and radiation, both of which can be complicated to analyze. If the experimental boundary conditions are not easily analyzed or if they do not produce the expected behavior in the specimen, additional experiments may be required to characterize the boundary conditions. The measured characteristics of the boundary conditions can be input into the analytical model to improve the correlation between predicted and measured behavior.

Two approaches to providing boundary conditions for thermal-structural experiments are often used. One approach is to use a specimen limited to the size of interest and to provide experimental boundary conditions that match, as closely as possible, those of the analysis. The second approach is to build a specimen larger than the region of interest so that there is transition structure surrounding the region of interest. The intent of the larger test specimen is to have any undesirable effects of imperfect boundary conditions to occur within the transition structure, so that the region of interest will have the desired boundary conditions. The first approach uses a smaller, less expensive specimen and test apparatus, however, the second approach may often provide a more accurate representation of the desired boundary conditions.

Thermal

Most thermal-structural tests of aerospace structures attempt to simulate aerodynamic heating of a high-speed vehicle. The most realistic heating can be obtained through flight tests or high-speed wind tunnel tests. Such tests, however, are expensive and place severe limitations on the specimen size and geometry and the distribution and duration of applied heating.

The most common method of simulating aerodynamic heating of aerospace structures is to use radiant heat. For temperatures up to 3000°F and heat fluxes up to 110 Btu/ft^{2}s, air-cooled quartz lamps[14] may be used. Graphite heaters[15] may be used to provide temperatures up to 4000°F and heat fluxes up to 350

Btu/ft^{2}s. Much higher heat fluxes, up to 2600 Btu/ft^{2}s over a 0.2 in. x 4 in. area, may be obtained using an arc lamp heater.[15] Quartz lamps are, by far, the most versatile and commonly used method of radiant heating. Arrays of quartz lamps may be configured to heat large or small, flat or contoured surfaces without the need for a low oxygen environment. Individually controlled lamp zones may be used to impose steady-state or transient temperature gradients over the test specimen. Quartz lamps may be rapidly and precisely controlled by varying the applied electrical power to lamps to produce a wide range of heat output from a fixed lamp array. However, a supporting framework is required to position the quartz lamps next to the structural panel, an electrical power distribution and control system is required to control all of the zones, and a clear view of surface to be heated must be provided. For higher heat fluxes, graphite heaters have many similar features to quartz lamps. However, graphite heaters require a low oxygen environment to avoid oxidation of the graphite filaments, do not respond as quickly as quartz heaters, and require considerably more electrical power. Arc lamps provide much higher heat fluxes, but are much more expensive than either quartz or graphite heaters, and are limited to heating much smaller regions. Any type of radiant heater may have difficulty producing an even heating distribution on an uneven surface. Some control over the amount of radiant heat entering and leaving a specimen can be obtained by coating the heated surface to provide a desired surface emissivity and absorbtivity.

Other methods of heating specimens are also used. Ovens are commonly used for smaller, uniform temperature specimens which do not require rapid changes in temperature. Induction heating can be used to provide high levels of heating to smaller specimens. Lasers may also be used to provide very high, localized heating. Blanket and bond-on electric resistance heaters may occasionally be used for lower temperature tests. A remarkable use of convective heating is using hot air to heat the full-scale Concorde supersonic transport.[16] The convective heating system was also used to cool the Concorde to complete the desired thermal cycle. For many tests, the only thermal requirement is to generate a particular temperature distribution. For some tests, however, a particular mode of heat transfer must be used (e.g., a transpiration-cooled panel normally requires a hot flow test to verify its performance).

It is often desirable to limit heat loss at some boundaries of a test specimen to avoid unwanted temperature gradients. Flexible fibrous insulation, which is light and flexible enough not to affect structural performance, is often used to cover unheated surfaces on hot structures to reduce unwanted heat loss from natural convection or radiation. To limit heat loss where the specimen contacts fixturing, the fixturing may be made of a low-conductivity material or a guard heater may be employed to heat the fixture to the same temperature as the specimen. Reducing the contact area between the specimen and the fixture can also reduce the heat loss from the specimen.

For some thermal-structural tests, the specimen must be cooled. Structures which operate below room temperature, such as cryogenic tanks, may be cooled by immersing part of the structure in a coolant or circulating a coolant through passages attached to or embedded in the structure. Tests which require steady-state thermal gradients to be maintained in the structure may require cooling if natural heat loss modes are not sufficient. Cooling is also sometimes required to prevent test equipment from overheating.

Structural

Fixturing for structural tests usually either limits the deformation of the specimen or imposes an applied deformation or force on the specimen. Sometimes the same fixture may be used for both purposes. It is beyond the scope of the present paper to attempt to describe the wide variety of fixtures which have been used for room-temperature structural tests. However, a few examples of room-temperature fixtures will be described.

Two of the most common structural boundary conditions attempted are simply supported and clamped edges. One common technique for simulating simple supports is the use of knife edge supports. The fixture contacts the panel on the upper and lower surface along the edge to be simply supported. The fixture has a very small radius where it contacts the panel so that it prevents out-of-plane motion, but allows rotation about the edge so that no moment can develop about the edge. Perfectly clamped edge conditions can be very difficult to achieve using fixturing, especially for highly loaded structures. An edge to be clamped is often built in to a very stiff edge fixture. The fixture may have to be firmly attached over a significant area of the upper and lower surface of the edge of a panel to achieve a good approximation of a clamped condition.

One of the most common techniques for loading structures is the use of closed-loop-controlled hydraulic actuators. Such hydraulic actuators are often combined with specially designed load frames to make general purpose structures testing machines. Fixturing is required to transfer the load from the testing machine to the specimens. Mechanical or hydraulic grips are often used to grasp small tensile specimens. Bolted or pinned fixtures are often used for larger tensile tests. Compression specimens are often compressed between flat and parallel platens. Obtaining high-quality test data from compression tests of panel-sized specimens can be difficult. A thorough description of the boundary conditions and test techniques used to test several curved, composite panels is given in Ref. 17. Other loading conditions, such as shear or torsion, can also be applied with appropriate fixturing. Distributed loading may be applied by techniques such as pressurizing a surface, applying distributed dead weights, or using a whiffle-tree arrangement to apply a tensile load from an actuator to distributed pads bonded to a surface.

Interactions

There are significant challenges associated with both thermal and structural experimental boundary conditions. Additional concerns must be considered for thermal-structural tests, which have combined thermal and structural boundary conditions.

In thermal-structural tests, the requirements for thermal and structural boundary conditions often conflict or interact with each other. Structural boundary conditions usually require fixtures to contact the specimen to introduce load or restrain motion. These structural fixtures may conflict with thermal boundary conditions, or vice versa, in several ways.

1) Thermal expansion of the fixtures may not match that of the specimen.

2) Thermal and structural fixtures need to occupy the same physical location or have access to the same location on the specimen.

3) Thermal fixtures may reinforce or stiffen a specimen.

4) Heat transfer paths through the fixtures may affect the temperature distribution in the specimen.

5) Acceptable temperature limits for fixtures or surrounding test equipment may be exceeded because of thermal loading.

6) Deformation of fixtures and/or specimen may affect loading.

7) Changes in material properties with temperature may affect interaction of specimen and test apparatus

Differences in thermal expansion between the specimen and fixtures can cause large undesired thermal stresses. These thermal stresses can be a result of differences in thermal expansion coefficient and/or differences in temperature of the materials. It is also possible for these differences in thermal growth to cause unwanted loosening of the fixtures. Fixtures, therefore, should be designed to either closely match the thermal expansion of the specimen or to provide a means to allow the proper unrestrained thermal expansion of the specimen. In some cases it is possible to make use of the difference in thermal expansion to help load the specimen.

Thermal-structural tests require both a structural and thermal boundary condition to be met on every surface of the specimen. Many combinations of boundary conditions require thermal and structural fixtures or test equipment that must have access to the same surface on the specimen or physically occupy the same space. Examples are surfaces that must be heated and loaded or cooled and simply supported. For these situations the fixtures must be designed to serve both the thermal and structural functions. There are other combinations of boundary conditions which can be easily achieved, such as a mechanically free, thermally insulated boundary, which can be achieved by covering the surface with fibrous insulation.

Thermal fixtures which contact the specimen may reinforce or stiffen the specimen. Blanket heaters, bond-on heaters, or fibrous insulation which cover thin-gauge regions on a specimen may significantly affect the stiffness and mass of the specimen. Coolant passages attached to a specimen may also affect its stiffness and mass distribution.

Structural fixtures which directly contact the specimen provide heat transfer paths from the specimen and could potentially cause unwanted temperature gradients at the boundary. The unwanted thermal gradients may then cause additional thermal stresses or deformations. The fixtures can be designed to reduce the heat flow through them by using a material with low thermal conductivity, heating the fixture to the same temperature as the adjoining specimen, and reducing the area of contact if possible.

When heating specimens, especially with radiant heat, the surrounding fixtures and test equipment are also exposed to heating. The fixtures which must be hot to perform their function must be made from materials which maintain acceptable properties at the operating temperature. Fixtures and

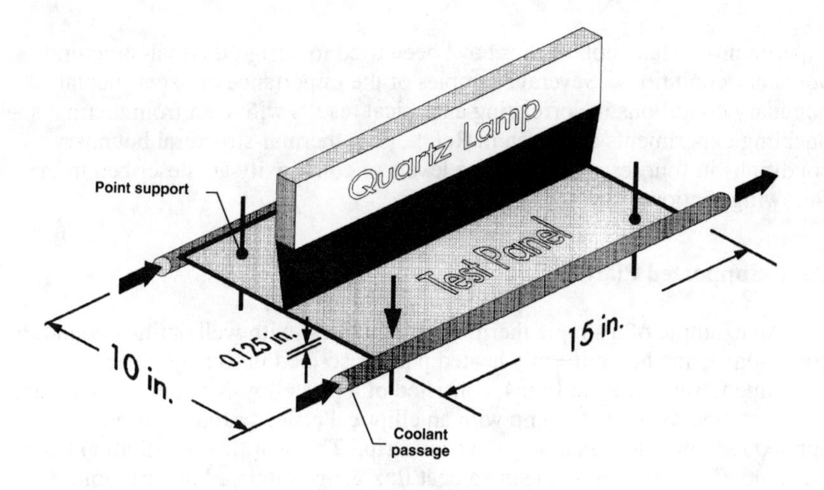

Fig. 4 Point-supported plate.

equipment which do not need to be heated to perform their function can be protected by insulation or cooling.

Both thermal and mechanical loading cause the specimen to deform. If such deformations are large, they may cause eccentricities in the mechanically applied loading or may move the specimen closer to or farther away from a heater. When such deformations are important, the experimental apparatus must be designed accordingly. When unwanted thermal gradients are produced in either the specimen or the fixtures, the specimen may undergo additional unwanted deformation.

Material properties of both the specimen and fixture may change with temperature. Usually a thermal-structural analysis will account for the variation of material properties in the test specimen. However, under some conditions material property variations with temperature may cause undesirable results in an experiment. Fixtures designed to restrain motion may become less stiff at higher temperatures and provide less restraint than desired. A portion of the specimen may be inadvertently overheated, resulting in premature yielding or melting of the specimen. Surface emissivity and absorbtivity may vary with temperature as a specimen is radiantly heated, introducing unwanted variation of applied heating.

In addition to the conflicting requirements for thermal and structural boundary conditions, instrumentation also requires access to the specimen. Optical, whole-field instrumentation techniques, for example, require a clear view of a relatively large surface area. This requirement for a view of the surface may conflict with a requirement to insulate or heat that surface.

Examples of Thermal-Structural Experiments

After considering the general issues involved in providing experimental boundary conditions for aerospace structures, it is helpful to examine particular experiments. Many approaches have been used to design thermal-structural boundary conditions. Several examples of the importance of experimental boundary conditions in correlating analytical results with data from thermal shell buckling experiments are given in Ref. 9. The thermal-structural boundary conditions in four tests with varying levels of complexity are described in the following section.

Point-Supported Plate

An example of a simple thermal-structural test with well defined boundary conditions is the nonuniformly heated plate described in Ref. 18. The experiment, illustrated in Fig. 4, consisted of a Hastelloy-X plate heated along the centerline by a quartz lamp with an elliptical reflector that focused the applied radiant heat flux along a narrow strip. The heat flux distribution from the lamp was characterized using a heat flux gauge mounted in a calibration apparatus. The edges of the plate parallel to the heater were maintained at a constant temperature by chilled water flowing through tubes attached to the edges. The tubes were slotted, fitted around the edges of the plate, and sealed so that the coolant directly contacted the plate. The temperature of the coolant was maintained constant using a chiller. The plastic tubes were not stiff enough to significantly restrict the deformation of the plate. To simulate an insulated boundary, fibrous insulation was used to cover the entire panel, except for the narrow heated strip and the region covered by the tubes. The insulation was light and flexible enough that it did not affect plate deformations. The heated strip was painted with high emissivity paint, thereby increasing the absorbtivity $\overline{\alpha}$ in Eqs. (4) and, consequently, the heat absorbed.

This experiment was intended to provide well understood experimental data for comparison with analysis, rather than to simulate a realistic portion of an aerospace structure. Therefore, to minimize uncertainty from the test, simple structural boundary conditions were chosen. The panel was supported at four points by rods which contacted both the upper and lower surfaces. At three of the points the rods contacted the panel with spherical tips to reduce contact area [A in Eqs. (6c) and (6d)] and thus minimize heat transfer from the panel, to provide as little bending restraint as possible, and to allow the plate to slide by the supports when there was differential thermal growth between the panel and support fixture. At the fourth point, the tips of the rods were conical and fit into a slight indentation on the panel. This point contact fixture minimized heat loss and moment restraint, yet prevented slippage due to thermal expansion, thereby providing a known reference point about which the movements due to thermal expansion occurred.

The interactions between the thermal and structural boundary conditions were kept at a minimum. The thermal boundary conditions were applied without introducing additional unwanted structural stiffness or loading. Structural restraints were designed to allow in-plane thermal expansion. The careful attention to designing simple, well understood, boundary conditions was one of the keys to this experiment successfully providing useful data for comparison with analysis.

Deflection is shown as a function of temperature for the center of the plate in Fig. 5. Little out-of-plane motion is indicated for lower temperatures. The sharp increase of out-of-plane deflection above 200°F indicates the onset of buckling behavior induced by the in-plane temperature variation.

Titanium Honeycomb Sandwich Panel

Tests of a titanium honeycomb sandwich panel, considered for use as a wing panel for a Mach 5 aircraft, are reported in Ref. 19. The panel was designed to withstand thermal stresses due to a thermal gradient through the honeycomb panel thickness with an outer skin temperature of 900°F. In the aircraft wing, the 23-in.-square panel would be surrounded by other, similar panels. The surrounding panels would impose a symmetry boundary condition along all four edges which would imply an insulated thermal boundary and a clamped structural boundary. In the experiment, however, the panel was allowed to expand in-plane and the edges of the panel were not forced to remain straight. Thermal stresses developed when the panel attempted to bend upward because of the temperature gradient through the thickness but was restrained from rotation at the edges by the test fixture. The preferred approach mentioned in Ref. 19 was to test the panel in the center of a multipanel array (transition structure approach). Because of budget constraints, however, two panels were tested individually.

The test fixture was designed to apply uniform temperatures across the upper surface, to prevent all four panel edges from rotating, and to allow the panel to expand in-plane. The upper surface of the panel was heated by an array of 12-in.-long quartz lamps, located 6 in. from the surface. The most challenging portion of this test was designing a fixture to restrain the edge rotation. Providing a perfectly clamped-edge boundary condition is extremely difficult even for a room temperature test. The fixture shown in Fig. 6 was used

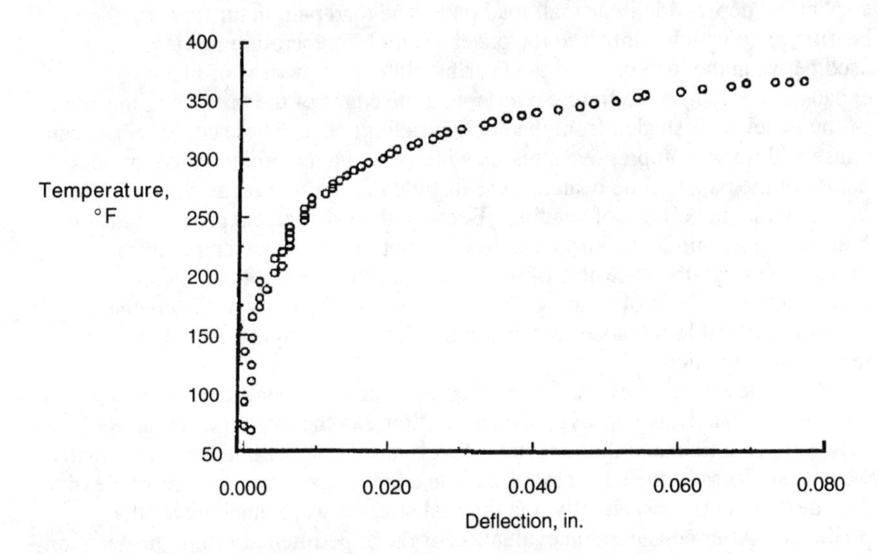

Fig. 5 Point-supported plate center temperature vs displacement.

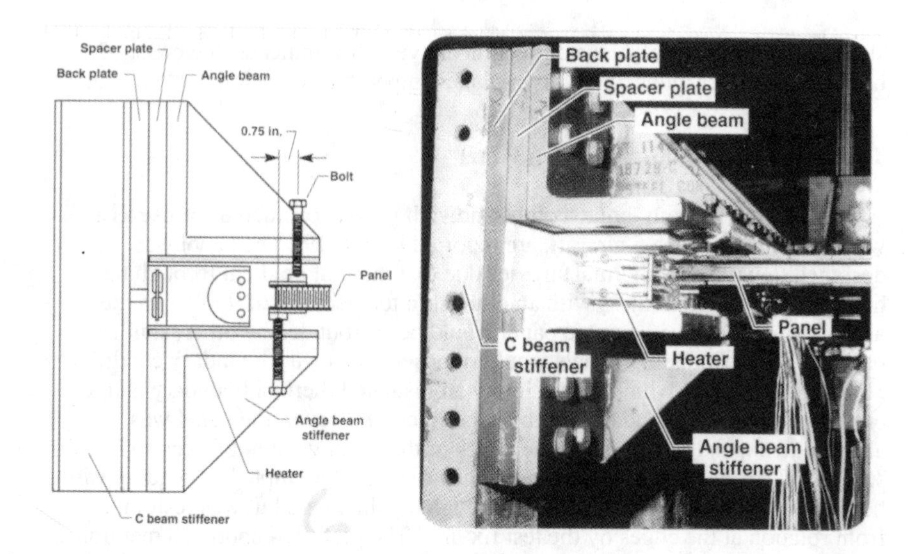

Fig. 6 Initial test fixture monent restraint for the titanium honeycomb sandwich panel.

initially. The angle beam, back plate, and C-beam stiffeners were intended to provide stiffness to react the thermal moment. Bolts were used to transfer the load from the panel to the stiffened fixture, thereby limiting the contact area [*A* in Eqs. (6c) and (6d)] and thus the heat loss into the fixture. There were 13 pairs of bolts along each panel edge. Each pair of bolts was offset, as shown, to provide a restraint to edge rotation. The spherical end of each bolt was seated in a spherical depression in a small load pad. The load pad, in turn, pressed on a bearing plate which reinforced the panel. A high-temperature lubricant was used between the load pad and the bearing plate to allow free in-plane expansion. Because the fixture overlapped the edges of the test panel, the edges of the panel were shaded from the applied radiant heat. These cooler edges can cause additional compressive stresses which reach a maximum value in the center of the panel. Side heaters were installed in the fixture, as shown in Fig. 6, to counteract the effects of shading. Because the side heaters provided heat to both the upper and lower surfaces, they did not maintain the temperature gradient through the thickness of the panel and, thus, did not completely counteract the effects of shading. The unheated surface was not insulated because the heat loss helped to develop the desired through-the-thickness temperature gradient.

Initial test results revealed large discrepancies between test and analysis. A more refined analysis improved the correlation, but the major discrepancy between the anticipated thermal stress levels and those obtained experimentally was the stiffness of the edge restraint. The edge fixture was much more flexible than desired and, consequently, the thermal stresses were much lower than predicted. After considerable evaluation of the experimental setup, the decision was made to simulate the design thermal stresses, rather than attempting to

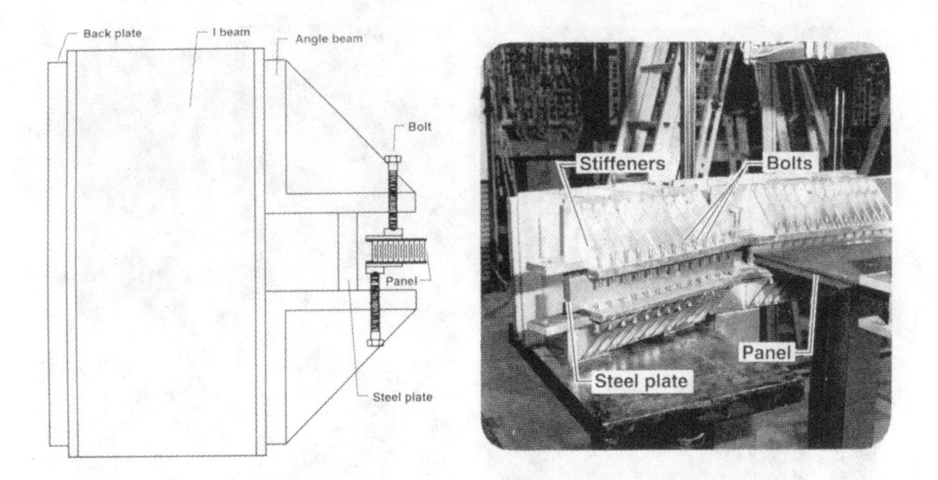

Fig. 7 Final test fixture moment restraint for the titanium honeycomb sandwich panel.

match the design environment of the panel. The edge fixture was modified, as shown in Fig. 7. The side heater was removed for two reasons: to allow the edges to be cooler so that the compressive stresses in the center of the panel would be higher and to allow the upper and lower angle beams to be more directly connected, which greatly stiffened the fixture. In addition, large I beams were used to further stiffen the fixture. The additional stiffening improved the edge fixity from 35 to 75% of a perfectly fixed restraint. The measured surface temperature distribution of the panel in the modified test fixture is shown in Fig. 8. The edges of the panel remained nearly at room

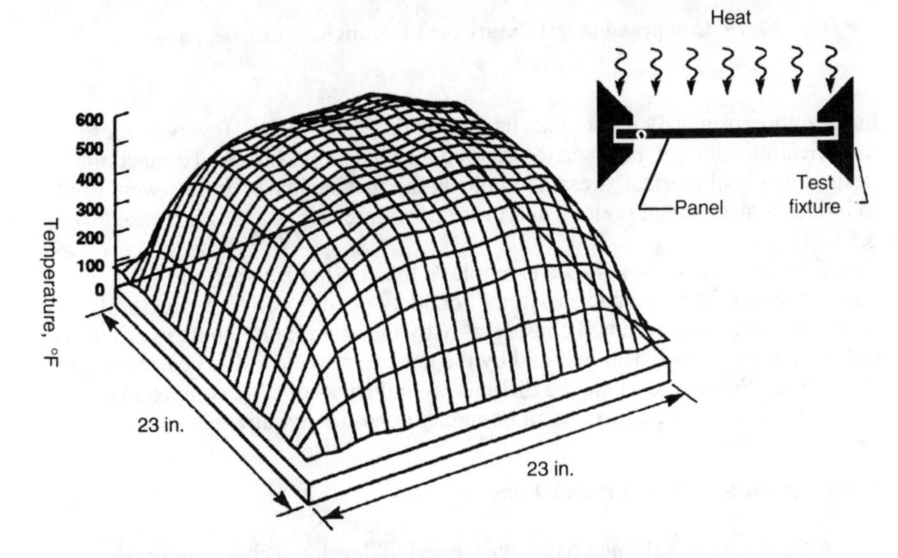

Fig. 8 Measured temperature distribution for titanium honeycomb sandwich panel.

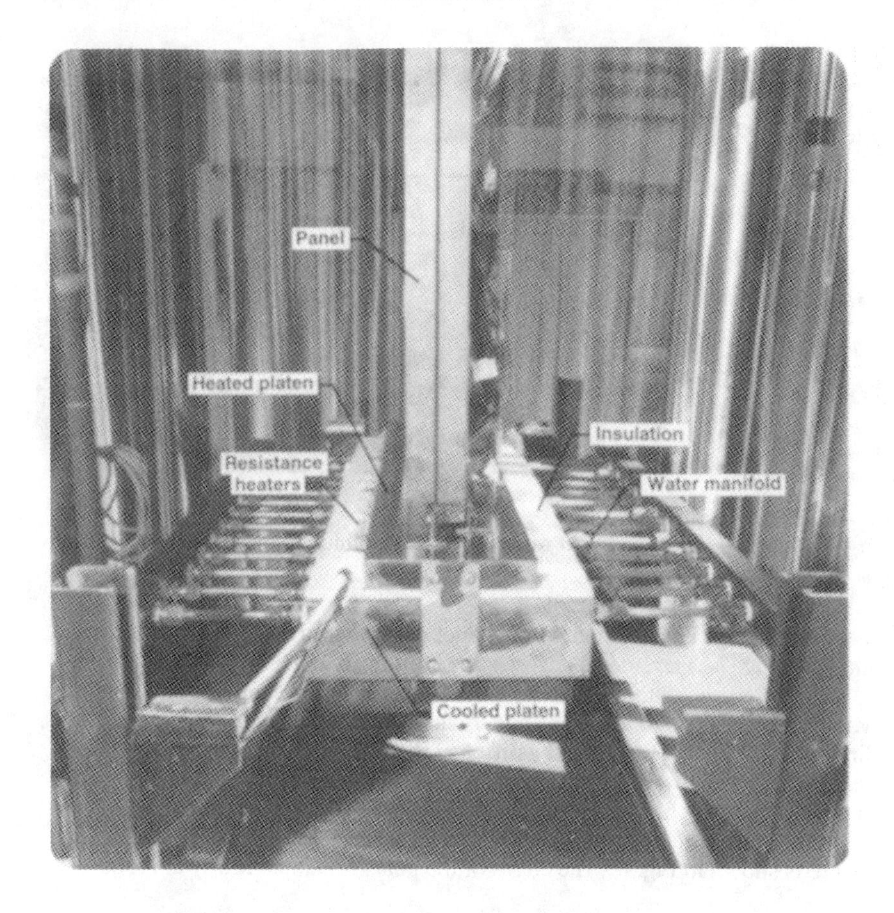

Fig. 9 Compression test fixture for titanium hat-stiffened panel.

temperature, primarily due to the shading by the test fixture. However, these modifications did not produce the desired thermal stress levels. To reach the desired levels of thermal stress, the bolts in the mechanical restraints were used to induce a mechanical preload and the temperature rise rate was increased from 3.5 to 15°F/s.

This experiment is a classic example of the conflict between the requirements of thermal and structural boundary conditions. The structural restraint fixture was softened considerably by thermal requirements to minimize contact area to reduce heat loss and by the need to provide a side heater inside the fixture. Sometimes it can be difficult to find an acceptable compromise between the thermal and structural boundary condition requirements.

Titanium Hat-Stiffened Panel Tests

A titanium hat-stiffened panel was tested to develop technology for the National Aero-Space Plane.[20] A 2-ft-square panel was tested in compression at

Table 2 Buckling load summary for titanium hat-stiffened panel

Test	Pre-Test Analysis Prediction	Test Results
Cross corrugation room temperature	P_{cr} = -6,500 lb	P_{cr} = -6,370 lb
Cross corrugation panel-frame gradient	T_{cr} = 360°F	T_{cr} = 410°F
Cross corrugation panel-frame gradient with load	P_{cr} = -4,220 lb	P_{cr} = -4,380 lb
Cross corrugation 500°F - uniform	P_{cr} = -5,500 lb	P_{cr} = -5,580 lb
Axial room temperature	P_{cr} = -39,700 lb	P_{cr} = -41,730 lb
Axial 500°F - uniform	P_{cr} = -36,900 lb	P_{cr} = -39,170 lb

temperatures up to 500°F. Part of the test fixture is shown in Fig. 9. The panel was mechanically compressed between flat and parallel load platens in a hydraulic loading machine. Because a large contact area was required to transfer the load, heat loss from the ends of the panel was reduced by using heated platens [equivalent to setting $T_1 = T_2$ in Eqs. (6c) and(6d)]. Cooled platens were used to prevent the heated platens from overheating the hydraulic load machine. A 0.25-in.-thick piece of insulation between the heated and cooled platens helped to prevent a large heat flow which may have caused warping of the platens. Titanium frames were bolted to all four edges of the panel to provide a means of introducing load into the loaded edges and stabilizing the unloaded edges. Table 2 lists some of the load cases and the predicted and measured buckling load. Panels were loaded parallel to the hat stiffeners for axial tests and transverse to the stiffeners for cross corrugation tests. For some of the tests, special load frames with embedded coolant passages were used on the unloaded edges (identified as panel-frame gradient test in Table 2). These frames were insulated, and water was circulated through the coolant passages to maintain the frames at room temperature. The cooled frames were thereby used to load the panel with thermal stresses. For another load case (not shown in Table 2), the load frames on the loaded ends were insulated to induce additional transverse thermal stresses.

The panel was enclosed in a quartz lamp oven with lamps grouped into four independently controlled zones on each side of the panel. The four zones gave the heaters the capability to compensate for effects such as natural convection up the sides of the panel. For some tests the panel was heated on both sides, and for other tests it was heated on only one side.

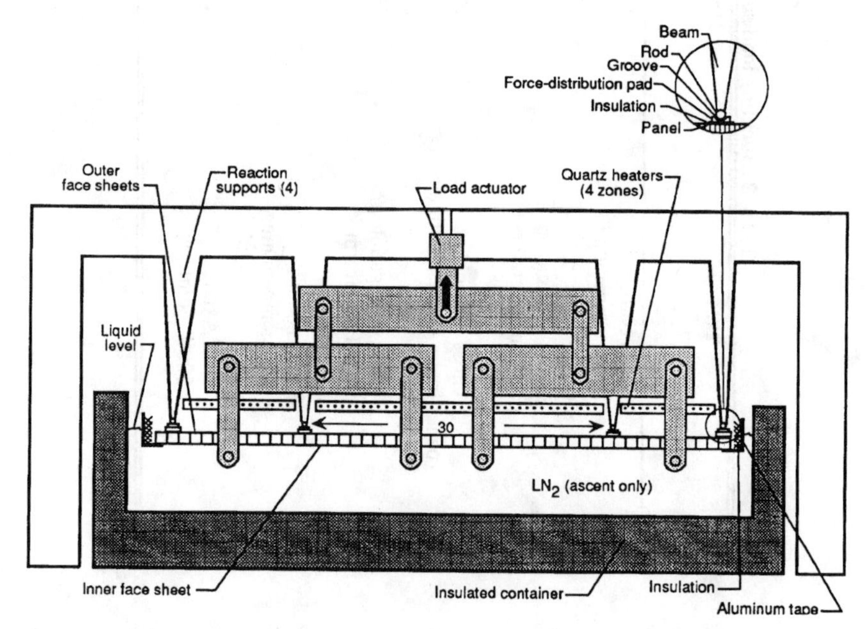

Fig. 10 Schematic of test apparatus for honeycomb sandwich cryotank structure.

Two examples of successfully combining thermal and structural functions for a test fixture were illustrated by the test series. The heated platens provided flat, rigid surfaces for introducing compressive mechanical load. They also minimized heat loss from the end of a uniformly heated panel by matching the temperature of the end of the panel. Use of heated platens, however, necessitated use of the insulation and cooled platens in the load train to prevent overheating the load machine or warping of the heated platens. Satisfying the thermal-structural boundary conditions of loaded, insulated ends introduced the additional complexity of cooled platens to protect the test machine. The load frames helped stabilize the edges of the panel, as well as providing a means of introducing load into the edges of the panel. Although there was some indication that the bolts holding the frames to the panel did not load evenly, there was no apparent effect on the buckling loads or modes as indicated by the close agreement between predicted and measured buckling loads shown in Table 2. The temperature of the load frames was varied to introduce thermal loading in the panel. Thus, the load frames were used to provide both thermal and structural boundary conditions.

Honeycomb Sandwich Cryotank Structure

Two Rene 41 honeycomb sandwich panels were tested under thermal-structural loads simulating those of an integral tank-and-fuselage structure for an

Fig. 11 Test apparatus and panel for honeycomb sandwich structure.

Earth-to-orbit vehicle.[21] A schematic representation of the test apparatus is shown in Fig. 10 and a photograph of the panel and test apparatus is shown in Fig. 11. The lower surface of the 12-in. x 72-in. panel was immersed in liquid nitrogen to simulate cryogenic hydrogen fuel in the vehicle. Simulated aerodynamic heating was applied to the upper surface of the panel by four zones of 10-in.-long lamps spaced on 0.875-in. centers and located 6 in. above the panel. Ceramic edge reflectors were attached around the periphery of the heaters to reduce heat loss at the edge of the panel and to reduce convective air currents. Mechanical bending loads were applied to the panel to simulate strains produced by internal pressure, and aerodynamic and thrust loads. Line loads were applied across the width of the panel at four locations on the cooled surface. The loads were reacted at four different locations across the width of the heated surface, as shown in Fig. 10. The load from a hydraulic actuator was distributed to the four load application points by a whiffle tree arrangement. At each load application point a stiff beam, attached to the whiffle tree at each end, spanned the width of the panel. The portion of the beam next to the panel had an 0.5-in. radius which nested in a semicircular groove in a steel load application pad. The load application block helped distribute the loading into the honeycomb panel without locally crushing the core or restraining bending rotation. A ceramic insulator was placed between the load application pad and the panel surface to limit heat transfer to the loading fixture [reducing k_2 in Eqs. (6c) and (6d)]. The load reaction system on the heated surface used the same

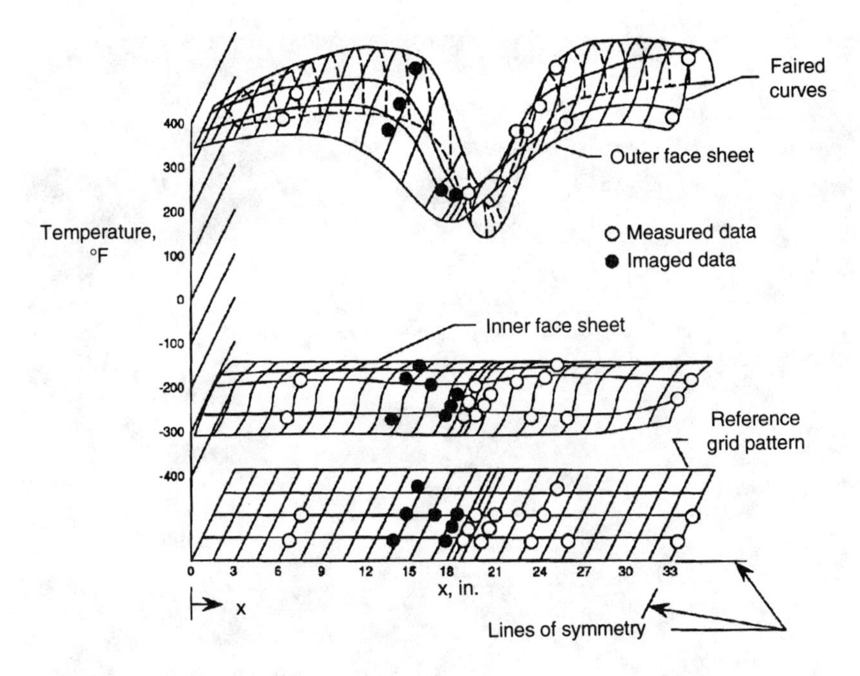

Fig. 12 Surface temperature distribution for honeycomb sandwich structure, first heating cycle.

load application pads and insulators. The structural boundary conditions allow the panel to freely expand in-plane but restrain thermal bowing due to the temperature difference through the panel thickness.

This was a well-designed experiment which illustrated the type of compromises which must be made to satisfy conflicting thermal and structural boundary conditions. The outer surface of the honeycomb sandwich panel was required to be heated to a uniform temperature, yet react mechanical load at four locations. The load reaction beams which traversed the panel at four locations on the heated surface, and their associated load application pads and insulation blocks, shaded significant regions of the surface from the radiant heating. This shading resulted in large variations of surface temperature (as shown in Fig. 12) and, consequently, variations in the deflections and stresses. The 10-in.-long lamps used to heat the panel provided a compact heating system which allowed the structural whiffle tree system, which extended down beside the heaters, to be located close to the edges of the panel, thereby providing a stiffer load train. However the 10-in.-long lamps did not provide a uniform heat flux distributions across the width of the panel. Several techniques were used in an attempt to compensate for this nonuniform heat flux. Aluminum tape was attached to the ceramic side reflectors to reflect more heat on the edges of the panel and a 1.5-in.-wide piece of fibrous insulation was suspended below the centerline of the heaters to reduce the radiation along the center of the panel. Although these modifications reduced the variations of temperature on the heated surface (as shown in Fig. 13), there remained a significant variation of temperature on the

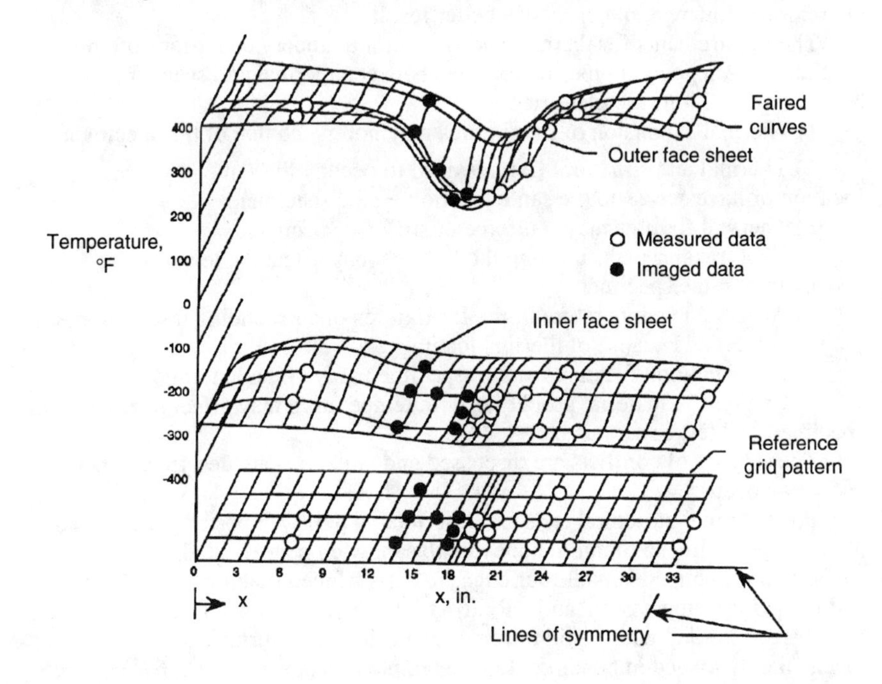

Fig. 13 Surface temperature distribution for honeycomb sandwich structure, 500th thermal cycle.

heated surface. Therefore, the measured temperature distribution had to be used in the structural analysis to obtain satisfactory correlation between test and analysis.

Summary

Development of current and future high-speed vehicles requires thermal-structural tests to evaluate structures designed for combined thermal and structural loading. Providing good boundary conditions for these tests is critical for obtaining reliable test data which can be used to verify vehicle designs. Test data are almost always compared with analysis to verify predicted behavior. In the current paper, some of the theoretical boundary conditions commonly used in thermal and structural analyses, as well as some experimental approaches to applying thermal and structural boundary conditions, are reviewed. Experimental boundary conditions that can be easily analyzed or that simplify analysis are preferred.

To predict the behavior of a thermally loaded structure, the temperature distribution throughout the entire structure must be known. Often, measured temperature distributions must be used in the analysis to obtain good correlation between predicted and measured structural behavior.

Small specimens, limited to the size of interest are often tested. However, when resources permit, larger specimens with transition structure surrounding the region of interest, may produce better results.

The requirements for thermal and structural boundary conditions often conflict. Seven general types of conflicts between thermal and structural boundary conditions are identified.

1) Thermal expansion of the fixtures may not match that of the specimen.

2) Thermal and structural fixtures need to occupy the same physical location or have access to the same location on the specimen.

3) Thermal fixtures may reinforce or stiffen a specimen.

4) Heat transfer paths through the fixtures may affect the temperature distribution in the specimen.

5) Acceptable temperature limits for fixtures or surrounding test equipment may be exceeded because of thermal loading.

6) Deformation of fixtures and/or specimen may affect loading.

7) Changes in material properties with temperature may affect interaction of specimen and test apparatus

These types of conflicts are discussed and situations are described where each may occur.

Four thermal-structural tests are described: a point-supported plate heated along its centerline, a titanium honeycomb sandwich panel heated on one surface and "clamped" on all four edges, a hat-stiffened titanium panel heated and loaded in compression, and a Rene 41 honeycomb sandwich panel heated on one surface, cooled with liquid nitrogen on the opposite surface, and mechanically loaded in bending. Each example illustrates approaches used to satisfy the often conflicting requirements of thermal and structural boundary conditions.

Some of the basic issues involved in providing boundary conditions for thermal-structural tests are discussed in this paper. The examples illustrate some of the difficulties that have been encountered, as well as some of the ingenious solutions that have been used to provide satisfactory compromises between the requirements of thermal and structural boundary conditions. Innovative approaches will be required to meet the challenge of designing boundary conditions for the thermal-structural tests needed to develop the next generation of high-speed aerospace vehicles.

References

[1] DeAngelis, V. M., "A Historical Overview of High-Temperature Structural Testing at the NASA Dryden Flight Research Facility," Structural Testing Technology at High Temperature, Proceedings of Conf., Society for Experimental Mechanics, Dayton, OH, Nov. 4-6, 1991.

[2] Lustig, S., "A History of High Temperature Testing at WPAFB, " Structural Testing Technology at High Temperature, Proceedings of Conf., Society for Experimental Mechanics, Dayton, OH, Nov. 4-6, 1991.

[3] Boggs, B. C., "The History of Static Test and Air Force Structures Testing," Wright-Patterson Air Force Base, AFFDL-TR-79-3071, Dayton, OH, June 1979.

[4] Heldenfelds, R. R., "Historical Perspectives on Thermalstructural Research at the NACA Langley Aeronautical Laboratory From 1948 to 1958," NASA TM 83266, Feb. 1982.

[5] Anon., "Workshop on Correlation of Hot Structures Test Data With Analysis - Vols. I and II," Proceedings of Workshop NASA Ames Research Center Dryden Flight Research Facility, NASA CP 3065, Edwards, CA, Nov. 15-17, 1988.

[6] Anon., "Structural Testing Technology at High Temperature," Structural Testing Technology at High Temperature, Proceedings of Conf., Society for Experimental Mechanics, Dayton, OH, Nov. 4-6, 1991.

[7] Thornton, E. A., "Thermal Buckling of Plates and Shells," *Applied Mechanics Reviews*, Vol. 46, No. 10, Oct. 1993, pp. 485-506.

[8] Kelly, H. N., and Blosser, M. L., "Active Cooling From the Sixties to NASP," NASA TM 109079, July 1994.

[9] Bushnell, D., and Smith, S., "Stress and Buckling of Nonuniformly Heated Cylindrical and Conical Shells," *AIAA Journal*, Vol. 9, No. 12, 1971, pp. 2314-2321.

[10] Boley, B. A., and Weiner, J. H., *Theory of Thermal Stresses.* Wiley, New York, 1960. reprint Krieger, Malabar, FL, 1985.

[11] Leissa, A. W., "Vibration of Shells," NASA SP-288, 1973.

[12] Tauchert, T. R., "Thermal Stresses in Plates - Dynamical Problems," *Thermal Stresses II*, edited by R. B. Hetnarski, Elsevier, Amsterdam, The Netherlands, 1987, pp. 1-56.

[13]Lukasiewicz, S. A., "Thermal Stresses in Shells," *Thermal Stresses III*, edited by R. B. Hetnarski, Elsevier, Amsterdam, The Netherlands, 1989, pp. 355-553.

[14]Fields, R. A., "Flight Vehicle Thermal Testing With Infrared Lamps," Structural Testing Technology at High Temperature, Proceedings of Conf., Society for Experimental Mechanics, Dayton, OH, Nov. 4-6, 1991.

[15]Sikora, T. P., and Leger, K. B., "Simulation of High Heat Flux Levels With Graphite Heating and Arc Lamps," Structural Testing Technology at High Temperature, Proceedings of Conf., Society for Experimental Mechanics, OH, Nov. 4-6, 1991.

[16]Harpur, N.F., "Concorde Structural Development," *Journal of Aircraft*, Vol. 5, No. 2, 1968, pp.176-183.

[17]Bushnell, D., Holmes, A. M. C., Flaggs, D.L., and McCormick, P.J., "Optimum Design, Fabrication and Test of Graphite-Epoxy, Curved, Stiffened, Locally Buckled Panels in Axial Compression," *Buckling of Structures*, edited by I. Elishakoff, J. Arbocz, and C.D. Babcock, Elsevier, Amsterdam, The Netherlands, 1988.

[18]Thornton, E. A., Coyle, M.F., and McLeod, R. N., "Experimental Study of Plate Buckling Induced by Spatial Temperature Gradients," *Journal of Thermal Stresses*, Vol. 17, No. 2, 1994, pp. 191-212.

[19]Richards, W. L., and Thompson, R. C., "Titanium Honeycomb Panel Testing," Structural Testing Technology at High Temperature, Proceedings of Conf., Society for Experimental Mechanics, Dayton, OH, Nov. 4-6, 1991.

[20]Teare, W. P., and Fields, R. A., "Buckling Analysis and Test Correlation of High Temperature Structural Panels," *Thermal Structures and Materials*, edited by E. A. Thornton, Vol.140, Progress in Astronautics and Aeronautics, AIAA, Washington, DC, 1992, pp. 337-352.

[21]Shideler, J. L, Fields, R. A., Reardon, L. F., and Gong, L., "Thermal and Structural Tests of Rene 41 Honeycomb Integral-Tank Concept for Future Space Transportation Systems," NASA TP 3145, May 1992.

Inverse Analysis for Structural Boundary Condition Characterization of a Panel Test Fixture

Sandra P. Polesky* and John L. Shideler*
NASA Langley Research Center, Hampton, Virginia 23665

Abstract

The applicability of a least squares inverse analysis technique to determine the structural boundary conditions of a panel test fixture is investigated. Experimental data of an aluminum beam and plate tested in the test fixture are utilized in estimating the test fixture stiffness. The effect of the magnitude of the sensitivity values for the displacement with respect to the stiffness parameters is discussed. Problems with ill-conditioning led to a reduced parameter model for an average boundary stiffness. Use of optimal experiment design criterion is proposed to improve estimates of the structural parameters. Recommendations for testing are made to improve the accuracy of analyses of test specimens using the test fixture.

Introduction

The development of advanced stiffened plate designs for high-speed aircraft wing panels has been an ongoing effort for improving aircraft technology. Wing structures designed for hypersonic aircraft must be lightweight and able to withstand high levels of aerodynamic heating. Both analytical and experimental studies are required to verify the structural response for advanced wing panel designs. An analytical study of two wing structures for Mach 5 cruise airplanes demonstrated an improved lightweight design that utilized titanium honeycomb core sandwich panels.[1] The honeycomb panels were designed to withstand thermal stresses caused by a through-the-thickness temperature gradient of 320°F during the ascent portion of flight.

Thermal-structural tests and analyses were conducted to investigate the structural response of a titanium honeycomb panel subject to a thermal gradient

---

* Aerospace Engineer, Thermal Structures Branch.

and appropriate boundary conditions.[2,3] Typically, an array of nine test panels would be necessary to obtain the proper boundary conditions on the center panel being studied. Because of the high cost for an array of titanium honeycomb panels, however, a test fixture was designed to thermally and structurally test a single panel with the appropriate edge boundary conditions representative of the actual wing.[2,3] The test goals included applying the thermal-structural loads determined in Ref. 1. The test fixture was designed to allow in-plane thermal expansion to simulate aerodynamic heating of the entire wing structure, to prevent out-of-plane bending at the panel edges, and to minimize the moment restraint attachment area so the test fixture would restrain the panel as if it were attached to spars and ribs as it would be in the actual wing structure.

Initial testing of the honeycomb sandwich panel in the stiffened panel test fixture, described in Refs. 2 and 3, led to large discrepancies between experimental data and analytically determined strains and panel deflections. The test fixture was designed to simulate a clamped boundary condition along all four panel edges. The clamped boundary condition would require a perfectly rigid test fixture. In reality, due to the flexibility of the test fixture, the actual boundary condition obtained was somewhere between a clamped and a simply supported boundary condition.

The complexity of the analysis required to model the appropriate boundary conditions for the panel was not anticipated prior to testing the honeycomb stiffened panel. Possible unknown boundary condition parameters that affected the experimental result included test fixture stiffness, preloading due to bolt restraints, and friction forces due to reaction forces in the bolts. The trial-and-error procedure for obtaining more realistic boundary conditions to correlate experimental data with analysis results did not necessarily provide the correct boundary conditions.[2] When the effect of varying a certain parameter produced an analysis result that better correlated with the experiment, it was assumed that the parameter was at a correct value and the effect of other possible parameters was neglected. However, the actual experimental results can depend on a combination of the various possible boundary conditions. When there is more than one unknown parameter affecting the experimental result, the ability to decipher the correct value for each parameter can become difficult, if not impossible.

In addition to complex loading and boundary conditions, a honeycomb sandwich panel adds a geometric complexity to the structural response. To gain a better understanding of the boundary conditions the test fixture imposes on panels and to maintain simplicity, an aluminum beam and plate were subsequently tested in the test fixture with simplified loading conditions. The beam and plate were tested with a concentrated mechanical load applied at the center and with a "finger-tight" bolt loading. The plate was also tested with only a bolt preload applied.

The objective of the study presented in this paper is to better characterize and understand the boundary conditions the panel test fixture imposes on the structural panels being tested. The use of an inverse analysis method to obtain the appropriate structural boundary conditions is investigated. A least squares parameter estimation technique is initially applied to a beam problem as a preliminary step in determining a structural stiffness boundary condition. The inverse analysis method is subsequently applied for correlating the aluminum plate experimental data with finite element analysis results. The ability to model the correct boundary conditions and, hence, analyze experiment designs poses certain requirements on experiment designs. Recommendations are made for designing future experiments which allow for optimum parameter estimation.

Test Fixture Description

The stiffened panel test fixture is located at NASA Dryden Flight Research Facility. A photograph of the upper surface of a honeycomb sandwich panel located in the test fixture is shown in Fig. 1. The test fixture was designed to prevent rotation at the panel edges while allowing in-plane translation. A schematic diagram of the test fixture is shown in Fig. 2. The upper and lower bolts shown in Fig. 2a were offset to provide a moment to restrain the panel from bending out-of-plane as the upper surface of the panel is heated. There were 52 moment restraint mechanisms (13 on each side) located around the perimeter of the panel as shown in Fig. 2b. The end of each steel bolt has a spherical radius that fits into a spherical socket in the load pad. A torque is applied to the restraining bolts to hold the panel in the test fixture prior to thermal loading. The bolts are supported by a large steel structure consisting of angle beam stiffeners and large I beams welded to the back plates. A high-

Fig. 1 Photograph of upper surface of honeycomb sandwich panel located in the test fixture.

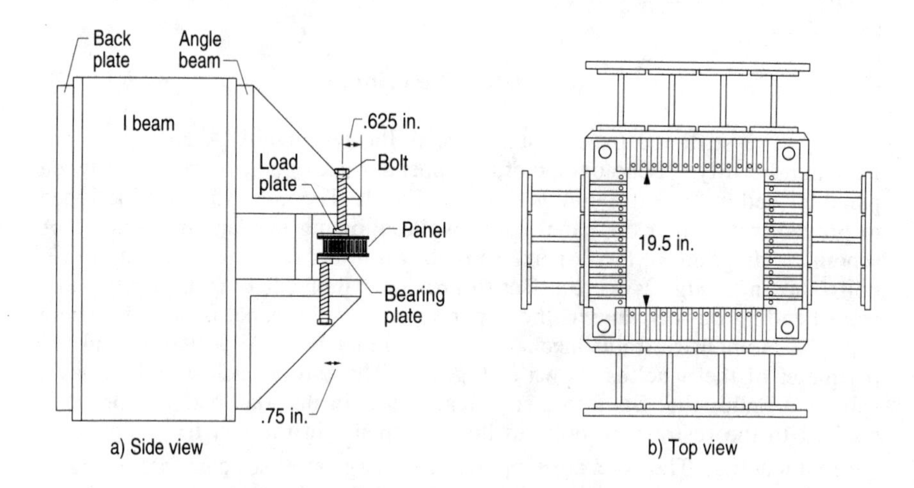

Fig. 2 Schematic diagram of test fixture.

temperature lubricant was used between the load pads and bearing plates to allow for in-plane thermal expansion of the panel.[3]

Analytical Modeling

Typically, structural analysis consists of determining the distribution of displacements in a geometrical model given the material properties, loading, and boundary conditions. The strains and stresses are subsequently determined from the displacement field. This is known as a direct field problem. When the geometry, material properties, loading, and/or boundary conditions are not known prior to testing, however, they become unknown parameters. An iterative procedure is generally required to solve the inverse problem of determining the unknown parameters that best correlate analysis results with experimental data. The sensitivity coefficients of the unknown parameters are a measure of the magnitude of change in the response with respect to a change in the parameter. The requirement that the sensitivity coefficients of the unknown parameters be linearly independent is necessary for existence of a unique solution to the inverse problem.[4]

The reaction loadings in the upper and lower restraining bolts can be represented as an applied moment and are analytically modeled as a rotational spring. Accordingly, the rotational springs are located at the center of the moment arm between the restraining bolts. Because of the geometry of the test fixture, it is expected that the rotational stiffness is higher at the corners of the test fixture and decreases toward the center of each panel edge. The effective stiffness of the test fixture can then be represented by values of rotational stiffness for each of the 52 restraining bolt mechanisms. From symmetry considerations on opposing sides of the test fixture, the number of unknown rotational stiffnesses reduces to 14. Preloading in the bolt mechanisms is modeled using an applied moment loading.

During thermal loading, in-plane expansion of the panel is reacted against by friction forces due to the vertical reactions in the restraining bolts. The friction force can be modeled by applying an in-plane force at the rotational spring locations. The in-plane force is a function of the normal force in the bolts which is a function of the applied moment preload and the additional moment reaction induced by bending of the plate during thermal loading. The effect of the friction force on a panel response needs to be determined when thermal loading is applied. Since only mechanical loading was applied to the aluminum beam and plate during testing, there is no need to account for friction forces due to the absence of in-plane thermal expansion.

Beam Analysis

A symmetric beam with rotational spring boundary conditions is modeled to resemble the aluminum beam tested in the stiffened panel test fixture. The beam was tested with a concentrated mechanical load applied at the center and a finger-tight bolt loading. One displacement potentiometer was located at the center of the beam to measure the beam deflection. A schematic diagram of the beam located in the test fixture and the corresponding analytical model are given in Fig. 3. A one-dimensional exact solution for the deflection of the beam with rotational spring boundary conditions is given by

$$y = \frac{PL^2\,(k\,x^2 + 2\,x\,EI\,)}{16\,EI\,(kL + 2\,EI)} - \frac{P\,x^3}{12\,EI} \qquad 0 \le x \le \frac{L}{2} \tag{1}$$

where y is the beam deflection, P the center applied load, L the beam length, x the spatial location along the length of the beam, E the modulus of elasticity, I the moment of inertia, and k the stiffness of the rotational spring boundary conditions.

The sensitivity of the deflection to changes in the rotational stiffness is obtained by differentiating Eq. (1) with respect to k,

$$X = \frac{\partial y}{\partial k} = \frac{P\,(L^2x^2 - L^3x)}{8\,(\,kL + 2EI)^2} \tag{2}$$

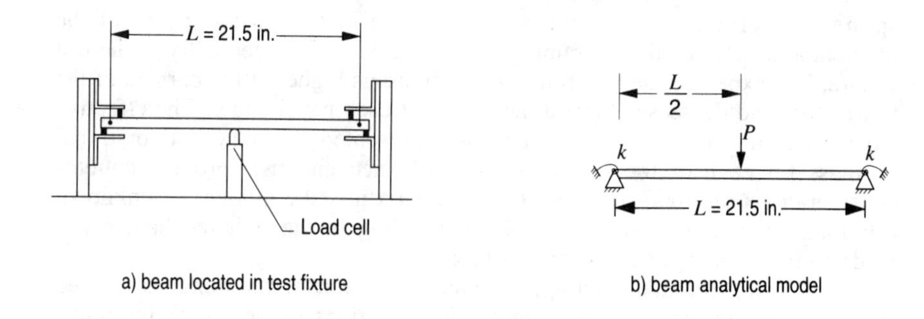

Fig. 3 Schematic diagram of boundary conditions on the beam.

where X is the sensitivity coefficient for k. The exact expression for the sensitivity is very valuable in determining optimum testing conditions for estimating the stiffness of the boundary conditions. One criterion for an optimum experiment design is to maximize the sensitivity value with respect to experimental variables. For this test, the experimental variables are the magnitude of the applied load and the location x where measurements of deflection are taken. Consequently, since the sensitivity is a linear function of the applied load, the maximum possible applied load, where the response is within the limits of elementary beam theory, should be used in estimating the boundary condition stiffness. Also, the sensitivity coefficient is maximized with respect to location x at the center of the beam where $x = L/2$. Hence, the center of the beam is the best deflection measurement location to use in estimating the rotational stiffness. It is also important to note that the sensitivity coefficient is nonlinear in terms of its parameter k.

The exact and finite element solution for the maximum deflection at the center of the beam is plotted as a function of spring stiffness in Fig. 4. The finite element model had 23 nodes where 21 nodes were equally spaced along the length of the beam and each end of the beam had an additional node for modeling the rotational springs. The displacements shown correspond to the maximum experimentally applied load of P = 2000 lb. Between the simply supported and fixed ends boundary condition solutions, one can observe a large variation in rotational spring stiffness over a relatively small range in maximum beam deflection by observing the range on the axes. The observation concurs with small sensitivities in the range of interest between the simply supported and fixed ends solutions, as can be computed from Eq. (2).

A least squares parameter estimation technique is applied to the beam finite element model to predict the stiffness value that best correlates the experimental data with analysis results. The sum of the squares of the

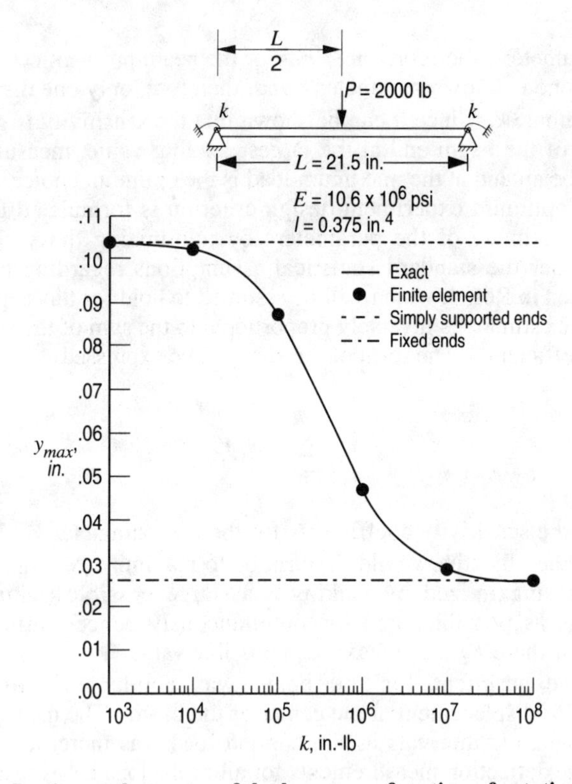

Fig. 4. Deflection at the center of the beam as a function of rotational stiffness.

differences can be expressed as

$$S = \sum_{i=1}^{n} (y_{ci} - y_{mi})^2 \tag{3}$$

where y_{mi} are measured values of displacement at a specific location and test time and y_{ci} are the corresponding calculated displacements at the same location and under the same load conditions. The object of the least square method is to minimize the sum of the squares of the differences with respect to each parameter. The resulting equations obtained are in the form

$$\frac{\partial S}{\partial p_j} = 0 = \sum_{i=1}^{n} \frac{\partial y_{ci}}{\partial p_j} (y_{ci} - y_{mi}) \qquad j = 1,m \tag{4}$$

where the derivative of S is taken with respect to each unknown parameter, p_j, and there are a total of m unknown parameters. This results in m equations for determining the m unknown parameters. For a solution to exist, the least squares method requires that there be at least as many measurements as there are unknown parameters, therefore, $m \leq n$. For the beam problem considered here, there is only one unknown parameter k and, therefore, only one displacement is needed to estimate k. Since it can be shown that the sensitivity is greatest for y at the center of the beam and at the largest loading value, measuring y at the center of the beam and at the maximum load is the optimum choice.

The optimum experiment design criterion is formulated to obtain the most precise estimate of the parameters by minimizing the variance of the estimate. Under the standard statistical assumptions regarding measurement errors described in Ref. 4, which will be assumed to hold for this experiment, the variance in the estimate is inversely proportional to the sum of the squares of the sensitivity coefficients. The form of the sum can be expressed

$$\Delta = \sum_{i=1}^{n} (X_i)^2 \tag{5}$$

where X_i are the sensitivity coefficients for the one parameter k. To minimize the variance, the objective would, in turn, be to maximize Δ. One can observe that Δ can be maximized by making n as large possible (taking as many measurements as possible) and by simultaneously concentrating all of the measurements where X_i is at a maximum possible value.

The experiment for the beam was conducted with only one measurement of displacement at the center of the beam. The measurement was taken at consecutive intervals as the applied load was increased. Therefore, using all of the deflection measurements for all of the load cases recorded would maximize Δ. If the optimum design criterion were considered prior to performing the experiment, however, more accuracy in the estimate could theoretically have been obtained by placing additional transducers near the center of the beam.

Considering the least squares minimization for the beam problem, Eq. (4) can be rewritten as

$$\sum_{i=1}^{n} \frac{\partial y_{ci}}{\partial k} (y_{ci} - y_{mi}) = 0 \tag{6}$$

where the summation is over the measurements taken as the loading is increased. A Taylor series expansion of y_{ci} can be used to linearize its relationship with the parameter k. A first-order approximation of y_{ci} is in given by

$$y_{ci} = y_{0i} + (k - k_0) \frac{\partial y_{ci}}{\partial k} \qquad i = 1,n \tag{7}$$

where the subscript 0 denotes some initial value. An initial guess of k_0 is used in a finite element analysis to compute y_{0i} , and then a 0.1% change in k_0 is used in the finite element analysis to compute y_{ci}. The sensitivity is then numerically computed as a finite difference approximation,

$$\frac{\partial y_{ci}}{\partial k} = \frac{y_{ci}(k_0 + \Delta k) - y_{0i}(k_0)}{\Delta k} \qquad i = 1,n \tag{8}$$

where Δk is the 0.1% change in k_0 Eq. (7) is substituted in Eq. (6) and solved for k to yield

$$k = k_0 + \frac{\sum_{i=1}^{n} \frac{\partial y_{ci}}{\partial k} (y_{mi} - y_{0i})}{\sum_{i=1}^{n} \left(\frac{\partial y_{ci}}{\partial k}\right)^2} \tag{9}$$

In the inverse analysis procedure, the sensitivities computed from Eq. (8) are used in Eq. (9) to predict k. The procedure is repeated with k_0 replaced by k until the change in k is within a 1% prescribed error tolerance.[5]

Experimental data from two separate tests for a finger-tight bolt loading condition were used for the least squares parameter estimation. For each test, the load was applied in increasing intervals and eight deflection measurements were recorded, resulting in 16 load-deflection data points for use in the parameter estimation routine. For a single restraining bolt mechanism, an average rotational stiffness of $k = 1.14 \times 10^5$ in.-lb was calculated from the parameter estimation analysis. The load-deflection plot for the calculated rotational stiffness is plotted along with the test data in Fig. 5. The calculated deflections for the applied loading correlate well with the test data. The results indicate that a rotational spring boundary condition is sufficient to characterize

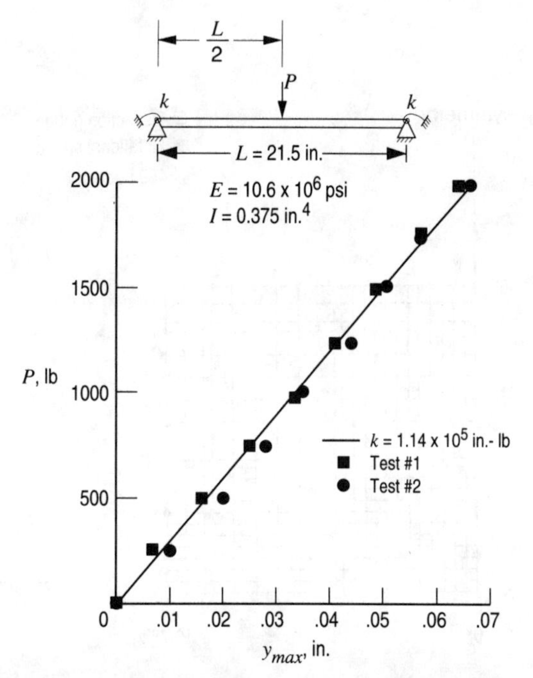

Fig. 5 Load-deflection diagram at the center of the beam.

the boundary conditions the test fixture imposes on the beam. Because of the fact that the sensitivity coefficients are small, however, the variance in the stiffness estimate is large and, consequently, there is not much accuracy in the value obtained.

Aluminum Plate Analysis

Because of the effect of stiffness coupling between the test fixture and the test article, the test fixture should possess a greater stiffness for a plate being tested as opposed to a beam located in the center of the test fixture. Therefore, it was necessary to independently determine the boundary condition the test fixture imposes on a plate being tested. An aluminum plate was fabricated and tested to determine the test fixture boundary conditions for any panel being tested in the test fixture. The plate was initially tested subject to a finger-tight bolt loading and a concentrated mechanical load applied at the center. Seven deflection gauges were located in one quadrant of the plate, as shown in Fig. 6. The plate was subject to a normal load applied at the center of the plate, and deflections were measured at approximately 1-lb intervals as the load was increased to 5000 lb. Subsequently, the plate was tested with only a bolt preload applied.

A schematic diagram of the finite element model of the aluminum plate is shown in Fig. 6. The nodes along the perimeter of the plate were fixed from out-of-plane displacements. Because of symmetry considerations, one-quarter of the plate was modeled with 14 rotational spring boundary conditions and 489 nodes. Although there are 14 unknown stiffness values associated with the springs, the size of the problem is reduced to 8 unknown parameters by initially

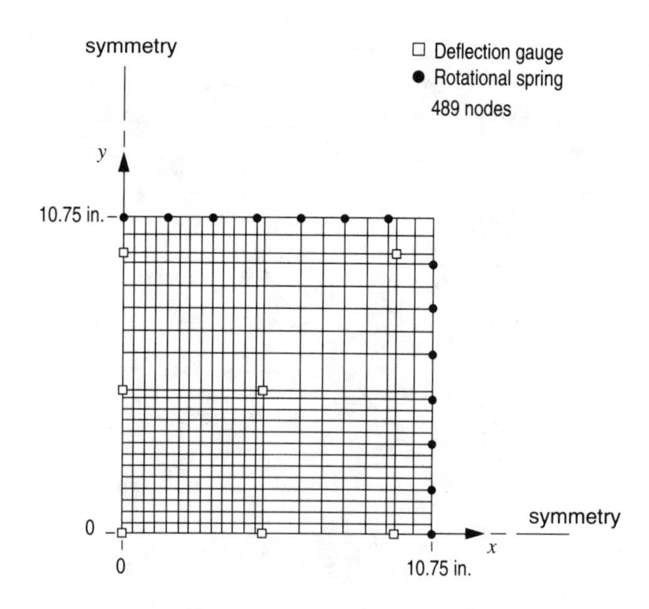

Fig. 6 Schematic diagram of plate finite element model.

assuming a cubic distribution of the stiffness along the panel edge. For the two plate edges, the stiffness is assumed in the form

$$k = k_1 + k_2\, y + k_3\, y^2 + k_4\, y^3 \quad (\text{ at } x = 10.75 \text{ in.}) \tag{10}$$
$$k = k_5 + k_6\, x + k_7\, x^2 + k_8\, x^3 \quad (\text{ at } y = 10.75 \text{ in.})$$

resulting in 8 unknown parameters, k_i, $i = 1,8$, for the two plate edges.

The requirement that the sensitivities of the parameters be linearly independent over the range of the observations was next considered. There are two experimental observation variables, the magnitude of the applied loading and the spatial location of the deflection gauges. The sensitivity coefficients were determined from finite element analyses. An initial estimate of an average constant test fixture stiffness of $k = 2.12 \times 10^5$ in.-lb was determined to be a good initial estimate by visually comparing the experimental and finite element deflection results. The parameters were each perturbed individually from the initial estimate, and the finite element results were used in conjunction with a first-order finite difference approximation of the sensitivity values. A graphical comparison of sensitivity coefficients over the range of the applied loading is shown in Fig. 7. The linear dependence of the sensitivity coefficients is shown in Fig. 7 (Ref. 4). It was determined that all of the sensitivity coefficients were linearly dependent over the range of the applied loading. Therefore, the stiffness coefficients cannot be uniquely determined from the results obtained by varying the applied loading. Next, the spatial observations were considered. Since there were only seven deflection gauges, a maximum of seven unknown coefficients

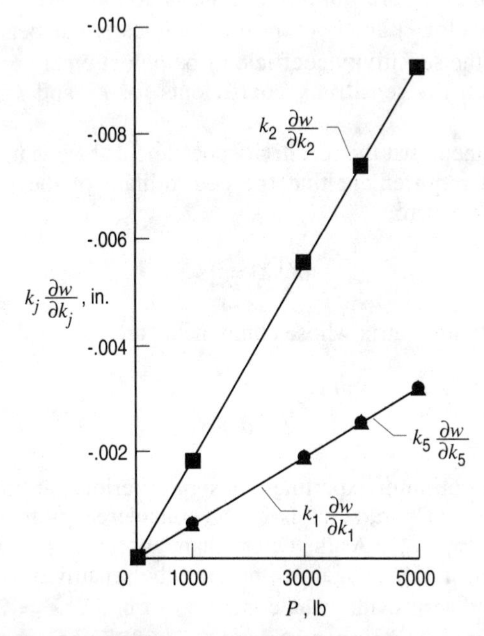

Fig. 7 Graphical comparison of sensitivity coefficients over the range of the loading.

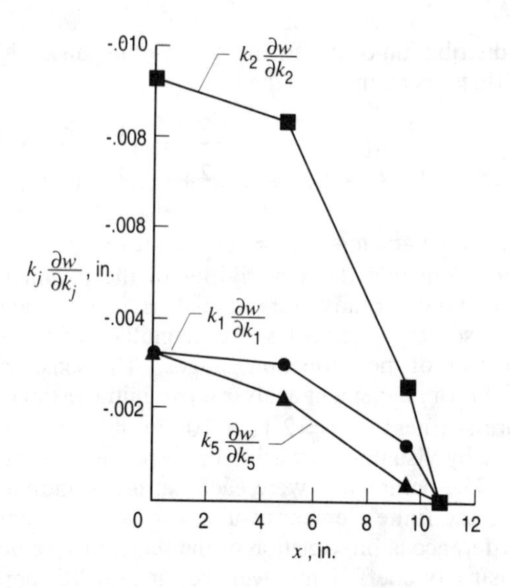

Fig. 8 Graphical comparison of sensitivity coefficients over the length of the plate at $y = 0$.

can only be considered. A graphical comparison of sensitivity coefficients over the length of a panel edge is shown in Fig. 8. Here, whereas the sensitivity coefficients for k_1 and k_2 are observed to be nearly linearly dependent, the sensitivity coefficients for k_1 and k_5 appear to be linearly independent. Because of the magnitude of the sensitivity coefficients being extremely small, however, the variation between the sensitivity coefficients for k_1 and k_5 is extremely small.

The requirement that the sensitivity coefficients be linearly independent is equivalent to the requirement that the determinant of the product of the sensitivity matrix be nonzero,

$$|\boldsymbol{X}^{\mathrm{T}}\boldsymbol{X}| \neq 0 \tag{11}$$

where $\boldsymbol{X}$ is the sensitivity matrix whose components are

$$X_{ij} = \frac{\partial w_i}{\partial k_j} \tag{12}$$

One approach to the optimum experiment design criterion, for problems where more than one unknown parameter is being considered, is to maximize the determinant given in Eq. (11). As the determinant approaches zero, the problem becomes ill-conditioned. Once again comparing the sensitivity coefficients for k_1 and k_2 only, the determinant was computed to be $|\boldsymbol{X}^{\mathrm{T}}\boldsymbol{X}| = 2.6 \times 10^{-32}$, which is approximated as being zero, and k_1 and k_2 are concluded to be approximately linearly dependent. Comparing the sensitivity coefficients for k_1

and k_5, the determinant is computed to be $|X^TX| = 4. \times 10^{-32}$, which is nearly as small as the determinant computed when k_1 and k_2 are considered. Although the coefficients appeared to be linearly independent graphically, the magnitude of difference between the sensitivity coefficients over the length of the plate is extremely small and results in a determinant near zero. Therefore, the sensitivity coefficients for k_1 and k_5 are assumed to be nearly linearly dependent. Linear dependence and even approximate linear dependence of the sensitivity coefficients results in an ill-conditioned matrix X^TX used in solving for the unknown stiffness parameters when utilizing the least squares technique. Not much accuracy can be expected in results obtained from an ill-conditioned matrix. Therefore, the problem of determining the stiffness of the test fixture should be reduced to a model with one unknown stiffness parameter, a constant average value of stiffness for the test fixture.

Beginning with the initial estimate of $k = 2.12 \times 10^5$ in.-lb used earlier, a one-parameter model for the test fixture stiffness is evaluated using the least squares approach described for the beam problem. Although the data obtained over the range in the loading does not yield information for determining additional parameters, it does aid in decreasing the variance in the estimate and, hence, in increasing the accuracy of the estimate. Three load cases were used in the least square estimate of a stiffness value of $k = 2.18 \times 10^5$ in.-lb for the test fixture rotational spring boundary conditions. The deflection distributions for the three load cases are shown in Figs. 9-11. The figures reveal good correlation between the experimental data and analysis results. All of the analysis results were within the ±0.002 in. accuracy range of the deflection gauge measurements. The experimentally and analytically obtained displacement distribution for the maximum applied load of P = 5020 lb is shown in Fig. 12 along with results for a simply supported and clamped boundary condition. As shown in Fig. 12, the test fixture stiffness is more closely represented by a clamped boundary condition than a simply supported boundary condition.

In addition to the concentrated mechanical load test of the aluminum plate, the plate was tested with only an applied bolt preload. The upper bolt of each edge restraint mechanism was torqued, where due to bolt losses the resultant forces in the bolts were unknown as well as the extent of variation in the bolt loading from mechanism to mechanism. The applied bolt preload was modeled analytically by applying a moment load at the center of each restraining bolt mechanism. A constant value of an applied moment is initially assumed to obtain an initial estimate of the applied moment load for further evaluation. Using the least squares technique with a single constant unknown moment parameter, an applied moment of M = 4842 in.-lb was calculated. The calculated displacement distribution is shown in Fig. 13 along with the experimental data. Large differences can be observed between the analytical results and experimental data near the outer perimeter of the plate (close to the supported boundaries). Linearly varying moment loads along each of the plate edges were next applied. The results of several linearly applied moment loads showed little change in the displacement field with large changes in the applied moment loading. A trial-and-error approach to arrive at a moment distribution that gave good correlation between analysis and experiment was unsuccessful. The displacement distribution proved to be very insensitive to variations in the

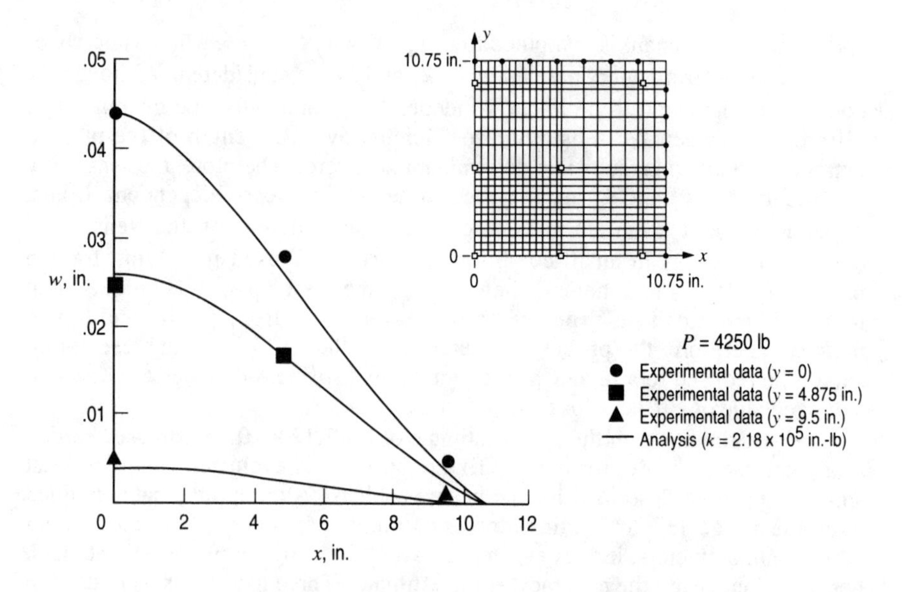

Fig. 9 Plate deflection curves for P = 4250 lb.

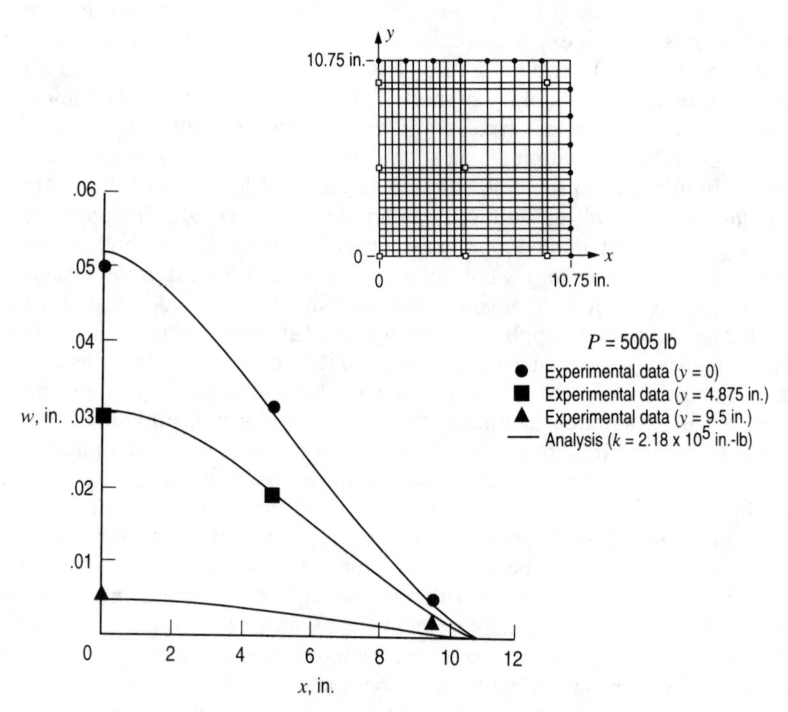

Fig. 10 Plate deflection curves for P = 5005 lb.

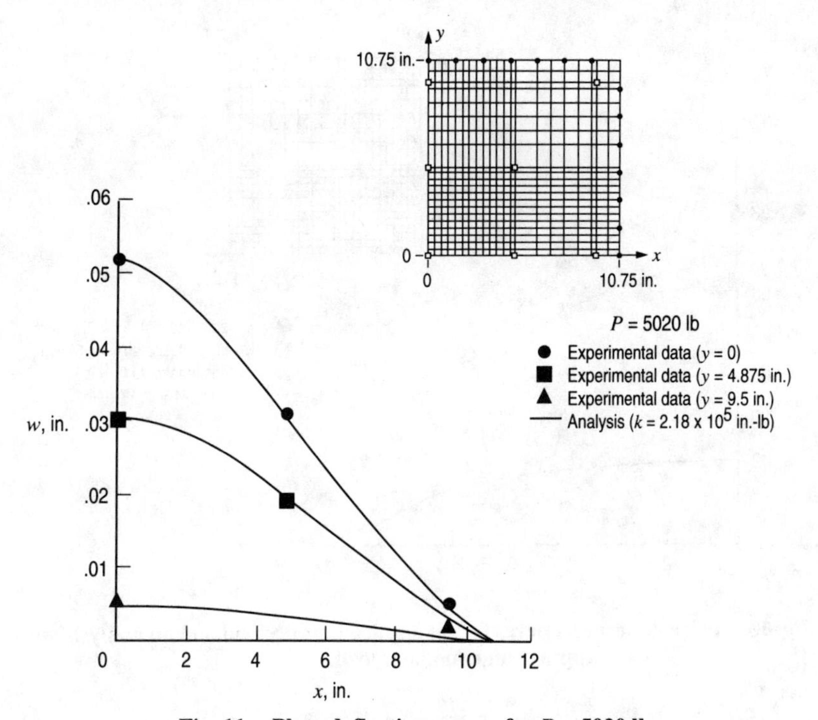

Fig. 11 Plate deflection curves for $P = 5020$ lb.

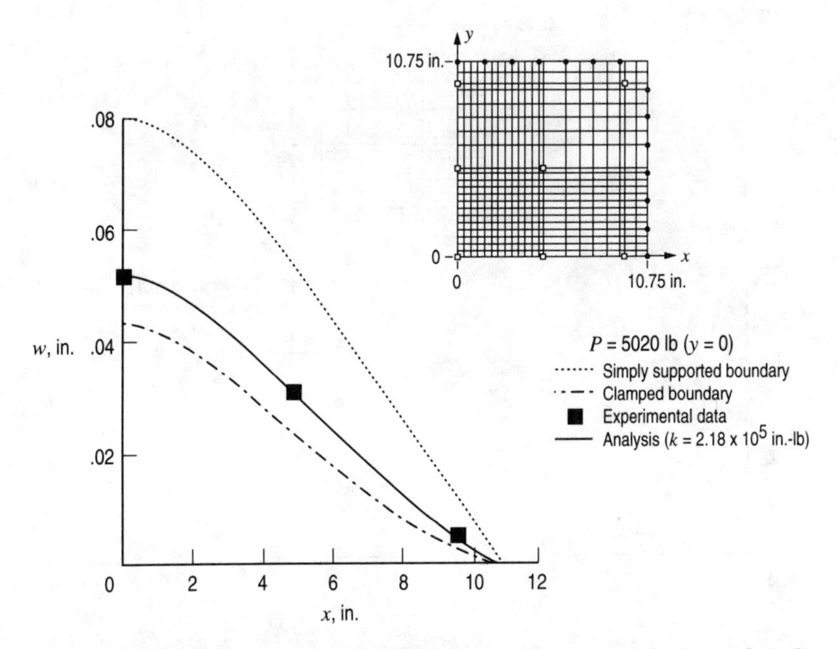

Fig. 12 Comparison of plate deflection for a simply supported, clamped, and rotational spring boundary condition.

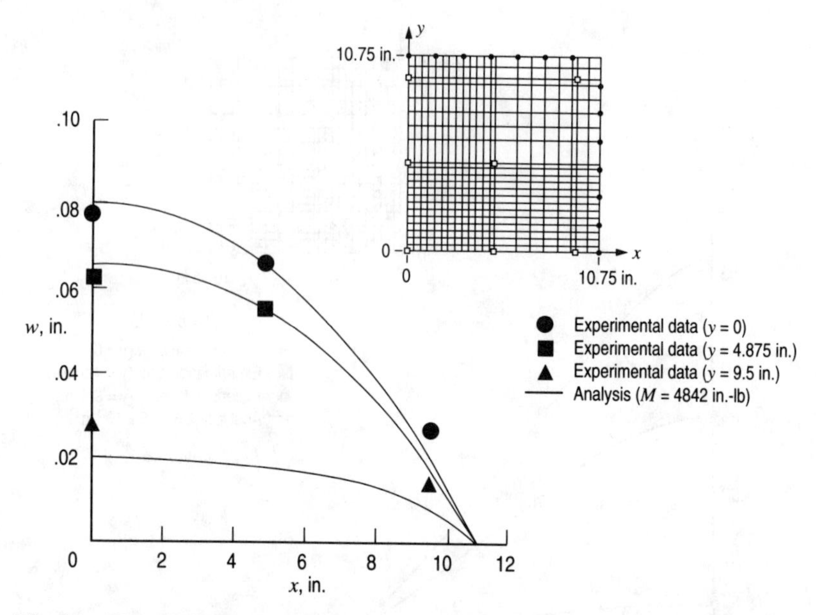

Fig. 13 Plate deflection curves for an applied bolt preload and an analysis with a uniformly applied edge moment load.

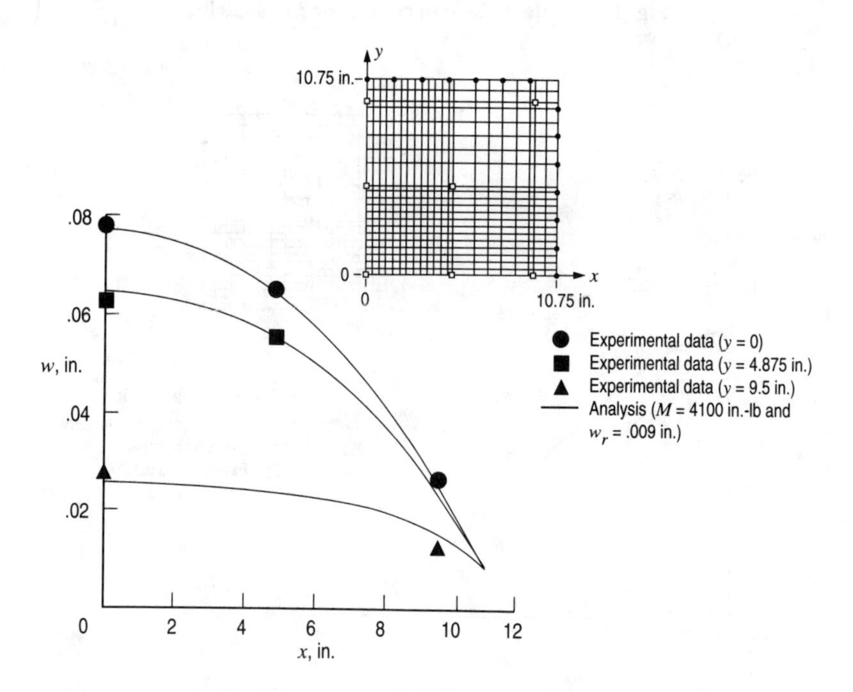

Fig. 14 Plate deflection curves for an applied bolt preload and an analysis with a uniformly applied edge moment load and rigid displacement.

applied moment loading. Therefore, a constant average moment load should be sufficient to characterize the test fixture loading on the plate.

Since there were large differences between the calculated deflections and experimental data near the outer edges of the plate, the possibility of a rigid displacement of the plate was investigated. When a rigid displacement of w_r = 0.009 in. is applied to the plate in addition to an applied moment of M = 4100 in.-lb, analytical results showed good correlation with the experiment, as shown in Fig. 14. However, the parameter estimation problem of computing the two unknown parameters, w_r and M, is ill-conditioned. Therefore, the solution which produced the desirable results displayed in Fig. 14 is not unique and does not necessarily prove that a rigid displacement exists. Hence, one must actually measure a rigid displacement, by locating a displacement potentiometer at the center of the moment arm in the bolt mechanisms, to assure its existence and magnitude.

Concluding Remarks

In the current study, a least squares parameter estimation technique is applied to determine the structural boundary conditions of a panel test fixture. An aluminum beam and plate were tested in the fixture to yield experimental data for characterizing the test fixture. The analytical models used rotational springs to represent the moment restraint provided by the offset bolts in the test fixture. In the inverse analyses, the unknown stiffness of the test fixture was represented by unknown rotational spring stiffnesses, and a bolt preload applied to the plate was represented by an unknown applied moment load.

The rotational spring stiffness became the single unknown parameter for the inverse beam analysis. Application of the least squares technique yielded a rotational spring stiffness that provided good correlation of experiment and analysis results. However, the variance in the stiffness estimate is expected to be large due to the sensitivity coefficient being extremely small. Optimal experiment design considerations can be used to plan future experiments to reduce the variance in the estimate.

For the inverse plate analysis, a cubic variation in rotational stiffness along the edge of the plate was initially assumed. The coefficients in the cubic expression were the eight unknown stiffness parameters. The linear dependence of the sensitivity coefficients for the parameters produced an ill-conditioned matrix for estimating parameters. Consequently, the parameter estimation problem reduced to single-parameter model for an average uniform boundary rotational stiffness coefficient. The least squares prediction of a constant test fixture stiffness boundary condition yielded analysis results that correlated well with the experimental deflection data. Hence, the use of constant stiffness rotational spring boundary conditions is sufficient for characterizing the boundary condition that the test fixture imposes on the aluminum plate when subjected to an applied concentrated mechanical load at the center of the plate. When a bolt preload was applied to the plate, a constant applied moment boundary condition in addition to a rigid displacement yielded analysis results that correlated well with experiment. However, to verify the existence and magnitude of the rigid displacement, additional experimentation would be required with a potentiometer located at the center of the moment arm in each restraining bolt mechanism.

References

[1]Taylor, A.H., Jackson, L.R., Cerro, J.A., and Scotti, S.J., "Analytical Comparison of Two Wing Structures for Mach 5 Cruise Airplanes," *Journal of Aircraft*, Vol. 21, No. 4, 1984, pp. 272–277.

[2]Jones, S.C., and Richards, L., "Titanium Honeycomb Panels Thermostructural Test and Analysis," Workshop on Correlation of Hot Structures Test Data With Analysis, NASA Ames-Dryden Flight Research Center, Edwards, CA, Nov. 15–17, 1988.

[3]Richards, W.L., and Thompson, R.C., "Titanium Honeycomb Panel Testing. Structural Testing Technology at High Temperature," Society for Experimental Mechanics, Bethel, CT, Nov. 1991, pp. 116–132.

[4]Beck, J.V., and Arnold, K.J., Parameter Estimation in Engineering and Science, Wiley, New York, 1977.

[5]Dorri, B., and Chandra, U.,"Determination of Thermal Contact Resistance Using inverse Heat Conduction Procedure," Numerical Methods in Thermal Problems, Vol. VII, Pt. 1, 1991.

Experimental Investigation of Thermally Induced Vibrations of Spacecraft Structures

Richard S. Foster* and Earl A. Thornton†
University of Virginia, Charlottesville, Virginia 22903

Abstract

Results of an experimental investigation of thermally induced vibrations of spacecraft structures now in progress at the University of Virginia's Thermal Structures Laboratory are presented. The primary objective of the research is to develop a fundamental understanding of thermally induced vibrations so that their adverse effects on spacecraft performance can be controlled or eliminated. Laboratory tests were conducted on a model spacecraft boom to investigate the character of the thermal-structural response of a representative flexible spacecraft appendage. Both stable and unstable thermally induced vibrations of the model spacecraft boom were studied. In addition, a new phenomenon not predicted by current theories has been observed: thermal strum.

I. Introduction

Spacecraft are typically designed with large flexible appendages such as solar panels for collecting the sun's radiant energy and long slender booms for positioning instrumentation or communications equipment away from the main spacecraft platform. Often, such appendages have low natural frequencies and are particularly susceptible to thermally induced vibrations. The vibrations are caused by sudden changes in the spacecraft's thermal loading, for example, during the night/day transitions in Earth orbit, and the concomitant changes in temperature throughout the structure. For large flexible appendages, the

---

*Graduate Student Researcher, Department of Mechanical, Aerospace and Nuclear Engineering. Member AIAA.

†Professor, Department of Mechanical, Aerospace and Nuclear Engineering. Associate Fellow AIAA.

internal moments resulting from temperature gradients can produce large-scale structural deformations. If the thermal and structural response times are of the same order of magnitude, the deformations may become vibratory, and under certain conditions, the vibrations may even become unstable.

Thermally induced vibrations were first posited by Boley[1] in the late 1950s. His analyses of long slender beams subjected to uneven heating revealed that if the effects of inertia were included in the thermoelastic equations, a vibratory response of the beam was possible. The response consists of a time varying displacement superimposed on a quasistatic deflection. For the case where inertia effects are ignored in the analysis, only the quasistatic deflection occurs. Boley further showed that the basic non-dimensional parameter of the problem is the ratio of the characteristic thermal response time to the natural period of vibration. This ratio will be denoted here as the Boley parameter B. For beams having a value of B greater than 10, thermally induced vibrations are of little consequence, and the response is given by the quasistatic solution. However, as the value of B approaches unity, inertia effects become increasingly important, and a dynamic analysis is required. The seminal analyses of Boley were followed by others who investigated thermally induced vibrations in plates and shells. More than a decade passed before the first experimental study confirmed the existence of such vibrations, but their effects became manifest much earlier.

During the 1960s, the United States launched and operated a series of six Earth-orbiting spacecraft called the Orbiting Geophysical Observatories (OGOs).[2] The OGO spacecraft were unique in that they were the first scientific satellites to be three-axis stabilized. Within the first few days of operation, OGO I-III (1964-1965) each experienced attitude control problems and had to resort to backup spin stabilized control. In addition, at various times during their orbits, the spacecraft were experiencing oscillations. A common feature among the OGO spacecraft was the use of long slender booms to position sensitive instrumentation away from onboard sources of electromagnetic radiation. By the spring of 1968, structural engineers began to conjecture that the oscillations were caused by thermally induced vibrations of the spacecraft booms during the night/day orbital transitions.

The first documented experimental demonstration of thermally induced vibrations was performed by Beam[3] at NASA Ames in June 1968. In his experiments, Beam used several photoflood lamps to heat a 3-ft-long boom. The boom, which supported a tip mass, consisted of a 1/2-in.-diameter beryllium-copper tube, slit and overlapped, having a wall thickness of 0.002 in. Beam's experiments confirmed the existence of stable thermally induced vibrations as predicted by theory, but in the process, he also discovered that the vibrations could become unstable. Beam and others soon began analytical studies of the spacecraft boom problem in an attempt to explain this new phenomenon. A more complete history of thermally induced vibrations and a summary of past research efforts are presented in a recent paper by Thornton and Foster.[4]

Even for contemporary spacecraft, thermally induced vibrations continue to cause operational problems. Perhaps the most highly publicized case is that of the Hubble Space Telescope (HST).[5,6] Launched in April of 1990, the HST has suffered from a number of problems, one of which has been termed thermal jitter. The jitter, which is most prevalent during the day/night and night/day orbital transitions, was caused by thermally driven bending of the solar arrays. The bending is produced by internal moments resulting from temperature gradients within the solar array's deployment booms. Because linear and angular momenta are conserved, motion of a flexible appendage, such as a solar array, results in a dynamic response of the entire spacecraft about the system mass center. For the HST, the dynamic response resulted in degradation of the telescope's image quality. The vibrating solar arrays were replaced with modified versions during the successful Space Shuttle mission to repair Hubble in December of 1993. Other recently launched spacecraft that have experienced thermally induced vibrations include Ulysses[7] in 1991, the Topography Experiment (TOPEX) satellite in 1992, and the Upper Atmosphere Research Satellite (UARS) in 1992.

The purpose of this paper is to present some results of an experimental investigation of thermally induced vibration research now in progress at the University of Virginia's Thermal Structures Laboratory. Both stable and unstable thermally induced vibrations have been studied in a model spacecraft boom. In addition, a new phenomenon not predicted by current theories has been observed: thermal strum. The remainder of this paper provides a description of thermally induced vibration experiments and the present experimental apparatus, example results of the tests, and a brief discussion of thermal strum.

II. Thermally Induced Vibration Experiments

To gain further understanding of thermally induced vibrations and their effects on spacecraft, it is useful to conduct experiments. Ideally, the experiments would be performed in orbit where the conditions identically match those expected for the candidate space structure. At present, however, space-based experiments are extremely expensive, thus alternative methods of investigation must be considered. Earth-based laboratory experiments are one such alternative.

In a laboratory experiment to study the behavior of spacecraft booms, a concentrated tip mass is suspended at the lower end of a long thin-walled steel tube.[8] The tube is rigidly fixed at its upper end and hangs vertically at rest under the influence of gravity. A heat lamp is turned on, thereby causing a sudden change in the boom's thermal loading. The incident heat flux gives rise to a circumferential temperature gradient within the tube which causes the boom to bend rapidly away from the lamp, and a thermally induced vibration ensues.

The spatial and temporal character of the applied thermal load play an important role in the response of the boom both in the laboratory and in space. In Earth orbit, the incident heat flux from the sun is essentially constant over length scales of spacecraft and, therefore, the spatial variation of the solar radiant heat flux may be ignored. To predict the thermal response of the boom, however, one must determine the rate at which heat energy is absorbed. Because the sun is a distant source, its rays are well collimated, and the rate at which energy is absorbed by the boom may be determined from a knowledge of the magnitude of the incident heat flux, the surface absorptivity of the boom, and the angle between the incident rays and the local surface normal vector of the boom.

When an orbiting spacecraft emerges from the Earth's umbra and penumbra into direct sunlight, there is a sudden change in its thermal loading. The side of the boom facing the sun begins to rise in temperature, and the boom bends away from the sun. As the boom bends, the angle between the local surface normal vector and the incident heat flux vector changes, thereby changing the energy absorption rate. The change in the energy absorption rate causes a change in the temperature distribution throughout the boom, which, in turn, causes a change in shape of the boom. In other words, the thermal and structural response of the boom are coupled.

Unlike the boom in space, the energy absorbed by the laboratory boom depends on the distance between the boom and the radiant heat source (heat lamp), thus the spatial variation of the heat flux must be considered. Shown in Figs. 1 and 2 are a schematic of the heat lamp geometry and a plot of the average heat flux at the center of the lamp as a function of distance from the front of the lamp reflector. As the boom nears the lamp, the heat flux increases, thereby increasing the temperature of the front side of the tube and

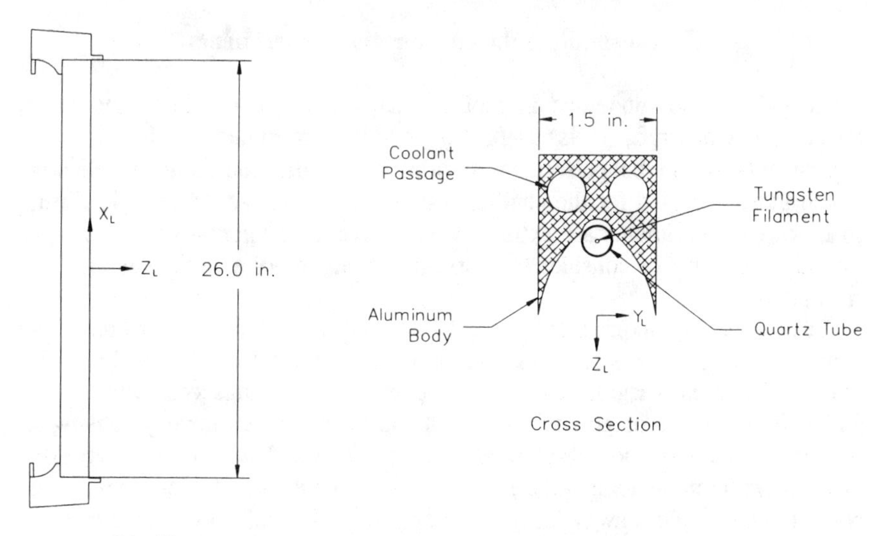

Fig. 1 Infrared heat lamp geometry.

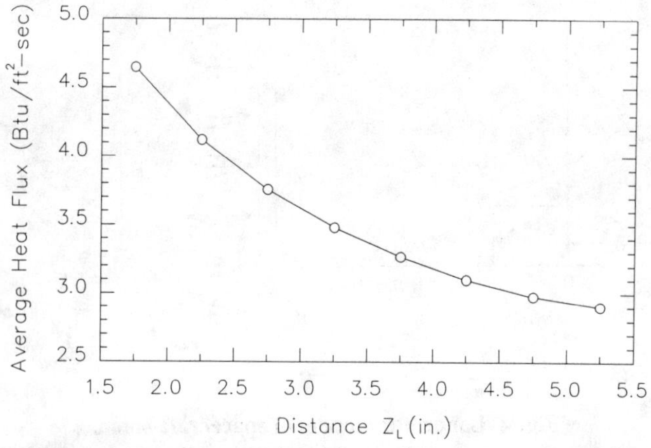

Fig. 2 Spatial dependence of average heat flux at lamp center, parabolic reflector, clear quartz tube, 100% power.

causing it to bend away from the lamp; as the boom moves away from the lamp, the applied heat flux decreases and the opposite effect occurs. Once again, boom deflections and temperatures are coupled. The coupling mechanisms just described are illustrated in Figs. 3 and 4. Both in space and in the laboratory, this thermal-structural coupling, under certain conditions, may cause the motion of the boom to become unstable, i.e., the amplitude of the oscillations continue to grow with increasing time. Such behavior is called thermal flutter.

III. Experimental Apparatus

In the experiments, which were conducted in air, a long, slender, model spacecraft boom was rapidly heated using an infrared heat lamp. As shown in

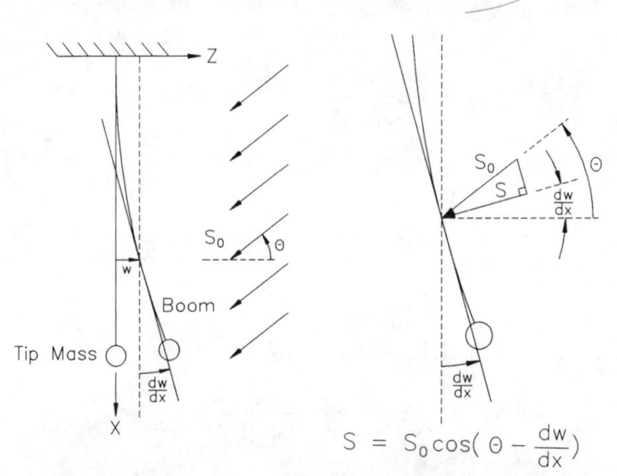

Fig. 3 Model of spacecraft boom with solar heat flux.

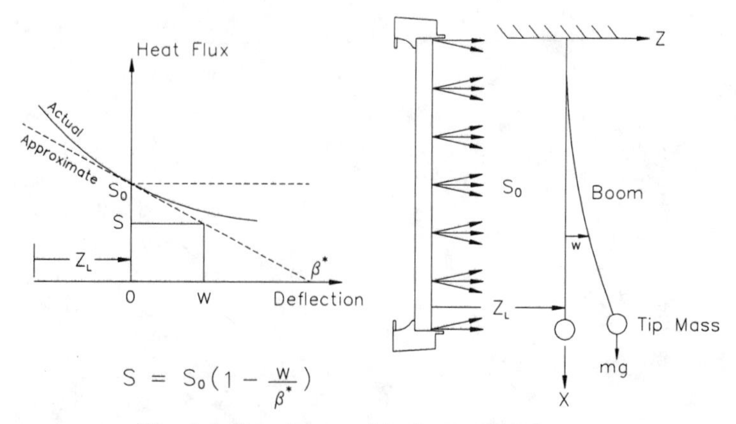

Fig. 4 Laboratory model of spacecraft boom.

Fig. 5, both the boom and lamp are mounted vertically within an aluminum frame. The upper end of the boom is firmly fixed, the lower end is free. The boom consists of a thin-walled stainless-steel tube, circular in cross section, and supports a brass tip mass at its free end.

A right-handed Cartesian coordinate system having three mutually orthogonal axes designated X, Y, and Z is used to describe the boom's geometry. The origin of the coordinate system is located at the fixed end of the boom, with the positive X axis collinear to the undeflected axis of the boom. The Y and Z axes of the coordinate system lie within the cross-sectional plane of the boom, with the positive Z axis in the direction opposite the heat lamp and the positive Y axis defined by the right-hand rule.

Two laser sensors were positioned in the proximity of the tip mass to simultaneously measure the Y and Z displacement components of the boom's

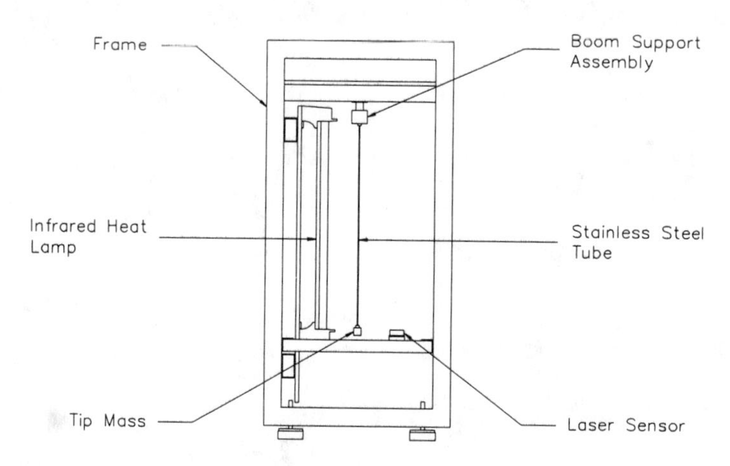

Fig. 5 Model spacecraft boom and test fixture.

free end. The displacement data was displayed and recorded using a personal computer based data acquisition and control system. Two types of test were conducted: 1) free vibration and 2) thermally induced vibration.

A. Test Fixture

The test fixture measures 2 ft × 2 ft × 4 ft and is fabricated of welded aluminum channel. Within the upper section of the fixture, a 1/2-in.-thick aluminum plate is secured, to which the boom support assembly is attached. The boom support assembly serves two purposes: 1) it provides a fixed support for the upper end of the boom and 2) it thermally isolates the boom from the test fixture. A schematic of the boom support assembly is presented in Fig. 6. As shown in the diagram, thermal isolation of the boom is provided by a silicon/glass composite grip. A steel insert and nipple, held within the grip, serve as a purchase for the boom, and the boom is secured in place by a stainless-steel collar. The undeformed boom is located at a distance of 3.75 in. from the front of the lamp's reflector.

Attached to the aluminum frame are a series of orthogonal slotted tracks which allow for the placement of calibration equipment and instrumentation. A pair of electrically isolated rectangular aluminum tubes provide the means for mounting and positioning the infrared heat lamp. The heat lamp employs a tungsten filament and parabolic reflector to produce a heated strip of dimensions 1.5 in. × 26 in. and was aligned parallel to the axis of the boom. An instrumentation deck was constructed at the level of the free end of the

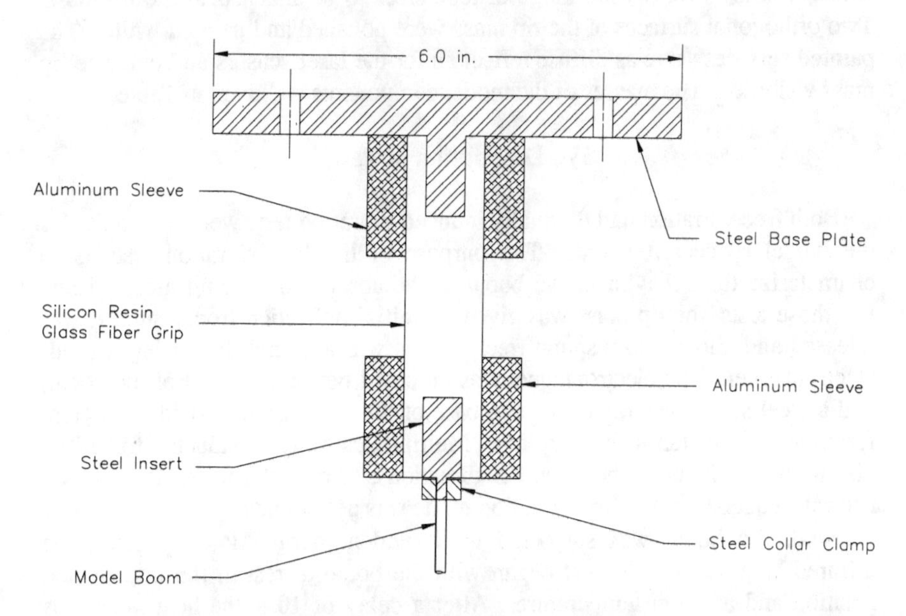

Fig. 6 Boom support assembly.

Table 1 Model boom masses

Component	Mass, g
Stainless-steel tube	7.4
Brass tip mass	124.5
Plug and pin	3.3
Collar clamp	5.2

boom for mounting the displacement sensors. Two laser sensors, one for each displacement component, were used to measure the displacement history of the boom's tip mass. The sensors measure displacement by focusing a spot of laser light on the tip mass surface and gauging the diffuse reflection. Heat shields were used to protect the sensors from excessive thermal transients during the tests.

B. Model Spacecraft Boom

The model spacecraft boom consists of a thin-walled stainless-steel (316) tube painted flat black with a brass tip mass affixed to the end. The tube is 25.5 in. long, 1/8 in. in diameter, and has a 0.006 in. wall thickness. Measuring 1 in. on an edge, the brass tip mass is cubic in shape and was machined to accept a composite plug. Centered in the plug is a steel pin which was inserted into the tube and secured with a stainless-steel collar clamp. The plug serves two purposes: 1) it provides thermal isolation between the tube and the tip mass and 2) it allows for a wide range of tube sizes to be attached to the tip mass. Two orthogonal surfaces of the tip mass were polished and painted white. The painted surfaces serve as diffuse reflectors for the laser sensors and enhance tip mass visibility. The masses of the model components are given in Table 1.

IV. Description of Tests

Both free vibration and thermally induced vibration tests were conducted for the model spacecraft boom. The purpose of the free vibration tests is to characterize the behavior of the boom in the absence of forcing mechanisms. For those tests, the tip mass was given an initial deflection from which it was released and allowed to respond freely. This was accomplished using a small electromagnet. The electromagnet was mounted beneath the end of the boom, and a steel screw was fitted into the base of the tip mass to provide sufficient ferromagnetic material for coupling. Multiple tests were conducted for initial displacements in both the Y and Z directions. From the tests, the damped natural frequency of the first vibration mode was determined.

Next, the boom was subjected to a sudden thermal loading using the infrared heat lamp. The test began with the boom at rest in the undeflected position and at room temperature. After a delay of 10 s, the heat lamp was turned on, and the boom was rapidly heated along its entire length. Thermally induced vibrations ensued. The vibrations were observed visually and by tracking the displacement history of the boom's tip mass with the laser sensors. Tests were conducted at various lamp power levels to determine the character of the boom's response for different thermal loadings.

V. Experimental Results

Typical test results are presented to demonstrate the characteristic behavior of the system. A more comprehensive investigation is anticipated in the coming months. The results of the free vibration tests are presented first, followed by those of the suddenly heated boom. The heat lamps used for these tests employed frosted quartz tubes. The heat lamp calibration data presented earlier in Fig. 2 was taken using a clear quartz tube and was included only to demonstrate the spatial dependence of the heat flux with respect to distance from the lamp, not for quantitative correlation with the thermally induced vibration tests. Calibration of heat lamps using frosted quartz tubes is forthcoming.

A. Free Vibration Tests

Free vibration tests were conducted for initial displacements in both the Y and Z directions. The displacement history for a duration of 500 s is presented in Fig. 7 for the case where the tip mass is given an initial deflection in the Z direction only. As seen in the figure, the overall motion of the boom is not rectilinear, and there clearly exists a coupling mechanism between the motions of the Y and Z directions. Shown in Figs. 8-10 are the tip mass's trajectory for

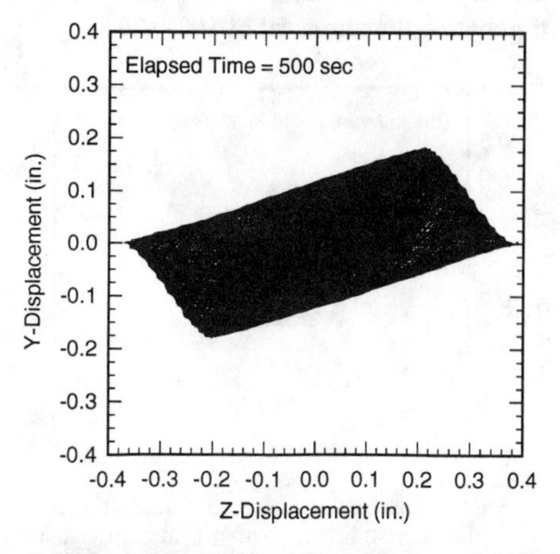

Fig. 7 Free vibration of model spacecraft boom, tip mass displacement history, *Y* and *Z* components, elapsed time of 500 s.

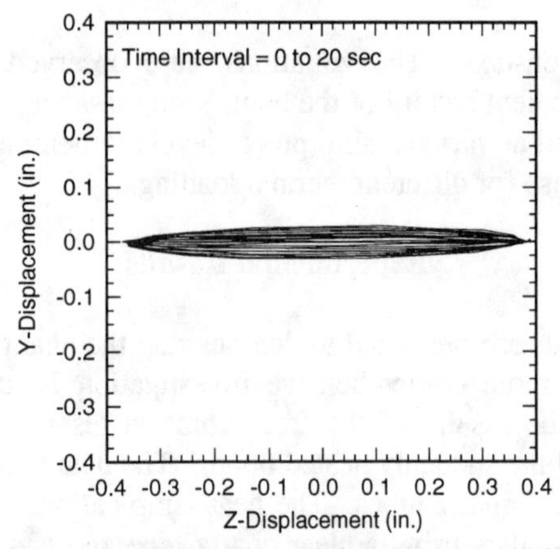

Fig. 8 Free vibration of model spacecraft boom, tip mass displacement history, *Y* and *Z* components, time interval of 0–20 s.

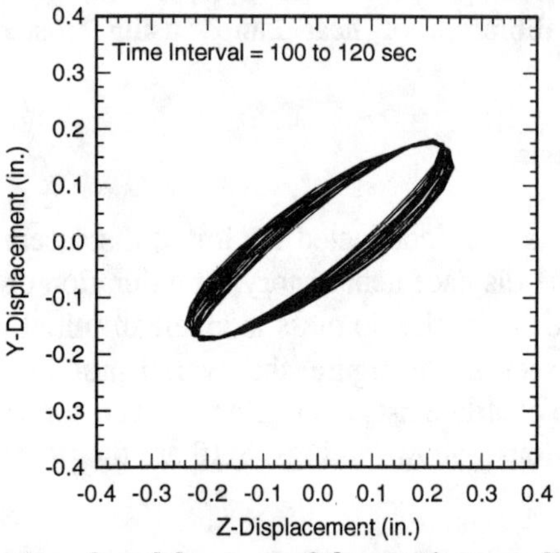

Fig. 9 Free vibration of model spacecraft boom, tip mass displacement history, *Y* and *Z* components, time interval of 100–120 s.

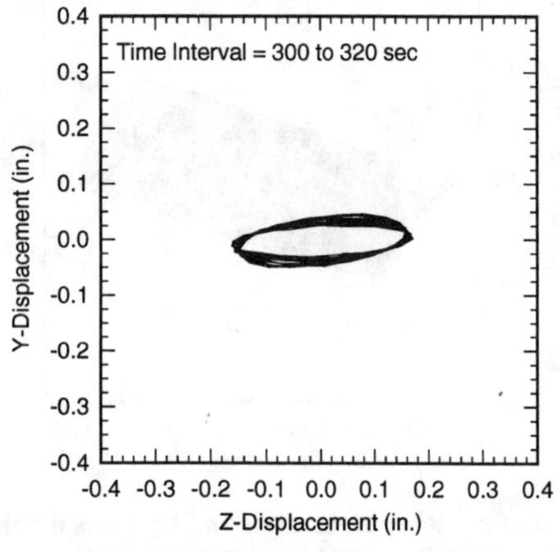

Fig. 10 Free vibration of model spacecraft boom, tip mass displacement history, *Y* and *Z* components, time interval of 300–320 s.

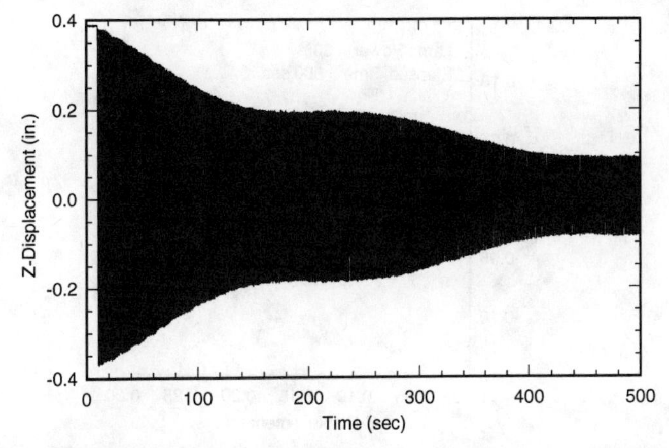

Fig. 11 Free vibration of model spacecraft boom, tip mass displacement history, *Z* component.

a 20 s time interval beginning at 0, 100, and 300 s, respectively. The displacement histories of the individual components for the same test are plotted in Figs. 11 and 12. From the response data, the damped natural frequency of the first vibration mode was determined to be 1.05 Hz.

Identification of the primary coupling mechanism between the *Y* and *Z* components has been difficult. Provided the boom is not handled, the test results are repeatable. If the boom is handled, however, for example, if it is removed from the support assembly and then reattached, the test results may be different. The thin-walled tubing is very delicate and may be easily deformed. This suggests that small changes in the shape or alignment of the

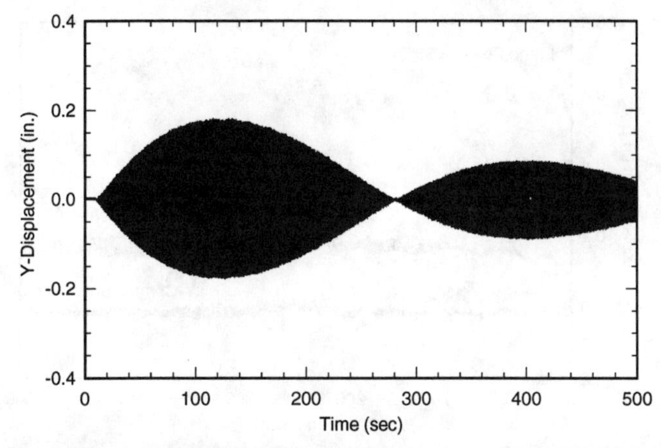

Fig. 12 Free vibration of model spacecraft boom, tip mass displacement history, *Y* component.

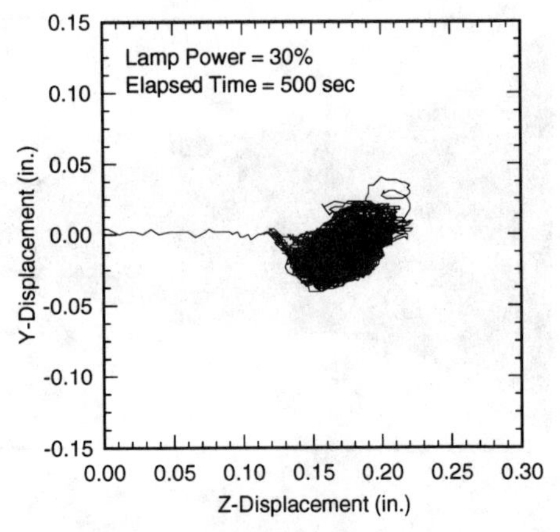

Fig. 13 Stable thermally induced vibration of model spacecraft boom, tip mass displacement, *Y* and *Z* components, elapsed time of 500 s.

boom may strongly influence its dynamic behavior. Instrumentation capable of addressing that issue is presently under investigation.

B. Thermally Induced Vibration Tests

Tests of the model spacecraft boom were conducted over a wide range of lamp power levels to explore the relationship between incident heat flux and the resulting thermally induced vibrations. Shown in Fig. 13 is a plot of the tip mass displacement history for a duration of 500 s for a lamp power level of 30% (958 W). Histories of the individual displacement components are presented in Fig. 14. As predicted by theory, the response is seen to consist of

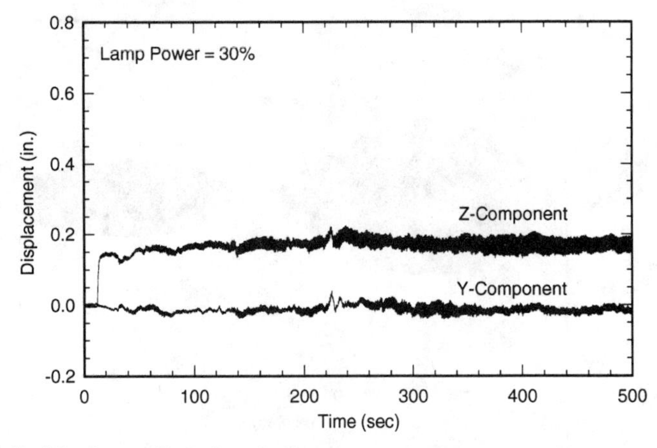

Fig. 14 Stable thermally induced vibration of model spacecraft boom, tip mass displacement history, *Y* and *Z* components.

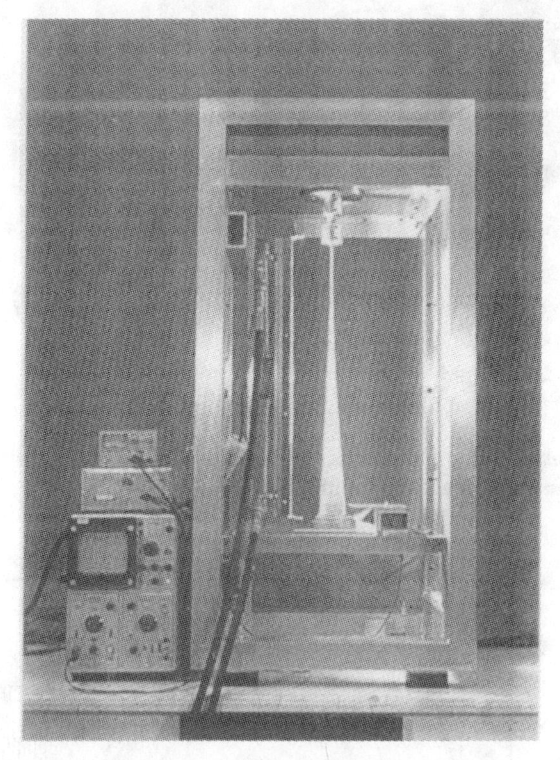

Fig. 15 Time lapsed photograph of thermally induced vibration.

two types of motion: an oscillatory displacement of the tip mass superimposed on a quasistatic deflection. Moreover, the response of the boom is shown to be stable, i.e., the amplitude of the oscillations does not continue to grow with time. A time lapsed photograph of the boom and test fixture during a typical test is presented in Fig. 15.

As the power level of the lamp and the incident heat flux increase, a stability boundary[4] is traversed and the response becomes unstable. Shown in Figs. 16 and 17 are the response of the boom for a lamp power level of 100% (2900 W). The unstable character of the vibrations is clearly demonstrated in the plots by the increasing amplitude of the oscillations.

What is not apparent in the plots, however, is the presence of a mode of vibration not predicted by current theories. The vibration appears to be the second natural mode of the boom, and it does not appear in the plots because, for the second mode, the tip mass acts like a node. The response resembles the behavior of a strummed guitar string, hence, the name thermal strum. To our knowledge, this phenomenon, that is, the simultaneous thermal excitation of both the fundamental and second natural bending modes of vibration for a fixed-free beam with a tip mass, has never before been reported and is not predicted by current theories. Methods of sensing and recording this behavior are currently being pursued.

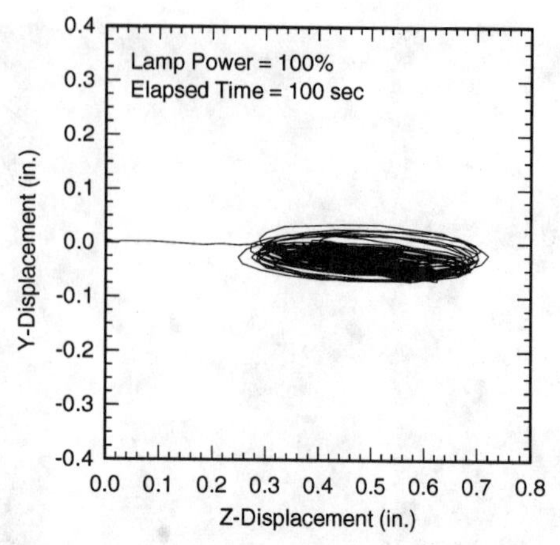

Fig. 16 Unstable thermally induced vibration of model spacecraft boom, tip mass displacement, *Y* and *Z* components.

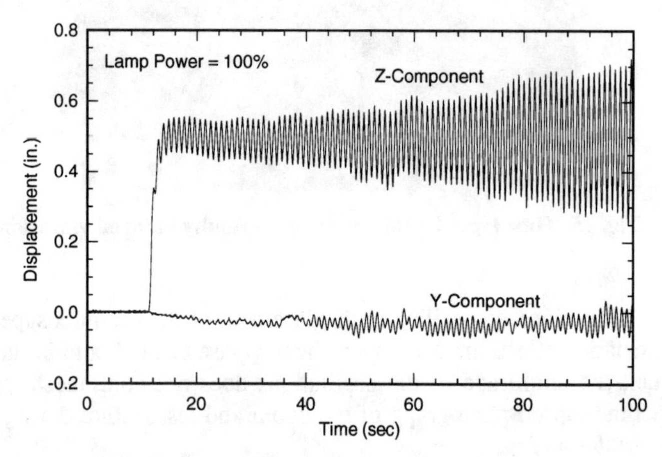

Fig. 17 Unstable thermally induced vibration of model spacecraft boom, tip mass displacement history, *Y* and *Z* components.

VI. Concluding Remarks

Results of an experimental investigation of thermally induced vibrations of a model spacecraft boom were presented. Two types of test were conducted: 1) free vibration and 2) thermally induced vibration. The results of the free vibration tests indicate that imperfections in the tube's shape and alignment may strongly influence its free vibration behavior. Methods for measuring the imperfections are presently under investigation. Both stable and unstable thermally induced vibrations were demonstrated in the laboratory for a model spacecraft boom in air. Similar tests will be conducted at low pressure when the laboratory's planned high vacuum test system becomes operational. Lastly, a new phenomenon was observed: thermal strum. Methods are presently being explored to quantitatively describe this new behavior and to explain its origins.

Acknowledgments

The research efforts of the authors were supported in part by the NASA Langley Research Center's Control Structures Interaction (CSI) Office, the University of Virginia's Academic Enhancement Program (AEP), and the Virginia Space Grant Consortium (VSGC).

References

[1]Boley, B. A., "Thermally Induced Vibrations of Beams," *Journal of Aeronautical Sciences,* Vol. 23, No. 2, 1956, pp. 179–181.

[2]Thornton, E. A., and Paul, D. B., "Thermal-Structural Analysis of Large Space Structures: An Assessment of Recent Advances," *Journal of Spacecraft and Rockets,* Vol. 22, No. 4, 1985, pp. 385–393.

[3]Beam, R. M., "On the Phenomenon of Thermoelastic Instability (Thermal Flutter) of Booms with Open Cross Section," NASA TN D–5222, June 1969.

[4]Thornton, E. A., and Foster, R. S., "Dynamic Response of Rapidly Heated Space Structures," Computational Nonlinear Mechanics in Aerospace Engineering, edited by S. N. Atluri, Vol. 146, Progress in Astronautics and Aeronautics, 1992, pp. 451–477.

[5]Sawyer, K., "Hubble Space Telescope's Flutter Will Require Solar-Panel Fix," *The Washington Post,* Nov. 11, 1990, p. A6.

[6]Thornton, E. A., and Kim, Y. A., "Thermally Induced Bending Vibrations of a Flexible Rolled-Up Solar Array," *Journal of Spacecraft and Rockets,* Vol. 30, No. 2, 1993, pp. 438–448.

[7]Anon., "Too Much Sun Gives Ulysses the Wobbles," *Electronics World and Wireless World,* March 1991.

[8]Sumi, S., Murozono, M., and Imoto, T., "Thermally Induced Bending Vibration of Thin-Walled Boom with Tip Mass Caused by Radiant Heating," *Technology Reports of Kyushu University,* Vol. 61, No. 4, 1988, pp. 449–455; Japanese with English Abstract.

Chapter 3. Analysis of High Temperature Composites

Recent Advances in the Mechanics of Functionally Graded Composites

Marek-Jerzy Pindera* and Jacob Aboudi†
University of Virginia, Charlottesville, Virginia, 22903
and
Steven M. Arnold‡
NASA Lewis Research Center, Cleveland, Ohio, 44135

Abstract

This paper summarizes recent developments of a new higher order theory for the thermomechanical response of functionally graded materials. In contrast to existing micromechanical theories that utilize classical (i.e., uncoupled) homogenization schemes to calculate microlevel and macrolevel stress and displacement fields in materials with nonuniform fiber spacing (i.e., functionally graded materials), the new theory explicitly couples the microstructural details with the macrostructure of the composite. Previous thermoelastic analysis has demonstrated that such coupling is necessary when the temperature gradient is large with respect to the dimension of the reinforcement, the characteristic dimension of the reinforcement is large relative to the global dimensions of the composite, and the number of reinforcing fibers or inclusions is small. In these circumstances, the standard micromechanical analyses based on the concept of the representative volume element used to determine average composite properties produce questionable results. Herein, examples are presented that illustrate further generalizations of the new higher order theory originally developed for the response of metal matrix composite plates with nonuniform fiber spacing in the through-the-thickness direction subjected to a thermal gradient. These generalizations include the incorporation of temperature-dependent inelastic models for the response of metallic constituents and the development of a two-

---

*Associate Professor, Civil Engineering and Applied Mechanics Department.

†Professor, currently on leave from the Faculty of Engineering, Tel-Aviv University, Israel.

‡Aerospace Research Engineer, Structural Fatigue Branch.

dimensional framework for modeling of composites functionally graded in two directions. The examples presented illustrate the shortcomings of the standard micromechanics approaches in analyzing functionally graded composites, the importance of including the inelastic effects, and the potential of the functional grading concept in reducing edge effects in laminated composite plates.

Introduction

The past 30 years have seen tremendous growth in the development and use of composite materials. The applications range from sporting and recreational accessories to advanced aerospace structural and engine components. Traditionally, the reinforcement phase in different types of composite materials is distributed uniformly such that the resulting mechanical, thermal, or physical properties do not vary spatially. Recently, a new concept involving tailoring or engineering the internal microstructure of a composite material to specific applications has taken root. This idea has been pursued vigorously by Japanese researchers who have coined the term functionally gradient materials (FGMs) to describe this newly emerging class of composites.[1,2] The idea involves spatially varying the microstructural details through nonuniform distribution of the reinforcement phase, by using reinforcement with different properties, sizes, and shapes, as well as by interchanging the roles of reinforcement and matrix phases in a continuous manner. The result is a microstructure that produces continuously changing thermal and mechanical properties at the macroscopic or continuum level.

Such an approach offers a number of advantages over the more traditional methods of tailoring the stiffness, thermal conductivity, and thermal expansion response of composite materials or structural elements and opens up new horizons for novel applications. Grading or tailoring the internal microstructure of a composite material or a structural component allows the designer to truly integrate both the material and structural considerations into the final design and final product. This brings the entire structural design process to the material level in the purest sense, thereby increasing the number of possible material configurations for specific design applications. This concept marks the beginning of a new revolution in the areas of engineering that deal explicitly with materials, such as materials science and mechanics of materials.

Functionally graded composites are ideal candidates for applications involving severe thermal gradients, ranging from thermal structures in advanced aircraft and aerospace engines to computer circuit boards. In such applications, a ceramic-rich region of a functionally graded composite is exposed to hot temperature while a metallic-rich region is exposed to cold temperature, with a gradual microstructural transition in the direction of the temperature gradient. By adjusting the microstructural transition appropriately, optimum temperature distribution can be realized. Microstructural grading can also be effectively used to reduce the mismatch in the thermomechanical properties between differently oriented, adjacent plies in a laminated plate. Thus, reduction of the interlaminar stresses at the free edge of a laminate that result from a large property mismatch between adjacent plies can be realized by using the functional grading concept to smooth out the transition between dissimilar plies. Along similar lines, joining of dissimilar materials can be made more efficient through the use of functionally graded joints. Other benefits to be realized from the use of functionally graded architectures include fracture toughness enhancement in ceramic matrix composites through tailored interfaces.

The potential benefits that may be derived from functionally graded composites have led to increased activities in the areas of processing, and materials science, of these materials. However, these activities are seriously handicapped by the lack of appropriate computational strategies for the response of functionally graded materials that explicitly couple the heterogeneous microstructure of the material with the global analysis. The standard micromechanics approach used to analyze the response of this class of materials is to decouple the local and global effects by assuming the existence of a representative volume element (RVE) at every point within the composite, Fig. 1 (cf., Wakashima and

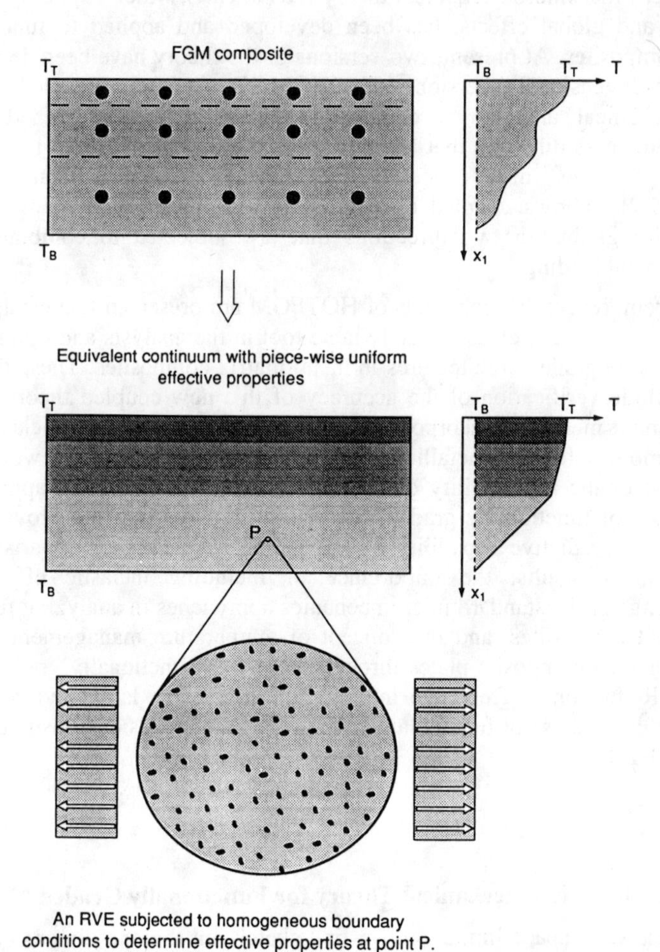

Fig. 1 Uncoupled micromechanics analysis of FGMs.

Tsukamoto[3]). This assumption, however, neglects the possibility of coupling between local and global effects, thus leading to potentially erroneous results in the presence of macroscopically nonuniform material properties and large field variable gradients. This is particularly true when the temperature gradient is large with respect to the dimension of the inclusion phase, the characteristic dimension of the inclusion phase is large relative to the global dimensions of the composite, and the number of uniformly or nonuniformly distributed inclusions is relatively small.[4] Perhaps the most important objection to using the standard RVE-based micromechanics approach in the analysis of FGMs is the lack of a theoretical basis for the definition of an RVE, which clearly cannot be unique in the presence of continuously changing properties due to nonuniform inclusion spacing.

As a result of the limitation of the standard micromechanics approches, a new higher order micromechanical theory (HOTFGM), which explicitly couples the local and global effects, has been developed and applied to functionally graded composites. At present, two versions of the theory have been developed. The one-dimensional version, HOTFGM-1D, allows coupled micro-macromechanical analysis of composite plates functionally graded in the through-thickness direction that are subjected to a thermal gradient in the same direction.[4–8] The most recently developed two-dimensional version, HOTFGM-2D, allows coupled micromechanical analysis of composite plates functionally graded in two directions that are subjected to combined thermomechanical loading.[9]

Herein, recent developments of HOTFGM are presented that establish the theory as an accurate, efficient, and viable tool in the analysis and optimization of functionally graded architectures in metal matrix composites. These developments include verification of the accuracy of this new coupled theory via the finite element method,[10] incorporation of temperature-dependent inelastic constitutive models for the metallic phase into the theoretical framework,[7] and assessment of the applicability of the uncoupled micromechanics approach in the analysis of functionally graded materials.[11–12] Examples are provided that illustrate the predictive capability of HOTFGM-1D through comparison with finite element results, the importance of including inelastic effects, the shortcomings of the standard micromechanics approaches in analyzing functionally graded composites, and the concept of temperature management in thermally protected composite plates through the use of functionally graded architectures. Reduction of free-edge interlaminar stresses in laminated composite plates through the use of functionally graded architectures is demonstrated using HOTFGM-2D.

Higher Order Micromechanical Theory for Functionally Graded Materials

Because of space limitation, we first sketch out the one-dimensional version of the coupled higher order theory, and then briefly discuss its extension to microstructures functionally graded in two directions.

One-Dimensional Theory

HOTFGM-1D is based on the geometric model of a heterogeneous composite, with a finite thickness H, extending to infinity in the x_2–x_3 plane and subjected to a temperature gradient produced by the temperature T_T and T_B applied to the top and bottom surfaces, respectively, see Fig. 2. The composite is reinforced by periodic arrays of fibers in the direction of the x_2 axis or the x_3 axis, or both. In the direction of the x_1 axis, called the functionally gradient (FG) direction, the fiber spacing between adjacent arrays may vary. The reinforcing fibers can be either continuous or finite length. The heterogeneous composite is constructed using a generic unit cell, which consists of either four or eight subcells, depending on whether continuously or discontinuously reinforced functionally graded composites are considered. The generic unit cell in the present framework is not taken to be an RVE whose effective properties can be obtained through homogenization. Rather, the RVE comprises an entire column of such cells spanning the plate's thickness. Thus, the response of each cell is explicitly coupled to the response of the entire column of cells in the FG direction, thereby directly coupling the microstructural details with the global analysis. This is in stark contrast with the standard uncoupled micromechanics approach commonly used in the analysis of FGMs.

The solution to the thermomechanical boundary-value problem outlined in the foregoing is solved in two steps. In the first step, the temperature distribution in a single column of cells, representative of the composite-at-large, spanning the FG dimension is determined by solving the heat equation under steady-state conditions in each cell subject to the appropriate continuity and compatibility conditions. The solution to the heat equation is obtained by approximating the temperature field in each ($\alpha\beta\gamma$) subcell of a generic unit cell using a quadratic expansion in the local coordinates $\bar{x}^{(\alpha)}$, $\bar{x}^{(\beta)}$, $\bar{x}^{(\gamma)}$, centered at the subcell's mid-

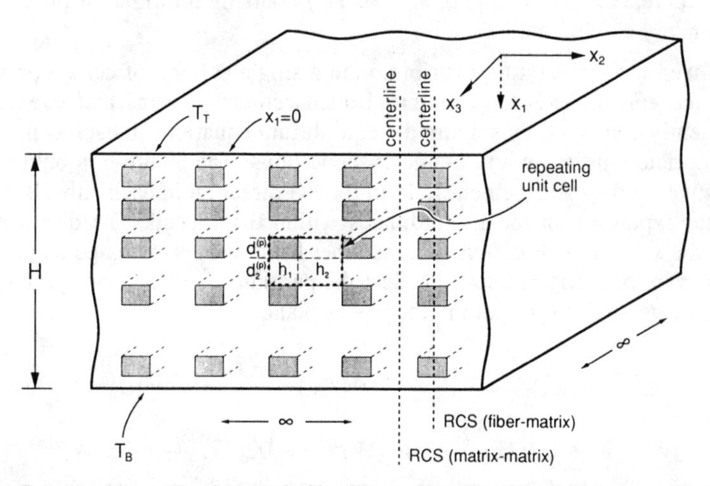

Fig. 2 Composite with nonperiodic fiber distribution in the x_1 direction. RCS denotes the representative cross-section.

point, that reflects the microstructure's symmetry and periodicity in the x_2–x_3 plane:

$$T^{(\alpha\beta\gamma)} = T_0^{(\alpha\beta\gamma)} + \bar{x}_1^{(\alpha)} T_1^{(\alpha\beta\gamma)} + 1/2(3\bar{x}_1^{(\alpha)2} - 1/4d_\alpha^{(p)2})T_2^{(\alpha\beta\gamma)}$$
$$+ 1/2(3\bar{x}_2^{(\beta)2} - 1/4h_\beta^2)T_3^{(\alpha\beta\gamma)} + 1/2(3\bar{x}_3^{(\gamma)2} - 1/4l_\gamma^2)T_4^{(\alpha\beta\gamma)} \tag{1}$$

A higher order representation of the temperature field is necessary in order to capture the local effects created by the thermomechanical field gradients, the microstructure of the composite, and the finite dimension in the FG direction, in contrast with previous treatments involving fully periodic composite media. The unknown coefficients associated with each term in the expansion are then obtained by constructing a system of equations that satisfies the requirements of a standard boundary-value problem for the given temperature field approximation. That is, the heat equation is satisfied in a volumetric sense, and the thermal and heat flux continuity conditions within a given cell, as well as between a given cell and adjacent cells, are imposed in an average sense. This yields $40M$ equations in the unknown $40M$ coefficients $T_i^{(\alpha\beta\gamma)}$ for a composite with M rows of fibers in the through-thickness direction of the form

$$\boldsymbol{\kappa}\,\boldsymbol{T} = \boldsymbol{t} \tag{2}$$

where the structural thermal conductivity matrix $\boldsymbol{\kappa}$ contains information on the geometry and thermal conductivities of the individual subcells $(\alpha\beta\gamma)$ in the M cells spanning the thickness of the FG plate, the thermal coefficient vector $\boldsymbol{T} = (\boldsymbol{T}_1^{(111)}, \ldots, \boldsymbol{T}_M^{(222)})$, where $\boldsymbol{T}_p^{(\alpha\beta\gamma)} = (T_0, T_1, T_2, T_3, T_4)_p^{(\alpha\beta\gamma)}$, contains the unknown coefficients that describe the thermal field in each subcell, and the thermal force vector $\boldsymbol{t} = (T_T, 0, \ldots, 0, T_B)$ contains information on the thermal boundary conditions.

Given the temperature distribution in a single column of cells representative of the composite-at-large, internal displacements, strains and stresses are subsequently generated by solving the equilibrium equations in each cell subject to appropriate continuity and boundary conditions. The solution is obtained by approximating the displacement field in the FG direction in each subcell using a quadratic expansion in local coordinates within the subcell. The displacement field in the x_2 and x_3 directions, on the other hand, is approximated using linear expansion in local coordinates to reflect the periodic character and symmetry of the composite's microstructure in the x_2–x_3 plane.

$$u_1^{(\alpha\beta\gamma)} = w_1^{(\alpha\beta\gamma)} + \bar{x}_1^{(\alpha)} \phi_1^{(\alpha\beta\gamma)} + 1/2(3\bar{x}_1^{(\alpha)2} - 1/4d_\alpha^{(p)2})U_1^{(\alpha\beta\gamma)}$$
$$+ 1/2(3\bar{x}_2^{(\beta)2} - 1/4h_\beta^2)V_1^{(\alpha\beta\gamma)} + 1/2(3\bar{x}_3^{(\gamma)2} - 1/4l_\gamma^2)W_1^{(\alpha\beta\gamma)} \tag{3}$$

$$u_2^{(\alpha\beta\gamma)} = \bar{x}_2^{(\beta)} \chi_2^{(\alpha\beta\gamma)} \tag{4}$$

$$u_3^{(\alpha\beta\gamma)} = \bar{x}_3^{(\gamma)} \psi_3^{(\alpha\beta\gamma)} \tag{5}$$

The unknown coefficients associated with each term in the expansion, i.e., $w_1^{(\alpha\beta\gamma)}$, $\phi_1^{(\alpha\beta\gamma)}$, $\chi_2^{(\alpha\beta\gamma)}$, $\psi_3^{(\alpha\beta\gamma)}$, $U_1^{(\alpha\beta\gamma)}$, $V_1^{(\alpha\beta\gamma)}$, $W_1^{(\alpha\beta\gamma)}$, are obtained by satisfying the appropriate field equations in a volumetric sense (zeroth, first, and second moment), together with the boundary conditions and continuity of displacements and tractions between individual subcells of a given cell and between adjacent cells. The continuity conditions are imposed in an average sense. This results in $56M$ equations in the unknown $56M$ coefficients in the displacement representation within each cell of a composite with M rows of fibers in the through-thickness direction of the form

$$\boldsymbol{K}\,\boldsymbol{U} = \boldsymbol{f} + \boldsymbol{g} \tag{6}$$

where the structural stiffness matrix $\boldsymbol{K}$ contains information on the geometry and thermomechanical properties of the individual subcells $(\alpha\beta\gamma)$ in the M cells spanning the thickness of the FG plate. The displacement coefficient vector $\boldsymbol{U}$ contains the unknown coefficients that describe the displacement field in each subcell, i.e.,

$$\boldsymbol{U} = (\,\boldsymbol{U}_1^{(111)}, \ldots, \boldsymbol{U}_M^{(222)}\,) \tag{7}$$

where $\boldsymbol{U}_p^{(\alpha\beta\gamma)} = (\,w_1, \phi_1, U_1, V_1, W_1, \chi_2, \psi_3\,)_p^{(\alpha\beta\gamma)}$, and the mechanical force vector $\boldsymbol{f}$ contains information on the mechanical boundary conditions and the thermal loading effects generated by the applied temperature. In addition, the inelastic force vector $\boldsymbol{g}$ appearing on the right–hand side of Eq. (6) contains elements, given in terms of the integrals of inelastic strain distributions $\boldsymbol{\varepsilon}^{p\,(\alpha\beta\gamma)}$ and Legendre polynomials $P_l(\cdot)$, $P_m(\cdot)$, $P_n(\cdot)$ of the orders l, m, and n, that are represented by the coefficients $R_{ij\,(l,m,n)}^{(\alpha\beta\gamma)}$, i.e.,

$$\boldsymbol{g} = \boldsymbol{g}(R_{ij\,(l,m,n)}^{(\alpha\beta\gamma)}) \tag{8}$$

where

$$\frac{R_{ij\,(l,m,n)}^{(\alpha\beta\gamma)}}{\mu_{(\alpha\beta\gamma)}} = \frac{p_{lmn}}{4} \int_{-1}^{1}\int_{-1}^{1}\int_{-1}^{1} \varepsilon_{ij}^{p\,(\alpha\beta\gamma)} P_l(\zeta_1^{(\alpha)}) P_m(\zeta_2^{(\beta)}) P_n(\zeta_3^{(\gamma)})\, \mathrm{d}V^{(\alpha\beta\gamma)} \tag{9}$$

with

$$p_{lmn} = \sqrt{(1+2l)(1+2m)(1+2n)}\,, \quad \mathrm{d}V^{(\alpha\beta\gamma)} = \mathrm{d}\zeta_1^{(\alpha)}\mathrm{d}\zeta_2^{(\beta)}\mathrm{d}\zeta_3^{(\gamma)}$$

These integrals depend implicitly on the elements of the displacement coefficient vector $\boldsymbol{U}$, requiring an incremental solution of Eq. (6) at each point along the loading path.

The choice of an appropriate technique for the solution of Eq. (6) depends on the inelastic constitutive model employed to calculate the inelastic strain distributions in each subcell from which the coefficients $R_{ij(l,m,n)}^{(\alpha\beta\gamma)}$ can be generated. In the present framework, two constitutive theories are employed to describe the inelastic response of the matrix phase, namely, the classical incremental plasticity theory (Prandtl–Reuss equations) and the Bodner–Partom unified viscoplasticity theory.[13] The plasticity theory is employed to efficiently model the inelastic constitutive response of the matrix phase when rate effects can be neglected, whereas the more computationally intensive Bodner–Partom theory is employed for those situations where rate-dependent deformation must be taken into account. It should be noted that the present formulation is sufficiently general to accommodate other types of unified viscoplasticity theories.

The solution of Eq. (6) based on the Prandtl–Reuss plasticity equations is carried out following Mendelson's iterative scheme briefly outlined next.[14] For the given thermal load increment, the plastic strain at any point within a subcell is expressed in terms of the plastic strain from the preceding loading state plus an increment that results from the imposed load increment,

$$\varepsilon_{ij}^{p}(\boldsymbol{x})\,|_{current} = \varepsilon_{ij}^{p}(\boldsymbol{x})\,|_{previous} + d\varepsilon_{ij}^{p}(\boldsymbol{x}) \tag{10}$$

The plastic strain distribution in each subcell is subsequently determined by calculating plastic strains at a number of equally spaced locations after updating the plastic strains at these locations using Eq. (10). The current values for the plastic strains at these stations are then used in determining the integrals $R_{ij(l,m,n)}^{(\alpha\beta\gamma)}$ given in Eq. (9), and thus the elements of the inelastic force vector $\boldsymbol{g}$ in Eq. (6). Updated values of the interfacial displacements are then obtained from these equations. With a knowledge of the current components of the displacement coefficient vector $\boldsymbol{U}$, solutions for the displacement components $u_1^{(\alpha\beta\gamma)}$, $u_2^{(\alpha\beta\gamma)}$, and $u_3^{(\alpha\beta\gamma)}$, Eqs. (3–5), at any point within each subcell are obtained, from which total strains and their corresponding stresses are calculated. These are then used to obtain new approximations for the plastic strain increments. The iterative process is terminated when the differences between two successive sets of plastic strain increments are less than some prescribed value. An advantage of this solution technique is its efficiency and relative quick convergence even for relatively large load increments.[15,16]

If, on the other hand, Bodner–Partom unified viscoplastic constitutive theory is employed to model the inelastic response of the matrix phase, then the solution of Eq. (6) is generated at each increment of the applied thermal load as follows. First, the viscoplastic strain increments that result from an imposed thermal load increment are determined at a number of points within each subcell using an explicit forward Euler integration scheme based on the known state of deformation. These increments are then used to calculate the current inelastic strain distributions in each subcell needed to determine the coefficients $R_{ij(l,m,n)}^{(\alpha\beta\gamma)}$, and, thus, the elements of the inelastic force vector $\boldsymbol{g}$. The knowledge of the current inelastic force vector allows one to determine the current values of the displacement field vector $\boldsymbol{U}$ by solving Eq. (6) and, thus, the current stress and strain states. The thermal load is incremented and the entire process repeated.

The magnitude of the applied thermal load increment is governed by the imposed rate of change of the temperature profile with respect to time and the magnitude of the time increment used to integrate the viscoplastic constitutive equations. The explicit forward Euler integration scheme presently employed requires the time increment to be sufficiently small so as to guarantee convergence of the integration process in view of the stiff behavior of this class of equations.

Two-Dimensional Theory

The analytical approach in the two-dimensional theory, as in the one-dimensional version, is based on volumetric averaging of the various field quantities together with the imposition of boundary and interfacial continuity conditions in an average sense. The previous restriction of periodicity in two orthogonal directions, however, is presently abondoned, thus allowing arbitrary distribution of one or more reinforcement phases in one plane. This leads to a significant generalization of the theory. As a result, composites with finite dimensions along the two functionally graded directions can be analyzed. The functionally graded architectures include rows of aligned inclusions (or continuous fibers) with variable spacing in the functionally graded directions and regular spacing in the periodic direction. Alternatively, completely random inclusion (or fiber) architectures in the plane defined by the functionally graded directions can be admitted. At present, the two-dimensional version of the theory is limited to the analysis of functionally graded composites in the linearly elastic range.

Illustrations

Finite Element Validation

As a first step, we present a finite element validation of the one-dimensional version of the coupled higher order theory. Figure 3 presents comparison between HOTFGM-1D and finite element results generated using the commercial ABAQUS code for the temperature and normal stress σ_{22} distributions in a uniformly graded SiC/TiAl plate with three through-thickness fibers subjected to a temperature gradient of 500°C. The results are given in the fiber-matrix representative cross section (RCS), see Fig. 2, and are similar to those obtained in the matrix-matrix RCS. For the purpose of this comparison, the SiC fibers and the titanium matrix were treated as elastic with temperature-independent properties. The thermal conductivity mismatch between the constituents was deliberately amplified (i.e., $\kappa_f / \kappa_m = 50$) in order to critically test the predictive capability of the new theory. Similar results have been obtained for composites with one and five through-thickness fibers, establishing HOTFGM-1D as an accurate and efficient method for the analysis of FGMs.[10]

Inelastic Constitutive Modeling Capability

The capability to model temperature-dependent inelastic response of the individual phases was incorporated into the theoretical framework of HOTFGM-1D in order to realistically model the response of functionally graded

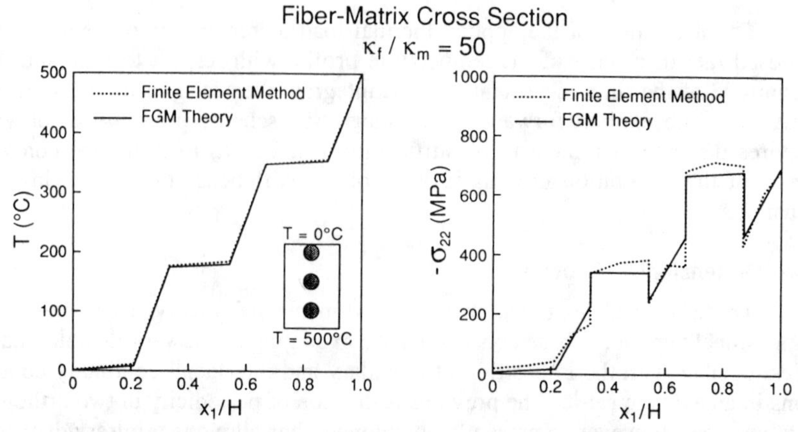

Fig. 3 Comparison between the coupled continuum theory and the finite element analysis for a SiC/Ti composite with three uniformly spaced fibers in the thickness direction: a) temperature profile across the fiber-matrix cross section and b) normal stress σ_{22} across the fiber-matrix cross section.

composites with metallic matrices in a wide temperature range. Examples that follow have been generated using the temperature-dependent inelastic response of Ti-6Al-4V illustrated in Fig. 4 that has been modeled with the classical plasticity and the Bodner–Partom unified viscoplasticity theories. In those situations where rate-dependent effects are not important, the Bodner–Partom unified viscoplasticity theory and the classical incremental plasticity theory with bilinear hardening produce nearly identical responses for the titanium alloy.

In order to set the previous results generated with the elastic formulation of the theory in perspective, Fig. 5 presents comparison between elastic and ine-

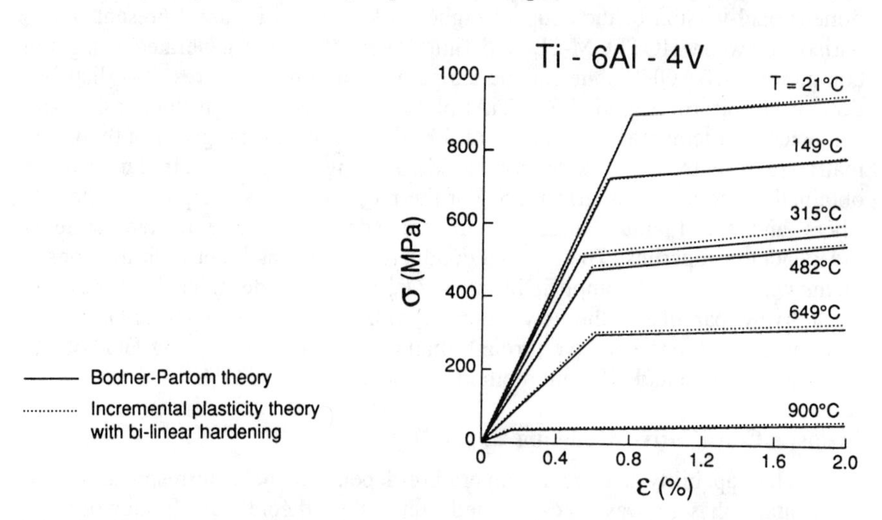

Fig. 4 Comparison of the titanium stress-strain response predicted by the Bodner–Partom unified viscoplasticity theory and the classical incremental plasticity theory.

lastic predictions for the normal stress σ_{22} distributions in a SiC/Ti-6Al-4V composite plate reinforced by 10 uniformly spaced fibers across the thickness. The inelastic results were generated using the Bodner–Partom unified viscoplasticity model for the titanium matrix. Two sets of results were obtained. In both instances, the applied temperature gradient was 500°C. In the first instance, the reference temperature at the bottom surface of the plate was 21°C whereas in the second instance this temperature was 400°C. For the lower reference temperature, the differences between the elastic and viscoplastic analyses are localized at the top surface of the plate exposed to the higher temperature. The differences are limited to the matrix region, with the elastic analysis overestimating the actual stress to an extent that decreases with increasing distance from the top surface. Virtually no difference is observed in the normal stress σ_{22} predicted by the elastic and inelastic analyses in the fiber phase in the immediate vicinity of the top surface and elsewhere. In contrast, the differences between elastic and inelastic analyses for the higher reference temperature are substantially more pronounced and propagate deeper into the plate. These differences are significant in both phases in the region exposed to the higher temperature. Comparison of the stress profiles presented at the two reference temperatures in Fig. 5 points to the importance of temperature-dependent properties of the constituent phases.

The in–plane force and moment resultants required to maintain the SiC/Ti-6Al-4V composite plate reinforced by 10 uniformly spaced fibers straight in the presence of the given temperature gradient are given in Fig. 6 for the two different reference temperatures. These quantities have been normalized with respect to the corresponding quantities obtained using elastic analysis. Included in the figure are the corresponding results for a SiC/Ti-6Al-4V composite plate reinforced by 10 exponentially spaced fibers with decreasing spacing toward the cold surface. As is observed, the error in the prediction of the

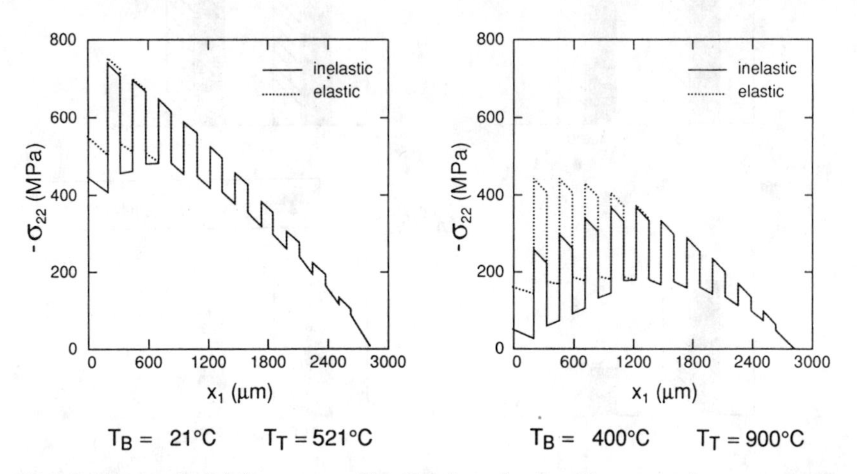

Fig. 5 Through-thickness σ_{22} distributions in the fiber-matrix cross section of a SiC/Ti-6Al-4V composite, with 10 uniformly spaced fibers in the thickness direction, subjected to a thermal gradient; effect of reference temperature T_B on the comparison of elastic and inelastic HOTFGM-1D analysis.

in–plane force and moment resultants introduced by neglecting the inelastic effects depends on the reference temperature, manner of functional grading, and the quantity of interest. In particular, the elastic analysis substantially overestimates the in–plane moment resultants M_2 and M_3 in the two configurations at the higher reference temperature. In the case of the exponentially spaced configuration, the elastic analysis even predicts an incorrect sign of the in–plane moment resultants. The incorporation of inelastic constitutive theories for the response of metallic phases into the HOTFGM-1D framework enables one to identify those regions where inelastic effects cannot be disregarded, especially in the presence of high thermal and stress gradients. As a result, the importance of the inelastic effects on the overall response of functionally graded composites can now be quantitatively assessed.[7]

Higher Order Theory Versus Uncoupled Micromechanics

An alternative solution methodology for the stress and displacement fields in functionally graded composites involves a combination of the continuum and standard micromechanics approaches in which local and global effects are decoupled. First, the continuum results are obtained by solving the given boundary-value problem of a homogeneous medium with equivalent or effective material properties. The effective properties are generated using a chosen micromechanics model without regard to whether a representative volume exists or not. That is, these effective properties are generated on the premise that no coupling exists between local and global responses. As stated before, this is the

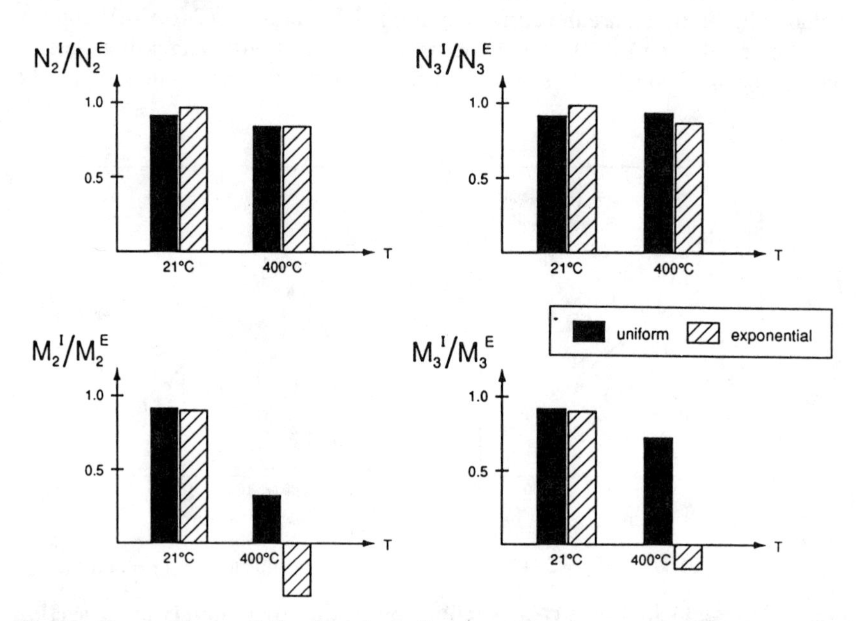

Fig. 6 Comparison of the in–plane force and moment resultants obtained with elastic and inelastic analyses for a SiC/Ti composite with 10 uniformly and exponentially spaced fibers in the thickness direction.

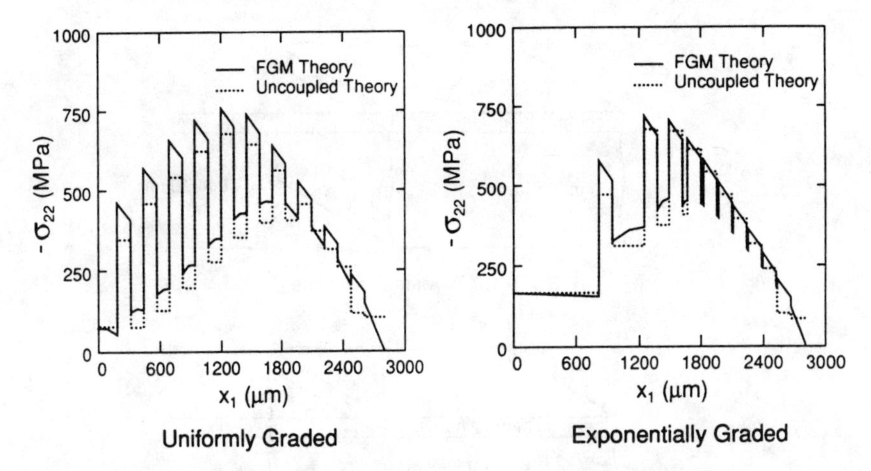

Fig. 7 Through-thickness σ_{22} distributions predicted by the uncoupled micromechanics analysis and HOTFGM-1D in the fiber-matrix cross section of a SiC/Ti-6Al-4V composite with 10 uniformly and exponentially spaced fibers in the thickness direction subjected to a thermal gradient, $T_B = 21°C$, $T_T = 900°C$.

standard micromechanical approach currently employed by researchers working in the area of functionally graded materials. These effective properties are subsequently used in the thermal boundary-value problem of an equivalent homogeneous composite subjected to the specified thermal loading. With the knowledge of the continuum or macroscopic thermal fields, the stresses in the individual phases of a repeating unit cell are then calculated by applying an average temperature over a given cell, treating it as an RVE within the framework of the chosen micromechanical model.

To assess the limits of applicability of the standard micromechanics approach in the analysis of FGMs, internal stress distributions in uniformly and nonuniformly graded composites subjected to a through-thickness temperature gradient were generated using HOTFGM-1D and the method of cells micromechanics model.[17] The method of cells was chosen for comparison purposes because it is a special case of the coupled theory and because it has been shown in previous investigations to be an accurate micromechanics model for the analysis of periodic composites. The normal stress distributions generated using the two approaches are compared in Fig. 7 for uniformly and exponentially graded SiC/Ti-6Al-4V composites with 10 through-thickness fibers, maintained at 21°C at the bottom surface and subjected to a temperature gradient of 879°C. The response of the titanium matrix was modeled using the Bodner–Partom unified viscoplasticity theory. As is observed, the normal stress may be substantially underestimated by the uncoupled micromechanics theory even in the case of composites with a relatively large number of through-thickness fibers. This is particularly true in the immediate vicinity of the boundary where coupling is important, as well as in regions of low-fiber concentrations. Substantially greater differences between the predictions of the two aproaches have been observed for fewer through-thickness fibers.[11–12]

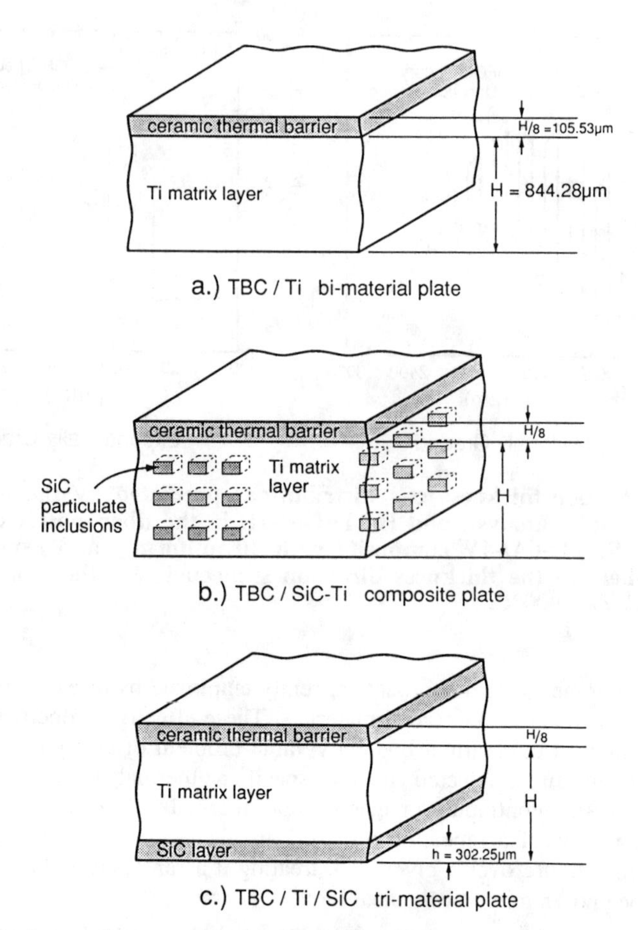

Fig. 8 Metal matrix composite plate configurations protected by a ceramic thermal barrier coating (TBC): a) TBC/Ti bimaterial plate, b) TBC/SiC-Ti composite plate, and c) TBC/Ti/SiC trimaterial plate.

Temperature Management via Functionally Graded Architectures

The last example involving HOTFGM-1D illustrates the potential of tailoring the microstructure of a SiC particulate-reinforced Ti-6Al-4V layer protected by a ceramic thermal barrier coating (TBC) at the top face which is exposed to an elevated temperature. The SiC cubical inclusions are distributed either uniformly or exponentially, with the exponential distributions characterized by the parameter δ. This parameter controls the SiC inclusion concentration at the bottom (cold) face of the plate, with increasing values of δ producing increasingly higher concentrations. The titanium matrix is modeled using the classical incremental plasticity theory. The TBC material is elastic with a low Young's modulus and thermal conductivity relative to those of the other constituents, and a thermal expansion coefficient comparable to that of the matrix. Its purpose, therefore, is to protect the portion of the Ti-SiC region subjected to an elevated temperature from excessively high temperatures.

The results are compared with the response of a bimaterial and trimaterial plate laminated with homogeneous plies under the same thermal gradient, see Fig. 8. This comparison is carried out in order to illustrate the advantages of employing uniformly and nonuniformly spaced two-phase microstructures to control the internal stress distributions with the objective of reducing the bending moment resultants in a plate subjected to a through-thickness temperature gradient.

Figure 9 presents the temperature distributions in the considered TBC-protected configurations subjected to a temperature of 900°C at the top surface and 21°C at the bottom surface. The temperature distributions in the configurations laminated with homogeneous plies have bilinear and trilinear appearance, whereas the temperature distribution in the SiC inclusion-reinforced plate exhibits a quasitrilinear appearance with slight changes in slope in the Ti-SiC region due to the different thermal conductivity between the titanium matrix and the SiC inclusions. The temperature profile of the TBC/Ti bimaterial configuration is bounded by the temperature profiles of the TBC/Ti/SiC trimaterial and the TBC/Ti-SiC composite configurations. When the SiC phase is in the form of a homogeneous plate bonded to the bottom surface of the titanium layer, the temperature distribution in such configuration is lower relative to the configuration consisting of the thermal coating bonded to the homogeneous titanium matrix. This is due to the higher thermal conductivity of the SiC layer relative to the titanium matrix in the low-temperature range at the cold surface of the plate. Alternatively, when the SiC phase is embedded directly in the titanium matrix in the form of inclusions, the temperature profile is now higher relative to the

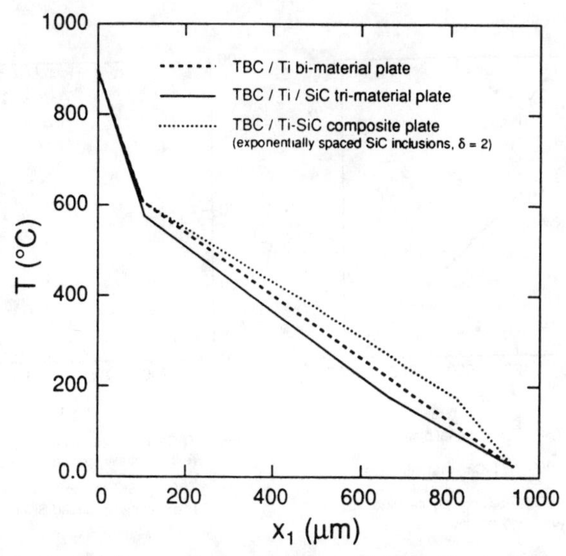

Fig. 9 Through-thickness temperature distributions in a TBC/Ti bimaterial plate, TBC/Ti-SiC composite plate with three exponentially spaced inclusions, and TBC/Ti/SiC tri-material plate (see Fig. 8) for T_{ref} = 21°C.

TBC/Ti bimaterial plate due to the higher effective thermal conductivity of the Ti-SiC composite region exposed to the elevated temperature. This is an important result that will have a direct bearing on the favorable redistribution of the internal stresses in the Ti-SiC region of the TBC/Ti-SiC composite plate when compared to the stress distribution in the Ti region of either the TBC/Ti bimaterial or TBC/Ti/SiC trimaterial plate.

The corresponding normal stress σ_{22} distributions and the resulting in–plane force and moment resultants are compared in Figs. 10 and 11, respectively. The normal stress distributions are presented in the bimaterial and trimaterial plates, and the uniformly and exponentially spaced ($\delta = 1$ and 2) configurations in the representative cross section (RCS) containing both phases. The results indicate that the presence of the SiC inclusions decreases the magnitude of the normal stress in the matrix phase directly adjacent to the ceramic TBC relative to the normal stress in the bimaterial and trimaterial plate. In regions farther away from the thermal barrier, however, the magnitude of the normal stress is now greater in the SiC particle-reinforced plate. The presence of the SiC particles, therefore, results in redistribution of the normal stress σ_{22}, suggesting that embedding the SiC particles in the titanium matrix in the manner indicated will increase the in–plane force resultants, while decreasing the moment resultants. This is illustrated in Fig. 11 where the in–plane force and moment resultants in the trimaterial and TBC/SiC-Ti configurations have been normalized by the in–plane force and moment resultants obtained for the TBC/Ti bimaterial plate in order to demonstrate the effect of the two-phase microstructure of the Ti-SiC region on these quantities. Since the SiC inclusions are stiffer relative to the titanium matrix, the in–plane resultant N_2 (and thus N_3) increases in the presence of the inclusion phase. More significantly, how-

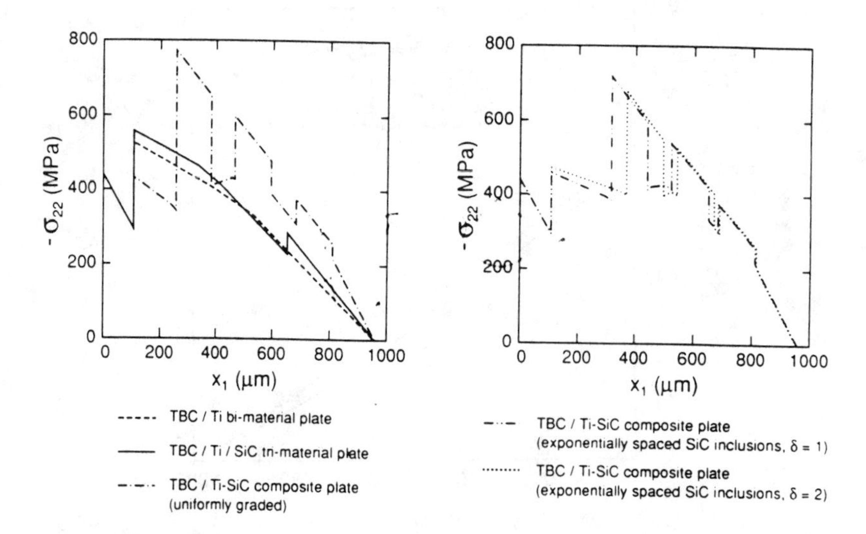

Fig. 10 Through-thickness σ_{22} distributions in thermally protected composite plate configurations with different substrate microstructures in the cross section containing both phases for $T_{ref} = 21°C$.

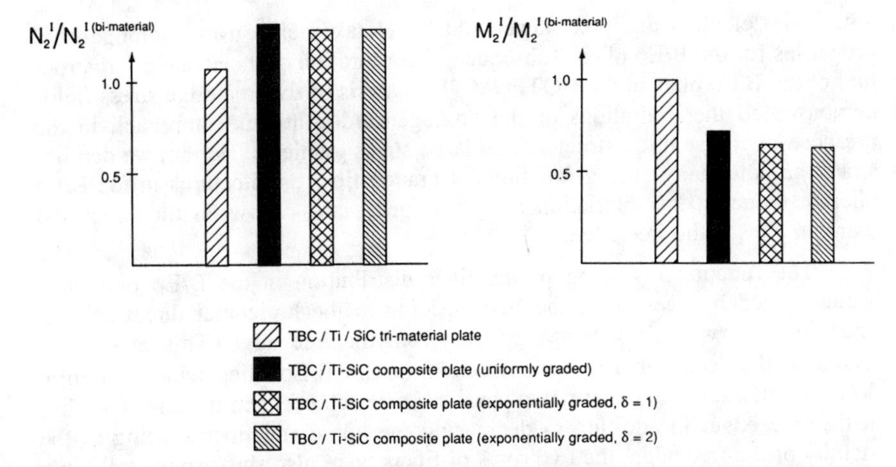

Fig. 11 In–plane force and moment resultants for the thermally protected composite plate configurations normalized by the corresponding quantities obtained for the TBC/Ti bimaterial plate.

ever, the presence of the SiC inclusions lowers the moment resultant M_2 (and thus M_3) relative to that of the TBC/Ti bimaterial plate. This reduction is a direct result of a more favorable (i.e., more uniform) redistribution of the normal stress σ_{22} which better utilizes the load carrying capability of the matrix phase in the regions exposed to lower temperatures (i.e., regions close to the bottom face of the plate). The greatest reduction in the moment resultant occurs for the exponentially spaced configuration with $\delta = 2$ since this configuration produces a more uniform distribution of σ_{22}. It is significant that the TBC/Ti/SiC trimaterial plate configuration offers a very modest reduction in the bending moment compared to that produced by both the uniformly and exponentially spaced TBC/Ti-SiC configurations.

The comparison of the temperature and stress distributions, and the resulting force and moment resultants within the three types of TBC-protected configurations just presented clearly illustrates the advantages that can be derived from the concept of internal temperature management through embedding of differently distributed particulate inclusions in metallic layers subjected to a temperature gradient.[7]

Interlaminar Stress Management via FG Architectures

We end the summary of the recent advances in the mechanics of functionally graded composites by presenting selected results obtained with the recently developed two-dimensional version of the coupled higher order theory.[9] This theory was employed to study the free-edge problem in a symmetrically laminated B/Ep-Ti composite plate (with two rows of boron fibers in the B/Ep plies parallel to the x_2 axis) subjected to a uniform temperature change, Fig. 12. This investigation established the capability of the theory to capture large stress gradients near a geometric discontinuity, such as the free edge, upon comparison with finite element analysis carried out by Herakovich[18] using homogenized properties for the B/Ep plies. Subsequent incorporation of the actual microstructure of the B/Ep plies in the HOTFGM-2D analysis of the free-edge stress fields demonstrated the limitations of the homogenized continuum approach in the presence of course microstructure and large-stress gradients. Herein, we demonstrate the potential of using functionally graded fiber architectures in the B/Ep plies in reducing the interlaminar normal (peel) stress between the B/Ep and titanium plies at the free edge.

The functional grading of the fiber distribution in the B/Ep plies was accomplished by decreasing the fiber spacing in the horizontal direction in a linear manner with decreasing distance from the free edge. This effectively increases the local fiber volume fraction in the B/Ep plies which, in turn, decreases the transverse thermal expansion mismatch between the adjacent plies at the free edge. In addition to decreasing the horizontal fiber spacing in the vicinity of the free edge, the two rows of fibers were also shifted vertically in a uniform manner in order to bring them closer to the B/Ep-Ti interface and thus decrease the local thermal expansion mismatch in the thickness direction.

Three different fiber architectures near the free edge in the B/Ep plies were generated using the combination of horizontal and vertical shifting of the

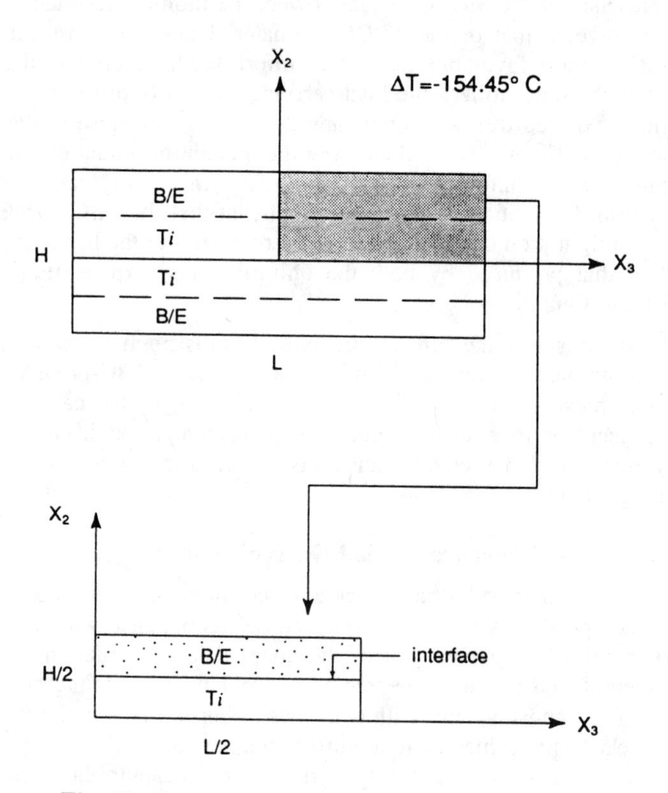

Fig. 12 Cross section of a [B/Ep - Ti]$_s$ laminate.

boron fibers. The first configuration was generated from the baseline, uniformly spaced configuration, by uniformly shifting the two rows of fibers in the vertical direction while preserving the uniform horizontal fiber spacing. The second configuration was generated by linearly decreasing the fiber spacing in the horizontal direction with decreasing distance from the free edge, in the manner described earlier, while preserving the vertical spacing of the baseline configuration. Finally, the third configuration was obtained from the second by vertically shifting the two rows of fibers in the same manner as was used to generate the first configuration.

The resulting peel stress distributions along the B/Ep-Ti interface in the baseline and functionally graded configurations caused by a uniformly applied temperature change of -154.45°C are illustrated in Fig. 13. Comparing the peel

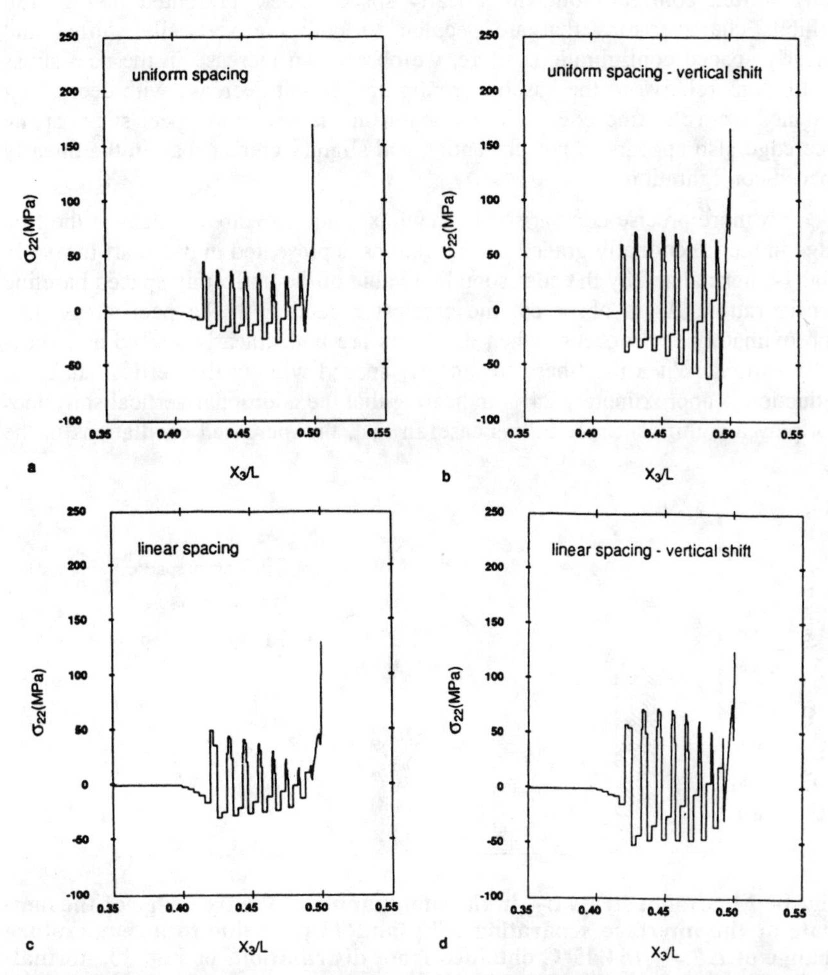

Fig. 13 Normal stress σ_{22} distributions in the titanium ply at the B/Ep-Ti interface near the free edge for the uniformly spaced and functionally graded fiber distributions.

stress distribution in the uniformly spaced baseline configuration, Fig. 13a, with the stress distribution in the vertically shifted configuration, Fig. 13b, we observe increased peel stress oscillations in the regions directly below the fiber locations. The relative increase in the oscillations is caused by the reduced distance between the interface and the bottom row of fibers produced by the vertical shift. The vertical shift also produces a reduction in the maximum peel stress at the free edge relative to the baseline configuration. The peel stress distribution in the configuration with the linearly spaced fibers, Fig. 13c, exhibits decreasing oscillations with decreasing distance from the free edge relative to the baseline configuration. A substantially greater decrease in the maximum peel stress at the free edge relative to the baseline configuration is also observed, as compared to the free-edge peel stress reduction observed in the vertically shifted configuration. Finally, the peel stress distribution in the vertically shifted configuration with linearly spaced fibers presented in Fig. 13d exhibits characteristics that are peculiar to both the vertically shifted and linearly spaced configurations. Here, we observe an increase in the peel stress oscillations relative to the baseline configuration that decrease with decreasing distance from the free edge. The reduction in the maximum peel stress at the free edge also appears to be substantial, and slightly greater than in the linearly spaced configuration.

A more precise comparison of the maximum peel stress values at the free edge in the functionally graded configurations is presented in bar chart format in Fig. 14, normalized by the corresponding value of the uniformly spaced baseline configuration. As is observed, the greatest reduction in the peel stress (i.e., approximately 25%) occurs when the fibers are both linearly spaced and vertically shifted. When the fibers are linearly spaced without the vertical shift, the reduction is approximately 24%, indicating that the additional vertical shift does not play a significant role in this case. In fact, the increased oscillations in the

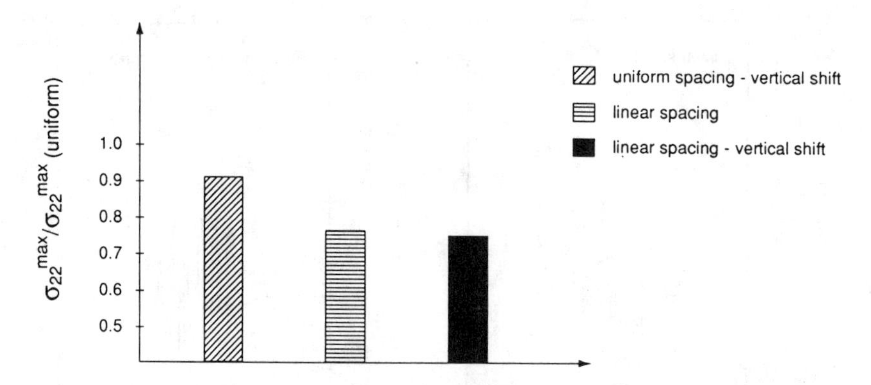

Fig. 14 Maximum stress σ_{22} in the titanium ply at the free edge of the laminate at the interface separating B/Ep and Ti plies due to a temperature change of ΔT = - 154.45°C obtained from distributions of Fig. 13, normalized with respect to the corresponding maximum stress obtained from the configuration with uniformly spaced fibers.

peel stress distribution caused by the vertical shift may not be desirable in some situations. Alternatively, when the fibers are vertically shifted without decreased horizontal spacing, the peel stress is reduced by a modest 9%. Thus, the major contribution to the reduction of the free-edge peel stress comes from functional grading in the horizontal direction.

Conclusions

This paper summarizes the recent advances in the thermomechanical analysis of functionally graded composites using a newly developed coupled higher order theory. In this new approach, the microstructural and macrostructural details are explicitly coupled when analyzing the response of a functionally graded composite subjected to a uniform or nonuniform loading, such as a through-thickness temperature gradient. Coupling of the local and global analyses allows one to rationally analyze the response of materials with continuously changing properties due either to nonuniform fiber spacing or the presence of several phases, as well as polymeric and metal matrix composites such as B/Ep, B/Al and SiC/TiAl that contain relatively few through-thickness fibers. For such materials, it is difficult, if not impossible, to define the representative volume element used in the traditional micromechanical analyses of macroscopically homogeneous composites.[19] Nevertheless, the traditional micromechanical analyses are currently applied to functionally graded composites by the majority of researchers.

The presented applications of the coupled higher order theory demonstrate that it is an accurate and efficient method for investigating the effect of fiber distribution on internal temperature and stress fields in the newly emerging class of functionally graded composites. The development of the theory has been justified by comparison with the results obtained using the standard micromechanics approach which neglects the micro–macrostructural coupling effects explicitly taken into account in the new theory. Because of the absence of such coupling, the standard micromechanics approach often underestimates actual stress distributions in composites with a finite number of uniformly or nonuniformly distributed fibers across the thickness dimension subjected to a thermal gradient.

The recent enhancements of the higher order theory's capabilities through the addition of inelastic constitutive models make possible the investigation of functionally graded composite plates subjected to through-thickness thermal gradients in a wide temperature range. In particular, functional grading of metallic layers protected by ceramic thermal barrier coatings has been shown to reduce the resulting thermally induced warping tendency in such configurations. Such applications of the new theory open up new areas of research dealing with the concept of optimum thermal management through functionally graded material architectures. Along similar lines, the extension of the one-dimensional version of the theory to materials functionally graded in two directions makes possible the investigation of the potential of functionally graded fiber architectures in reducing edge effects in laminated composites. The presented example indicates that substantial reductions in the thermally induced free-edge peel stress can be achieved for the considered B/Ep-Ti laminate by a relatively modest increase in the local fiber volume fraction directly at the free edge.

The authors have demonstrated that inclusion of inelastic effects in the analysis of functionally graded materials may be important in the presence of large temperature gradients in applications involving the one-dimensional version of the higher order theory.[7,12] Others have demonstrated the importance of inelastic effects within the context of the free-edge problem (cf., Williamson et al.,[20] Drake et al.,[21] and Suresh et al.[22]). Therefore, in subsequent investigations inelasticity effects will be incorporated into the theoretical framework of the two-dimensional version of the higher order theory.

Acknowledgments

The first two authors gratefully acknowledge the support provided by the NASA Lewis Research Center through the Grant NASA NAG 3-1377. The authors are grateful to Furman W. Barton, Chair of the Civil Engineering and Applied Mechanics Department at the University of Virginia for providing partial support during the second author's sabbatical leave. Fruitful discussions with Carl T. Herakovich about the thermal free-edge problem are also much appreciated.

References

[1]Yamanouchi, M., Koizumi, M., Hirai, T., and Shiota, I., *Proceedings of the First International Symposium on Functionally Gradient Materials* (Sendai, Japan), Oct. 8–9, 1990.

[2]Fukushima, T., *Proceedings of the International Workshop on Functionally Gradient Composites* (San Francisco, CA), Japan International Science and Technology Exchange Center, Nov. 5–6, 1992.

[3]Wakashima, K. and Tsukamoto, H., "Micromechanical Approach to the Thermomechanics of Ceramic-Metal Gradient Materials," *Proceedings of the First International Symposium on Functionally Gradient Materials* (Sendai, Japan), Oct. 8–9, 1990, pp. 19–26.

[4]Aboudi, J., Pindera, M.-J., and Arnold, S. M., "Thermoelastic Response of Metal Matrix Composites with Large-Diameter Fibers Subjected to Thermal Gradients," NASA Lewis, Cleveland, OH, *NASA TM 106344*, Oct. 1993.

[5]Aboudi, J., Arnold, S. M., and Pindera, M.-J., "Response of Functionally Graded Composites to Thermal Gradients," *Composites Engineering*, Vol. 4, No. 1, 1994, pp. 1–18.

[6]Aboudi, J., Pindera, M.-J., and Arnold, S. M., "Elastic Response of Metal Matrix Composites with Tailored Microstructures to Thermal Gradients," *International Journal of Solids and Structures*, Vol. 31, No. 10, 1994, pp. 1393–1428.

[7]Aboudi, J., Pindera, M.-J., and Arnold, S. M., "Thermo-Inelastic Response of Functionally Graded Composites," to be published, *International Journal of Solids and Structures*.

[8]Aboudi, J., Pindera, M.-J., and Arnold, S. M., "A Coupled Higher-Order Theory for Functionally Graded Composites with Partial Homogenization," to be published, *Composites Engineering*.

[9]Aboudi, J., Pindera, M.-J., and Arnold, S. M., "Thermoelastic Theory for the Response of Materials Functionally Graded in Two Directions," to be published, *International Journal of Solids and Structures.*

[10]Pindera, M.-J. and Dunn, P., "An Evaluation of a Coupled Microstructural Approach for the Analysis of Functionally-Graded Composites via Finite-Element Method," NASA Lewis, Cleveland, OH, *NASA CR 195455*, April 1995.

[11]Pindera, M.-J., Aboudi, J., and Arnold, S. M., "Limitations of the Uncoupled, RVE-Based Micromechanical Approach in the Analysis of Functionally Graded Composites," *Mechanics of Materials*, Vol. 20, No. 1, 1995, pp. 77–94.

[12]Pindera, M.-J., Aboudi, J., and Arnold, S. M., "Thermo-Inelastic Analysis of Functionally Graded Materials: Inapplicability of the Classical Micromechanics Approach," *Inelasticity and Micromechanics of Metal Matrix Composites*, edited by G. Z. Voyiadjis and J. W. Ju, Elsevier, New York, 1994, pp. 273–305.

[13]Bodner, S. R., "Review of a Unified Elastic-Viscoplastic Theory," *Unified Constitutive Equations for Creep and Plasticity*, edited by A. K. Miller, Elsevier, Amsterdam, The Netherlands, 1987, pp. 273–301.

[14]Mendelson, A., *Plasticity: Theory and Applications*, Krieger, Malabar, FL, reprint edition, 1983.

[15]Pindera, M.-J., Salzar, R. S., and Williams, T. O., "An Evaluation of a New Approach for the Thermoplastic Response of Metal-Matrix Composites," *Composites Engineering*, Vol. 3, No. 12, 1993, pp. 1185–1201.

[16]Williams, T. O. and Pindera, M.-J., "Convergence Rates of the Method of Successive Elastic Solutions in Thermoplastic Problems of a Layered Concentric Cylinder," *Engineering, Construction, and Operations in SPACE IV*, edited by R. G. Galloway and S. Lokaj, American Society of Civil Engineers, New York, 1994, pp. 348–357.

[17]Aboudi, J., *Mechanics of Composite Materials: A Unified Micromechanical Approach*, Elsevier, New York, 1991.

[18]Herakovich, C. T., "On Thermal Edge Effects in Composite Laminates," *International Journal of Mechanical Science*, Vol. 18, 1976, pp. 129–134.

[19]Hill, R., "Elastic Properties of Reinforced Solids: Some Theoretical Principles," *Journal of the Mechanics and Physis of Solids*, Vol. 11, 1963, pp. 357–372.

[20]Williamson, R. L., Rabin, B. H., and Drake, J. T., "Finite Element Analysis of Thermal Residual Stresses at Graded Ceramic-Metal Interfaces. Part I. Model Description and Geometrical Effects," *Journal of Applied Physics*, Vol. 74, No. 2, 1993, pp. 1310–1320.

[21]Drake, J. T., Williamson, R. L., and Rabin, B. H., "Finite Element Analysis of Thermal Residual Stresses at Graded Ceramic-Metal Interfaces. Part II. Interface Optimization for Residual Stress Reduction," *Journal of Applied Physics*, Vol. 74, No. 2, 1993, pp. 1321–1326.

[22]Suresh, S., Giannakopoulos, A. E., and Olsson, M., "Elastoplastic Analysis of Thermal Cycling: Layered Materials with Sharp Interfaces," *Journal of the Mechanics and Physics of Solids*, Vol. 42, No. 6, 1994, pp. 979–1018.

Micromechanical Analysis of Thermal Response in Textile-Based Composites

E. H. Glaessgen* and O. H. Griffin Jr.†
Virginia Polytechnic Institute and State University
Blacksburg, Virginia 24061

Abstract

Textile composites have the advantage over laminated composites of a significantly greater damage tolerance and resistance to delamination. Currently, a disadvantage of textile composites is the inability to examine the details of the internal response of these materials under load, including thermal loading. Traditional approaches to the study of textile based composite materials neglect many of the geometric details that affect the performance of the material. The present three dimensional analysis, based on the representative volume element (RVE) of a plain weave, allows prediction of the internal details of displacement, strain, and stress. Through this analysis, the effect of geometric and material parameters on the aforementioned quantities are studied.

Introduction

Textile based composite materials have received considerable attention in the literature in recent years.[1,2] These traditional approaches to the study of textiles are typically an extension of proven techniques for the study of laminated materials. They tend to be based on strength of materials and classical lamination theory (CLT), a homogenized finite element approach that uses classical micromechanics or CLT as a basis for material properties, or a beam in a matrix approach that violates compatibility of the constituent finite elements at all locations except for the nodes.[3] The result is a lack of ability to determine details of the internal response of these materials under load. A macro finite element approach was developed by Whitcomb[4] and Woo and Whitcomb[5] as a compromise between the previously mentioned models and the type of detailed finite element models developed for the present study. Although the analysis by Whitcomb shows some details of the load distribution within the composite, it neglects many of the geometric details that affect the performance of the mate-

* Graduate Research Assistant, Department of Engineering Science and Mechanics.
† Professor and Associate Dean, Department of Engineering Science and Mechanics.

rial. Techniques employing conventional finite elements have also been reported.[6-8] These have been developed for woven composites under mechanical loads and provide varying degrees of accuracy and details of results.

The present finite element based technique allows the interrogation of the internal response of textiles. This approach greatly reduces the geometric simplifications required for modeling, since the finite element models are taken directly from the textile geometry model (TGM).[9] Although originally developed for the analysis of textiles under mechanical loads,[6] this technique is easily extended to include response to thermal loading. The present analysis, based on the representative volume element (RVE) of a textile composite allows prediction of the load, mode, and location of failure initiation within the RVE. Through these models, not only is gross characterization possible, but internal details of displacement, strain, stress, and failure parameters can be studied. Specifically, this discussion focuses on the analysis of a plain weave textile composite. A uniform temperature change is considered.

Geometries Studied

Although it is desirable to avoid more geometric simplifications than necessary, certain simplifications are altogether unavoidable. Sectioning of manufactured woven textile composites shows that material variances exist throughout the textile. Accounting for the differences between RVEs in a given architecture must be treated as a statistical problem and is beyond the scope of this paper; instead, a typical geometry is studied. The yarn centerline is assumed to be a smooth (order O2) b-spline whereas the cross section is an O2 b-spline of an elliptical shape. The RVE is assumed to repeat both its geometry and response throughout the entire material. Details of the construction of the finite element model are given in Ref. 6.

In the present study, the geometry is determined by sampling points on the yarn centerline and cross-sectional perimeter of a fabricated material. Details of the RVE size and yarn dimensions are given in Table 1. As shown in Fig. 1, the geometric rendering of the RVE, the x and z directions are the in-plane directions, whereas y is the out-of-plane direction.

Material Selection

An unfortunate limitation of the finite element method is that the geometry and material properties must be selected a priori. As with the geometric parameters,

Table 1: Model geometry

RVE size, in.	*a*/*b*, axial yarn, in./in.	*a*/*b*, transverse yarn, in./in.
0.275 * 0.050 * 0.1960	0.0787 / 0.0098	0.0866 / 0.00866

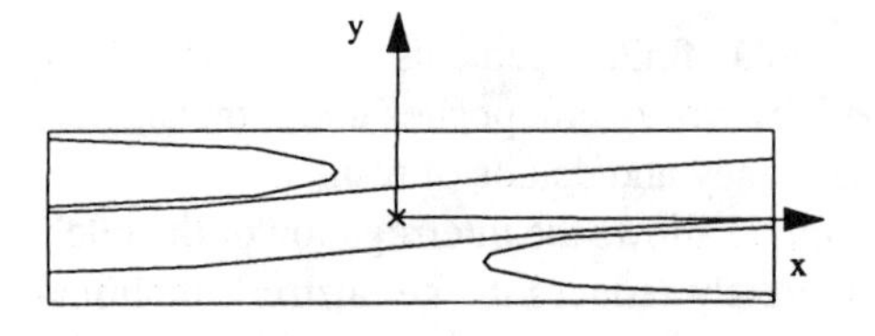

Fig. 1a Side view of RVE.

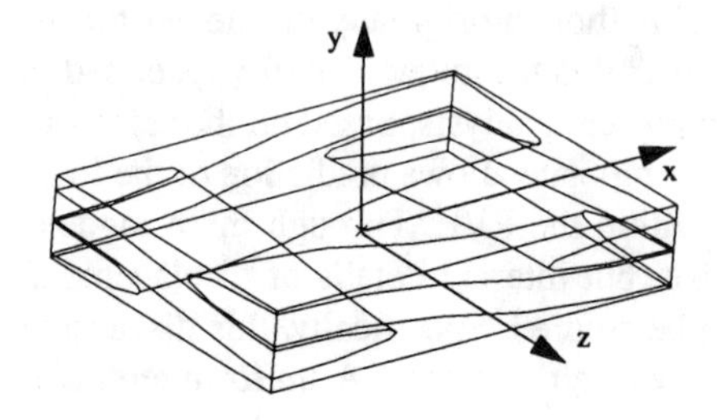

Fig. 1b Isometric view of RVE.

representative values of the material properties commonly found in textile composites are used. The values have been chosen to represent typical glass/epoxy and carbon/epoxy materials. The material properties are given in Table 2.

Finite Element Model

Choice of proper discretization for volumes as complicated as those found in textiles leads to a trade off between mesh convergence and model size. A typical global element size is approximately 1/20 of the model characteristic dimension, whereas local refinements are made to ensure convergence in known regions of high-stress gradient. Once the finite element discretization of both the matrix volumes and the yarn volumes has been completed and checked for element distortion, a concern considering the complex shape of the mesh volumes, the element coordinate systems of the yarn elements are aligned with the b-spline defining the orientation of the yarn volume. This alignment insures that the yarn finite elements will behave as subdomains within a homogeneous piecewise transversely isotropic material with the properties given in Table 2. Both materials were modeled with 10-node tetrahedron (ABAQUS C3D10) elements.

A uniform temperature change is considered as it is a fundamental case when considering the structural behavior of a material under thermal loading. Typically, textile preforms are layered and are many RVEs wide, thus it is appropriate to consider that a randomly selected RVE may be subjected to virtually any degree of constraint. This becomes important when accounting for the relative position of a RVE within a textile preform. To bound the limits of constraint, five boundary conditions are considered: 1) periodic, 2) out-of-plane symmetry, 3) out-of-plane free, 4) in-plane symmetry, and 5) in-plane free boundary conditions.

The first of these boundary conditions, periodic, represents the upper bound of constraint, a RVE far from any boundary. To simulate this situation, the opposing

Table 2: Material properties

Material Properties	Epoxy	Glass/Epoxy	Carbon/Epoxy
E_x, E6 psi	0.5	5.5	23.0
E_y, E6 psi	0.5	1.43	1.58
E_z, E6 psi	0.5	1.43	1.58
G_{xy}, E6 psi	0.1875	0.59	1.00
G_{xz}, E6 psi	0.1875	0.59	1.00
G_{yz}, E6 psi	0.1875	0.55	0.60
υ_{xy}	0.33	0.27	0.27
υ_{xz}	0.33	0.27	0.27
υ_{yz}	0.33	0.30	0.30
α_{xy}, E-6 in./in.°F	25.0	4.83	0.010
α_{xz}, E-6 in./in.°F	25.0	19.6	11.50
α_{yz}, E-6 in./in.°F	25.0	19.6	11.50

faces are restrained with multipoint constraints (MPC) to remain uniformly straight and parallel to one another. The second, out-of-plane symmetry, is an intermediate case similar to the behavior of a two-layer un-nested textile. Out-of-plane free boundary conditions are representative of the least constrained RVE, that is, a composite that is one RVE thick. Whereas the out-of-plane symmetry case has one free face, the out-of-plane free case has both top and bottom surfaces free to deform from their original planar state.

Of the two in-plane variations, the in-plane symmetry case is an intermediate degree of constraint representing a narrow composite two RVEs wide. The final case, in-plane free, is the lower bound of restraint, a composite one RVE wide. In this last case, the material is assumed to have two edges that are allowed to deform freely.

In all cases, a uniform temperature change of 1°F is applied to each element within the finite element model. Although this represents the steady-state loading condition, geometric complexities and material dissimilarities are sufficient to induce the complex behavior shown in the results portion of this paper.

Finite Element Results

The results that follow are based on the finite element models discussed in the preceding section. The results are divided into two main sections: 1) displacements and 2) strain energy density. The latter section is discussed because it is the

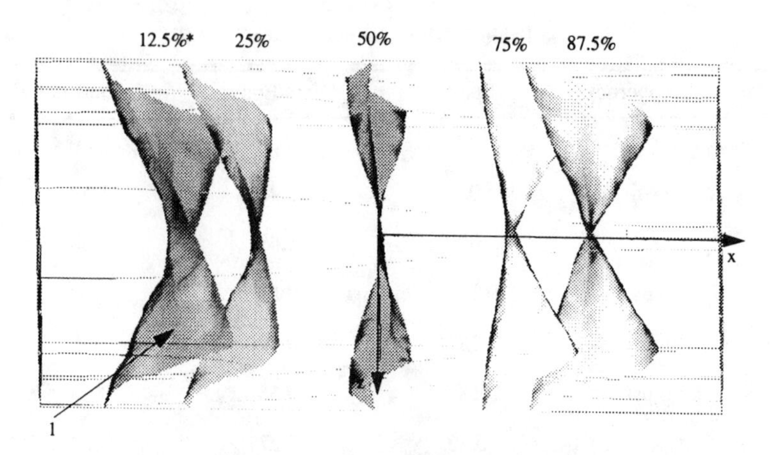

Fig. 2a Axial displacement U_x, periodic boundary conditions $U_{x,\,max}$ = 8.89E-7 in./°F, * % of total displacement represented by isosurface.

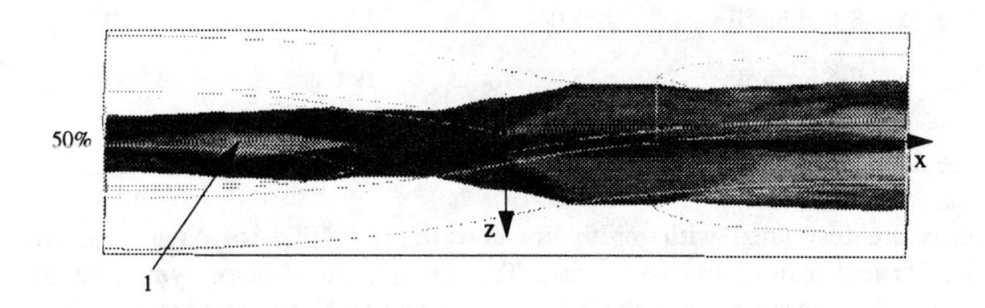

Fig. 2b Out-of-plane displacement U_y, periodic boundary conditions $U_{y,\,max}$ = 1.63E-6 in./°F.

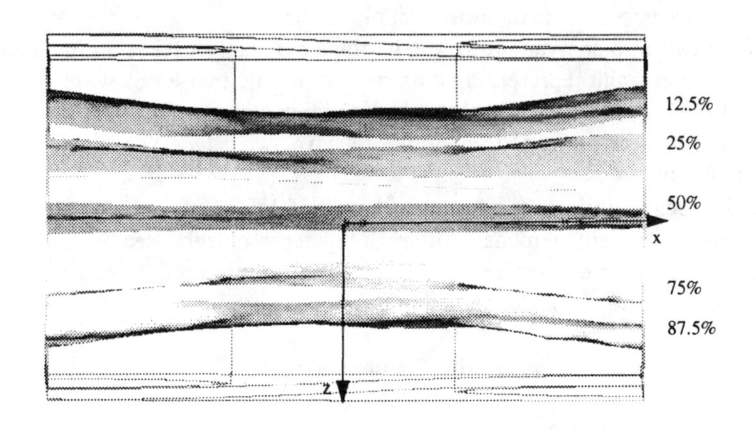

Fig. 2c Transverse displacement U_z, periodic boundary conditions $U_{z,\,max}$ = 9.19E-7 in./°F.

basis of many energy-based failure criteria and also provides a useful scalar quantity for visualization independent of material coordinate system.

Note that familiar terms are used in describing the geometry of the RVE. For example, lower and upper refer to the surface having most negative y coordinate (y= -0.025 in.) and the most positive y coordinate (y= +0.025 in.), respectively.

The models have been checked for convergence, and for the idealized geometries considered are converged to within a few percent. In the figures that follow, however, it is more beneficial to consider the qualitative nature of the isosurfaces than it is to compare exact numbers represented by the isosurfaces.

Displacement Results

Periodic Boundary Conditions

The isosurfaces illustrated in Fig. 2a represent the computed internal axial displacements of a well-restrained region within plain weave textile composite under a uniform temperature change. As prescribed by the boundary conditions, the exterior boundaries remain planar and parallel. Displacements within this RVE are, as expected, the most uniform and are similar to the case of uniform axial displacement.[6] Although not planar, every point on the surface 1 is 12.5% of the maximum value. It can be thought of as representing the average axial strain in that part of the model. Thus, the axial yarn is seen to strain more than the transverse yarn that is just above it in the RVE. Further, recalling elementary elasticity, the angular change in the displacement isosurface 1 can be compared directly to a local shear strain. The sharper the angle, the greater the shear strain. This shear strain is seen in both the x-y and x-z planes.

Figure 2b represents the out-of-plane deformation associated with periodic boundary conditions. Note that the maximum value of the out-of-plane displacement is 83% larger than the corresponding in-plane displacement for these boundary conditions. This occurs for two reasons: 1) the coefficient of thermal expansion of the yarn's transverse direction is much greater than the CTE in the yarn axial direction, and 2) the in-plane stiffness of the axial and transverse yarns restrains the in-plane deformation.

As with the in-plane deformation for this fully restrained case, the top and bottom boundaries are forced to remain straight and parallel from the applied boundary conditions. However, internally, the deformations are as shown and indicated by 1.

Figure 2c is the baseline of transverse displacement (fully constrained) for the RVE models. Again, the outer boundaries are allowed to deform in a planar fashion only. Although the temperature change is uniform in the x and z directions, as indicated in Table1, the RVE geometry is not uniform. Thus, it is reasonable to expect that the internal deformation behavior is not uniform in the two in-plane directions.

Out-of-Plane Boundary Conditions

When the upper surface of the RVE is released to deform out of plane, the reduced constraint results in the axial deformation shown in Fig. 3a. This case is

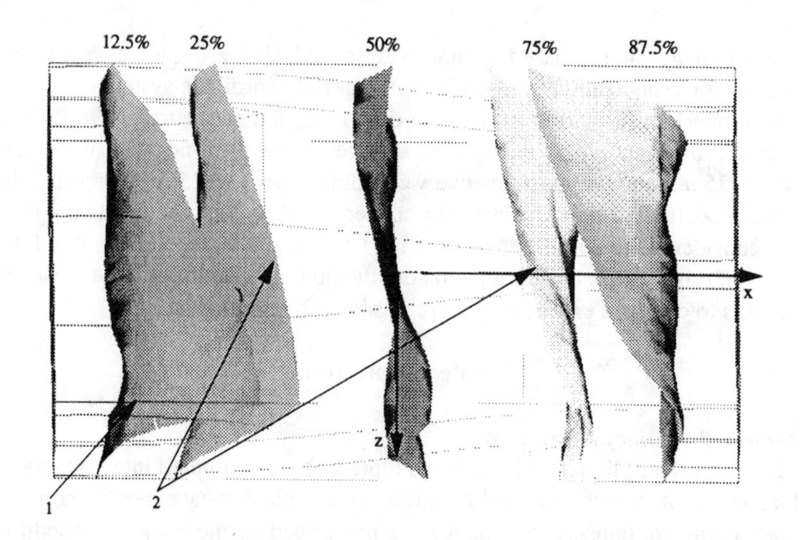

Fig. 3a Axial displacement U_x, free top surface $U_{x,\,max}$ = 9.08E-7 in./°F.

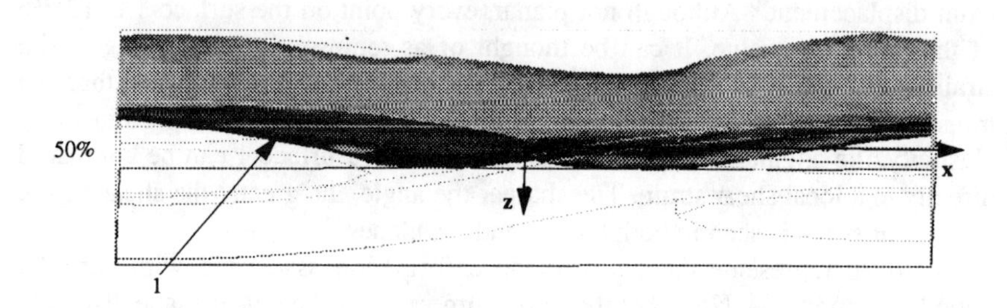

Fig. 3b Out-of-plane displacement U_y, free top surface $U_{y,\,max}$ = 2.12E-6 in./°F.

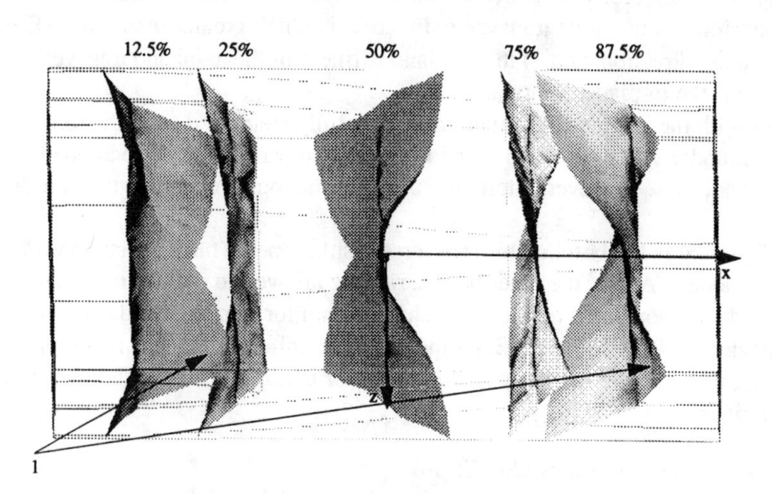

Fig. 3c Axial displacement U_x, free top/bottom surface $U_{x,\,max}$ = 9.92E-7 in./°F.

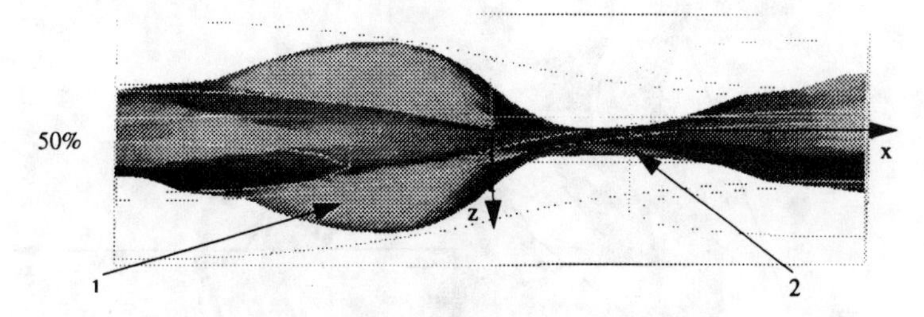

Fig. 3d Out-of-plane displacement U_y, free top/bottom surface $U_{y,\,max}$ = 2.39E-6 in./°F.

representative of a layer of a two-layer un-nested composite and provides an intermediate condition to the fully restrained case shown in Figs. 2, and the unconstrained case that will be seen as Figs. 4. Note here that the lower portion of isosurface at 1 shows a similar deformation as the lower portion of isosurface 1 in Fig. 2a, whereas the free condition on the top surface localizes the axial deformation to the region between the transverse yarns. As before, the local strains associated with isosurface at 1 can be visualized as the ratio of the value of displacement represented by the isosurface (12.5% of $U_{x,\max}$) to the distance from the isosurface to the left-side boundary. Thus, a disproportionate percentage of the axial deformation occurs in the most axially compliant region of the RVE 2.

Out-of-plane deformations, Fig. 3b, are dish shaped. Now, the restraint on out-of-plane deformation is relaxed, and the upper surface is allowed to deform freely. Note the isosurface shown is at 50% (1),whereas the bottom surface is the fixed, or 0%, surface. This surface is concave, and the maximum displacement occurs above the point of maximum concavity. For clarity, isosurfaces other than the one at 50% have been omitted from the plot; however, it is important to note that the maximum value of displacement occurs along a surface that is a function of the local weave geometry.

By further decreasing the constraint on the model to allow for both top and bottom surfaces to freely deform, a situation as shown in Fig. 3c results. In the regions of maximum constraint (those regions near the yarns), the axial displacement isosurfaces at 1 are similar to those in the fully restrained case (Fig. 2a). However, differences are greatest in the more compliant regions away from the yarns. Here, the angular change in isosurface and resulting shear strains are reduced because of the absence of the planar constraint.

An interesting artifact in Figure 3d is at 1, the isosurface of 50% of the maximum out-of-plane displacement. Because of the angle from which this image was taken, it appears that considerable out-of-plane deformation occurs at 1, whereas little occurs at 2. Actually, because of the antisymmetric geometry, the two regions of the displacement are antisymmetric with respect to one another. Because of the absence of in-plane constraint, the maximum deformation is 47% larger than for a similar layer embedded within the composite (Fig. 2b). Thus, it is possible to infer that there is a nonuniform displacement gradient not only

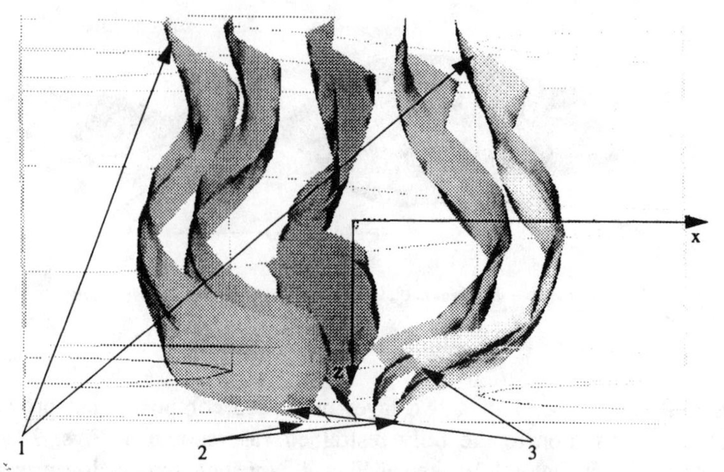

Fig. 4a Axial displacement U_x, free front surface $U_{x,\max}$ = 1.15E-6 in./°F.

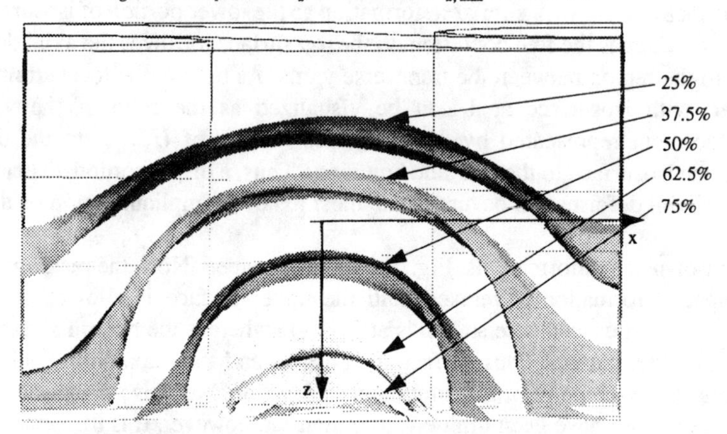

Fig. 4b Transverse displacement U_z, free front surface $U_{z,\max}$ = 3.16E-6 in./°F.

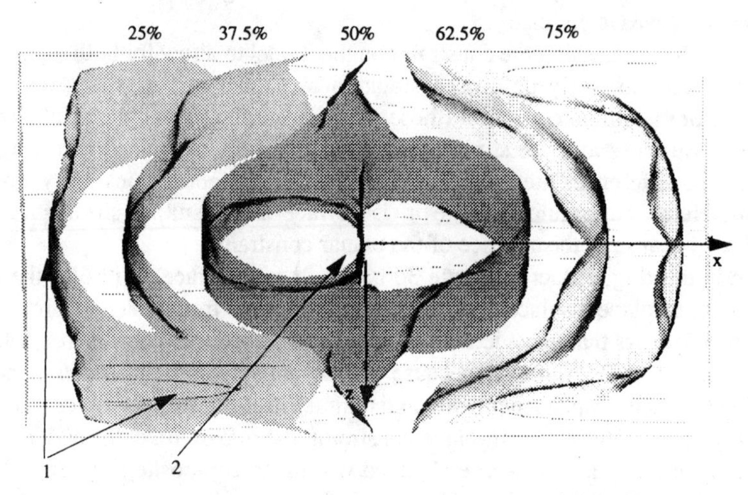

Fig. 4c Axial displacement U_x, free front/back surface $U_{x,\max}$ = 1.02E-6 in./°F.

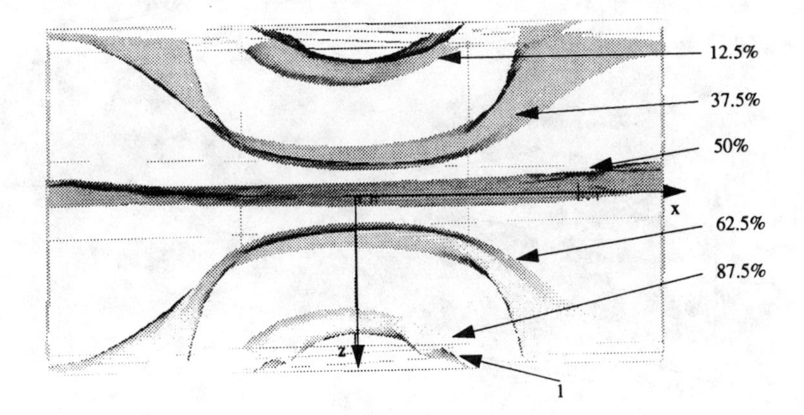

Fig. 4d Transverse displacement U_z, free front/back surface $U_{z,\,max}$ = 4.73E-6 in./°F.

within the individual RVEs (Fig. 3d), but among the several RVEs through the thickness of a typical textile composite and that it is a function of their position from the centerline of the composite.

In-Plane Boundary Conditions

To simulate a free edge, the front boundary of the RVE has been unconstrained in Fig. 4a to allow free deformation. Note that the axial deformation along the back, or constrained, surface 1 is similar to the axial deformation in the same region for the fully constrained case. However, a very different condition exists at the free edge. A disproportionately large percentage of the total axial deformation occurs near the compliant center of the front edge 2. Further, increased shear strains develop. For example, the x-y component, shown as an angular change in the isosurface, is prominent along the front edge at 3.

Figure 4b illustrates the transverse deformation, in particular, the transverse deformation along the free edge. Nominally, only slightly more than 25% of the total deformation in the RVE occurs in the back-half at 1. Also, the greatest transverse deformation occurs near the center of the RVE. This is the least constrained (transverse) region of the model both because of the domination of the region by axial yarns and by its distance from the stiffening effects of the restraints along the transverse edges at 2.

When both the front and back edges of the RVE are released, simulating a composite one RVE wide, the axial deformation is as shown in Fig. 4c. The internal deformation is redistributed such that the axial strain varies greatly with position from the x-axis at 1. Further, the center of the RVE, the matrix dominated region, is found to be in a state of near constant displacement at 2.

Corresponding transverse displacements are shown in Fig. 4d. Similar to the transverse displacements shown in Fig. 4b, the compliant region away from the front/back edge boundary conditions and the transverse yarns, has the greatest transverse displacement. The combined constraining effect of planar edges and

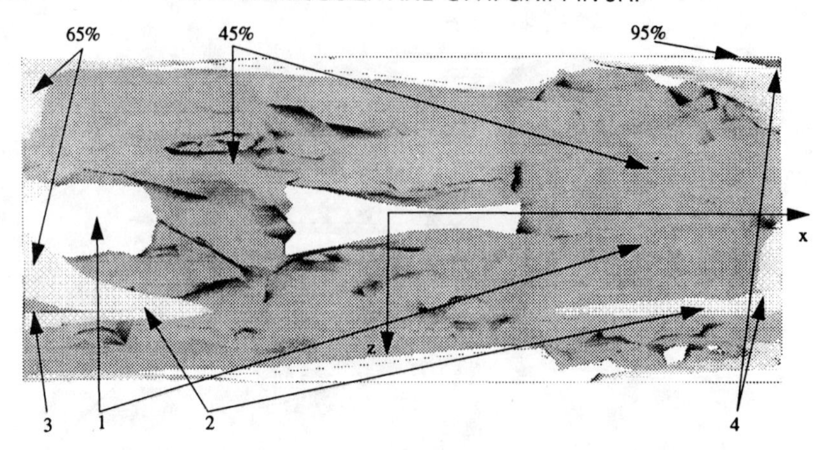

Fig. 5a Strain energy density, SED periodic boundary conditions 0.546E-4 < SED < 6.68E-4 (in.-lb/in.^{3})/°F.

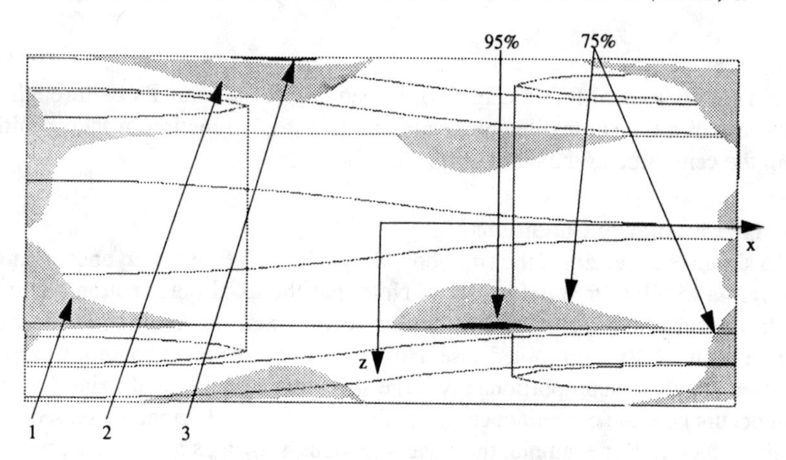

Fig. 5b Strain energy density, SED free top/bottom surface 0.532E-4 < SED < 5.06E-4 (in.-lb/in.^{3})/°F.

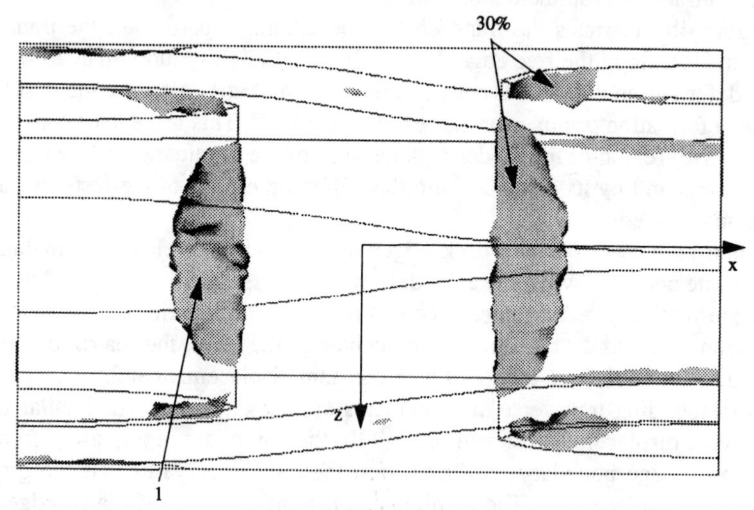

Fig. 5c Strain energy density, SED free front/back surface 0.248E-4 < SED < 17.90E-4 (in.-lb/in.^{3})/°F.

transverse yarns is similar to the thermal expansion case for a plate with cantilevered edges. The center is allowed to displace from the nominal centerline of the edge, however, the displacement is not as large as if the edges were allowed to rotate freely.

The largest transverse displacement is 50% larger than for the case of one free edge and over 400% larger than for the fully constrained case. A comparison can be made with the free edge case for laminated composites where the edge displacements are seen to vary as a function of through thickness y coordinate along the edge. In contrast to the laminate, the transverse displacements for the textile are also greatly dependent on axial coordinate at 1.

Strain Energy Density Results

Strain energy density (SED) results for the case of periodic boundary conditions are shown in Fig. 5a. The SED distributions in this model correspond to the displacement distributions shown in Figs. 2 discussed earlier. The left- and right-hand regions shown at 1 are similar. In the figure, the two regions of the isosurface are antisymmetric as a function of the yarn geometry.

Note that the yarns contain much of the SED in the model at 1, whereas the largest values of SED are found in the regions of the matrix immediately above and below the axial/transverse yarn crossovers at 2. Because this architecture is not quite balanced, the absolute maximum values occur at the alternate corners on the top and bottom of the model at 3. Checking the stresses and strains in the several regions of interest indicates that a matrix failure primarily due to through-thickness tension at region 3 is the likely first failure.

The centerline paths of the axial yarns are not fully antisymmetric for this geometry. An important result of this geometric artifact is that the distance from the axial yarn outer surface to the horizontal boundary of the RVE is different for adjacent pairs of interstices. Thus, of the eight corner locations in the RVE, there are two pair of four distinct values of SED. A typical pair is shown in at 4.

Since SED is a scalar product of strain and stress components in the model, it is always zero or positive for conservative systems. As shown in Fig. 5a, it is greater than zero for all locations within the model and reaches a maximum of 6.68E-4 in.-lb/in.^{3} in regions 3.

When the out-of-plane restraints are removed from the top and bottom surfaces, the strain energy distribution is as shown in Fig. 5b. This figure corresponds to the displacements shown in Figs. 3 that have been discussed earlier. Large values of SED are found in the corners of the matrix similar to the fully restrained case indicated by 1. However, the largest values of SED in the model are found in alternate inboard locations 2 and 3 and result primarily from the combined effect of through-thickness and axial tension. Thus, whereas the material in which probable first failure occurs has not changed, the location has been altered significantly.

When the in-plane restraints are removed from the front and back surfaces, the strain energy distribution is similar to that in Fig. 5c and corresponds to the dis-

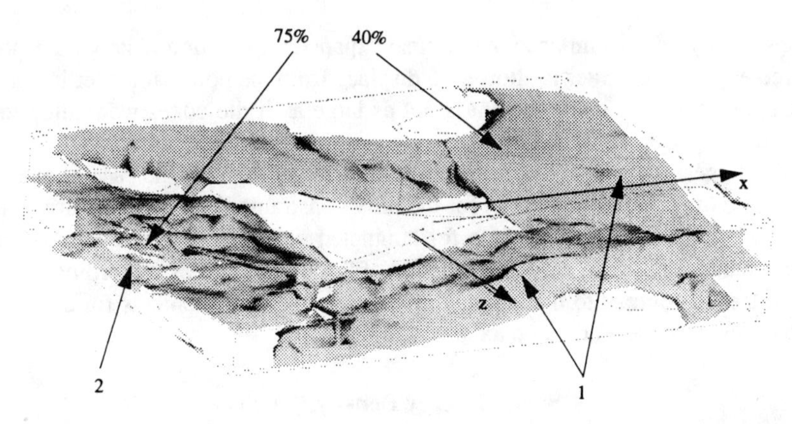

Fig. 5d Strain energy density, SED periodic boundary conditions glass/epoxy yarns 0.281E-4 < SED < 4.08E-4 (in.-lb/in.^{3})

placements in Figs. 4. The largest values of SED are very localized such that the interior cusp regions of the transverse yarns contain all of the SED with values above 30% of the maximum at 1. The maximum value of SED is increased by 165% over the fully restrained case. Now, the mode has changed to axial tension in the transverse yarns as determined in the local coordinate system.

In contrast to the mode and location shown in Fig. 5a for a carbon/epoxy system, if the material parameters are changed to simulate glass/epoxy, strain energy densities such as those in Fig. 5d result. The yarns contain all of the SED above 40% of the maximum value. However, in the glass/epoxy material system, the thermal and mechanical mismatches are not as severe as with carbon/epoxy. The largest values of SED are found near the centerline of the yarns at 2 and are primarily the result of axial loads in the local coordinate systems.

Discussion of Analysis

Several general conclusions may be drawn from the results that have been presented. Among these are the following.

1) The response of even simple textile composites, such as plain weaves, is fully three dimensional as shown in the figures in this paper.

2) Stacking of layers restrains in-plane deformation of the weave.

3) Severe dimpling is characteristic of an (unrestrained) single layer.

4) Free edge localizes axial/transverse displacement between transverse yarns.

5) Low levels of SED for ΔT show similar distribution to mechanical load.

6) Maximum SED for fully restrained carbon/epoxy is located in the corners and is dependant on corner geometry.

7) Decreased mechanical mismatch in glass/epoxy changes the location of maximum SED to the transverse yarns (fully restrained case).

8) Removing the out-of-plane restraint initiates the greatest SED away from corners.

9) The free-edge condition redistributes the SED into transverse yarns.

The technique that has been used for this analysis allows consideration of the three-dimensional geometry of textiles, interrogation of the details of the behavior of complex textile architectures, inclusion of general boundary conditions, and extraction of components of material response.

The method does require a large model size and execution time and several simplifications that are unavoidable.

References

[1] Dexter, H.B., Camponeschi, E.T., and Peebles, L., "3D Composite Materials," NASA CP 24020, Hampton, VA, 1985.

[2] Raju, I.S., Foye, R.L., and Avva, V.S., "A Review of Analytical Methods for Fabric and Textile Composites," Proceedings of Indo-US Workshop on Composite Materials for Aerospace Applications, July 23-July 27, 1990, Bangalore, India.

[3] Carter, W.C., Cox, B.N., Dadkhah, M.S., and Morris, W.L., "An Engineering Model of Woven Composites Based on Mircromechanics," Acta Metallurgica, accepted.

[4] Whitcomb, J.D., "Three-Dimensional Stress Analysis of Plain Weave Composites," NASA TM-101672, Nov. 1989.

[5] Woo, K., and Whitcomb, J.D., "Global/Local Finite Element Analysis for Textile Composites," 34th AIAA/ASME/ASCE/AHS/ACS Structures, Structural Dynamics and Materials Conference, La Jolla, CA, pp. 1721-1731, AIAA, Washington, DC, 1993, AIAA Paper 93-1506-CP.

[6] Glaessgen, E.H., Pastore, C.M., Griffin, Jr., O.H., and Birger, A., "Modeling of Textile Composites," submitted for publication, Composites Engineering.

[7] Lene, F., and Paumelle, P., "Micromechanisms of Damage in Woven Composite," Composite Material Technology, PD-Vol 45, 1992, pp. 97-105.

[8] Blacketter, D.M., Walrath, D.E., and Hansen, A.C., "Modeling Damage in a Plain Weave Fabric-Reinforced Composite Material," Journal of Composites Technology and Research, Vol. 15, No. 2, 1993, pp. 136-142.

[9] Pastore, C.M., Gowayed, Y.A., and Cai, Y., "Application of Computer Aided Geometric Modeling for Textile Structural Composites," Computer Aided Design in Composite Material Technology. Computational Mechanics, Southampton, UK, 1990, pp. 45-53

Recent Advances in the Sensitivity Analysis for the Thermomechanical Postbuckling of Composite Panels

Ahmed K. Noor*
University of Virginia, NASA Langley Research Center, Hampton, Virginia 23681

Abstract

Three recent developments in the sensitivity analysis for the thermomechanical postbuckling response of composite panels are reviewed. The three developments are: effective computational procedure for evaluating hierarchical sensitivity coefficients of the various response quantities with respect to the different laminate, layer, and micromechanical characteristics; application of reduction methods to the sensitivity analysis of the postbuckling response; and accurate evaluation of the sensitivity coefficients of transverse shear stresses. Sample numerical results are presented to demonstrate the effectiveness of the computational procedures presented. Some of the future directions for research on sensitivity analysis for the thermomechanical postbuckling response of composite and smart structures are outlined.

Introduction

Significant advances have been made in the development of computational models and strategies for the numerical simulation of the thermomechanical buckling and postbuckling responses of composite panels (see, for example, Refs. 1–5 and the review article, Ref. 6). More recently, attempts have been made to extend the domain of sensitivity analysis to the thermomechanical postbuckling response and to evaluate the sensitivity of the response to variations in the panel characteristics (see Refs. 5 and 7). The sensitivity coefficients (derivatives of the response quantities with respect to design variables) can be used to 1) determine a search direction in the direct application of nonlinear mathematical programming

---

* Ferman W. Perry Professor of Aerospace Structures and Applied Mechanics and Director, Center for Computational Structures Technology.

algorithms; 2) generate an approximation for the postbuckling response of a modified panel (along with a reanalysis technique); 3) assess the effects of uncertainties, in the material and geometric parameters of the computational model, on the postbuckling response; and 4) predict the changes in the postbuckling response due to changes in these parameters.

A number of techniques have been developed for evaluating the sensitivity coefficients using either the governing discrete (e.g., finite element) equations or the continuum equations of the structure. The techniques used with the discrete equations can be grouped into three categories: analytical direct-differentiation methods, finite difference methods, and semianalytical or quasianalytical methods. In analytical methods, the finite element equations of the structure are differentiated analytically, and the exact sensitivity coefficients are calculated. In the finite difference methods, approximate values of the sensitivity coefficients are generated based on the difference between the response quantities associated with the original and the perturbed parameters. The semianalytical methods use the factorized stiffness matrix of the finite element equations and a finite difference approximation to the right-hand side to compute the sensitivity coefficients.

In the present paper, some of the recent advances in the sensitivity analysis for thermomechanical postbuckling of composite panels, aimed at increasing its utility and improving the computational efficiency, are reviewed. Discussion focuses on three recent developments; namely, 1) an effective computational procedure for evaluating hierarchical sensitivity coefficients of the various response quantities with respect to the different laminate, layer, and micromechanical characteristics; 2) the application of reduction methods to the sensitivity analysis of the postbuckling response; and 3) the accurate evaluation of sensitivity coefficients of transverse shear stresses. Sample numerical results are presented which demonstrate the effectiveness of the computational procedures presented. Also, a number of research areas which have high potential for increasing the utility of sensitivity analysis are identified.

Mathematical Formulation

The panel is modeled by using a two-dimensional, first-order shear deformation shallow shell theory with the effects of large displacements, moderate rotations, average transverse shear deformation through-the-thickness, and laminated anisotropic material behavior included. A linear Duhamel–Neumann type, constitutive model is used, and the material properties are assumed to be independent of temperature. The thermoelastic constitutive relations used in the present study are given in Appendix A. A total Lagrangian formulation is used and the panel deformations, at different values of the applied loading, are referred to the original undeformed configuration. The panel is discretized by using two-field mixed finite element models, with the fundamental unknowns consisting of the generalized displacements at the nodes, and the stress resultant parameters. The stress resultants are allowed to be discontinuous at interelement boundaries. The sign convention for generalized displacements and stress resultants for the model is shown in Fig. 1.

The external loading consists of a uniform temperature change (independent of the coordinates x_1, x_2, and x_3) and an applied edge displacement q_e. The sensitivity coefficients are obtained by using the analytical direct-differentiation approach.

The governing finite element equations for the postbuckling response and the first-order sensitivity coefficients of the panel can be written in the following compact form:

$$[K]\{Z\} + \{G(Z)\} - q_1 \{Q^{(1)}\} - q_2 \{Q^{(2)}\} = 0 \tag{1}$$

$$\left[[K] + \left[\frac{\partial G_i}{\partial Z_j}\right]\right]\left\{\frac{\partial Z}{\partial \lambda}\right\} = -\left[\frac{\partial K}{\partial \lambda}\right]\{Z\} + q_1 \left\{\frac{\partial Q^{(1)}}{\partial \lambda}\right\} \tag{2}$$

where $[K]$ is the global linear structure matrix which includes the flexibility and the linear strain-displacement matrices; $\{Z\}$ is the response vector which includes both the unknown (free) nodal displacements and the stress-resultant parameters; $\{G(Z)\}$

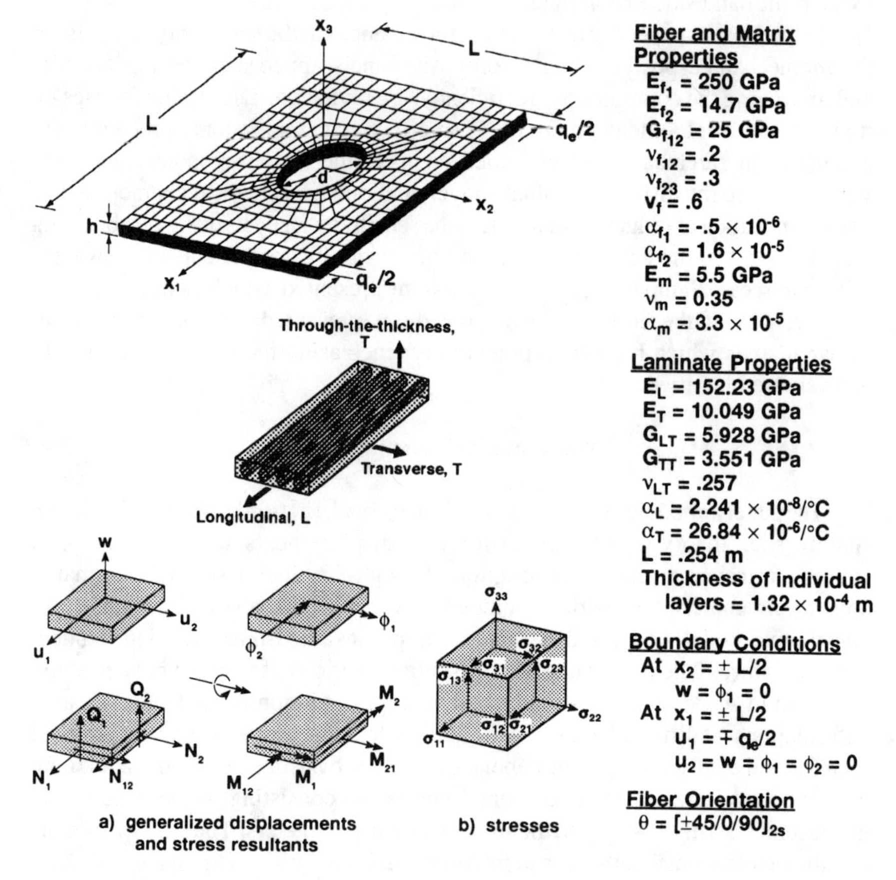

Fig. 1 Panels considered in the present study and sign convention for generalized displacements, stress resultants, and stresses.

is the vector of nonlinear terms; q_1 and q_2 are thermal strain and edge displacement parameters; $\{Q^{(1)}\}$ is the vector of normalized thermal strains; $\{Q^{(2)}\}$ is the vector of normalized mechanical strains; and λ refers to a typical material, lamination, or geometric parameter of the panel. The form of the arrays $[K]$, $\{G(Z)\}$, $\{Q^{(1)}\}$, $\{\partial Q^{(1)}/\partial\lambda\}$ and $\{Q^{(2)}\}$ is described in Appendix B.

Note that Eqs. (1) are nonlinear, but Eqs. (2) are linear. The matrix on the left-hand side of Eqs. (2) is identical to that used in the Newton–Raphson iterative process. Therefore, if the Newton–Raphson technique is used in generating the postbuckling response, the evaluation of each sensitivity coefficient requires the generation of the right-hand side of Eqs. (2) and a forward-reduction/back substitution operation only (no decomposition of the left-hand side matrix is required).

Hierarchical Sensitivity Coefficients

The postbuckling response characteristics of laminated composite structures are dependent on a hierarchy of interrelated parameters including laminate, layer, and micromechanical (fiber, matrix, and interface/interphase) parameters. A study of the sensitivity of the response to variations in each of these parameters provides an insight into the importance of the parameter and helps in the development of new materials to meet certain performance requirements.

Herein, a computational procedure is presented for evaluating the sensitivity coefficients of the postbuckling response with respect to three sets of parameters, namely, panel, layer, and micromechanical parameters. Henceforth, the three sets of parameters will be referred to as $\lambda_i^{(p)}$, $\lambda_j^{(\ell)}$ and $\lambda_k^{(m)}$ where superscripts (p), (ℓ), and (m) refer to the panel, layer, and micromechanical parameters, respectively; and the indices i, j, and k range from 1 to the number of parameters in each category.

The panel parameters $\lambda_i^{(p)}$ include the extensional stiffnesses A_{IJ}, bending stiffnesses D_{IJ}, bending-extensional coupling stiffnesses B_{IJ}, transverse shear stiffnesses $A_{I'J'}$ and the thermal effects appearing in the laminate constitutive relations [see Eqs. (A1) in Appendix A] where $I, J = 1, 2, 6$ and $I', J' = 4, 5$. The layer parameters $\lambda_j^{(\ell)}$ include the individual layer properties: Young's moduli E_L and E_T, shear moduli G_{LT} and G_{TT}, major Poisson's ratio ν_{LT}, coefficients of thermal expansion α_L and α_T, fiber-orientation angle θ, and layer thickness $h^{(\ell)}$, where subscripts L and T refer to the longitudinal (fiber) and transverse directions, respectively. The parameters $\lambda_k^{(m)}$ refer to the fiber, matrix, and interface/interphase moduli $E_f, E_m, E_p, G_f, G_m, G_p$ and fiber and matrix volume fractions v_f and v_m. The subscripts f, m, and p denote the fiber, matrix, and interface/interphase property, respectively.

The computational procedure consists of evaluating the sensitivity coefficients with respect to each of the panel (or laminate) stiffnesses $\{\partial Z/\partial\lambda_i^{(p)}\}$ using Eqs. (2). The sensitivity coefficients with respect to the layer and micromechanical parameters are then obtained by forming the following linear combinations:

$$\left\{\frac{\partial Z}{\partial\lambda_j^{(\ell)}}\right\} = \sum_i a_{ij}\left\{\frac{\partial Z}{\partial\lambda_i^{(p)}}\right\} \tag{3}$$

and

$$\left\{\frac{\partial \mathbf{Z}}{\partial \lambda_k^{(m)}}\right\} = \sum_j b_{jk} \left\{\frac{\partial \mathbf{Z}}{\partial \lambda_j^{(\ell)}}\right\} = \sum_i c_{ik} \left\{\frac{\partial \mathbf{Z}}{\partial \lambda_i^{(p)}}\right\} \tag{4}$$

where

$$a_{ij} = \left\{\frac{\partial \lambda_i^{(p)}}{\partial \lambda_j^{(\ell)}}\right\} \tag{5}$$

$$b_{jk} = \left\{\frac{\partial \lambda_j^{(\ell)}}{\partial \lambda_k^{(m)}}\right\} \tag{6}$$

$$c_{ik} = \left\{\frac{\partial \lambda_i^{(p)}}{\partial \lambda_k^{(m)}}\right\} = \sum_j a_{ij}\, b_{jk} \tag{7}$$

The a coefficients relate the panel (laminate) stiffnesses to the individual layer properties and are obtained from the lamination theory. The b coefficients relate the layer properties to the micromechanical (constituent) properties, and the c coefficients relate the laminate stiffnesses to the micromechanical properties (see Fig. 2). If the laminate stiffnesses are uniform and the constitutive relations of the

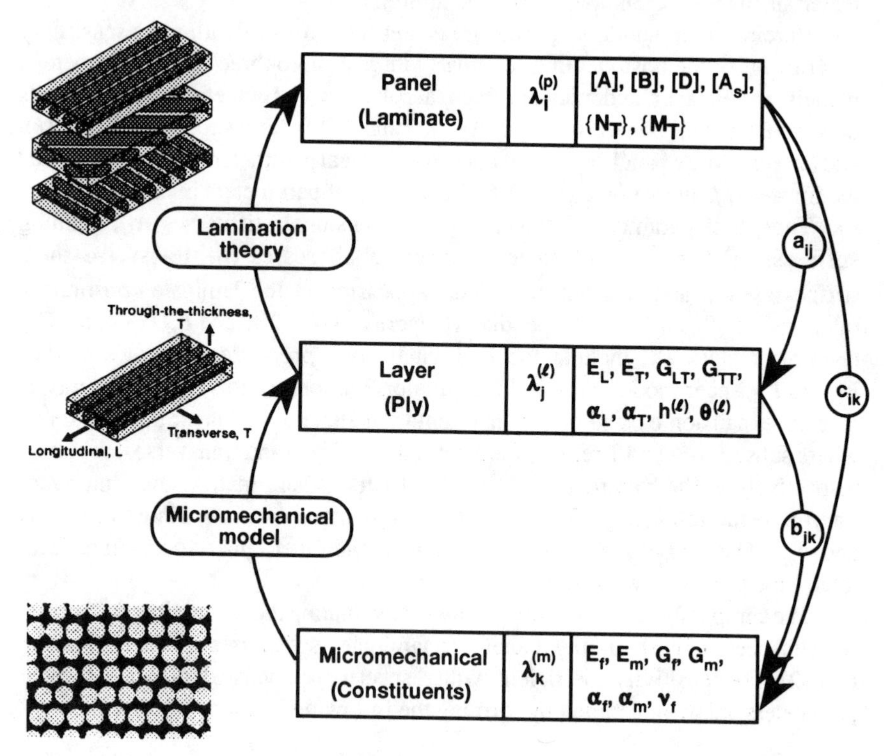

Fig. 2 Hierarchical sensitivity coefficients.

laminate, layer, and constituents are linear, then the *a*, *b*, and *c* coefficients are constants and need to be generated only once for each panel, even when the response is nonlinear.

Application of Reduction Methods

The basic idea of reduction methods is that of the condensation of a large system (algebraic and/or differential equations) to a similar (in some sense) much smaller substitute. The reduction methods which have been successfully applied to sensitivity analysis can be thought of as two-step hybrid analysis techniques combining a discretization method with a Bubnov–Galerkin (or a direct variational) technique. In the first step, a number of global approximation vectors (modes or basis vectors) for approximating both the postbuckling response of the panel and its sensitivity coefficients are generated using a discretization method with a perturbation technique. In the second step, the amplitudes of the global approximation vectors (or basis vectors) are determined via a Bubnov–Galerkin (or a direct variational) technique. The application of reduction methods to sensitivity analysis is described in Ref. 8 and is outlined subsequently.

Basis Reduction and Reduced System of Equations

The response vector $\{\mathbf{Z}\}$ and its first-order derivatives with respect to λ, $\{\partial\mathbf{Z}/\partial\lambda\}$, are each expressed as a linear combination of a few preselected global approximation vectors. The parameter λ refers to a typical laminate parameter from any of the three groups of parameters $\lambda^{(p)}$, $\lambda^{(\ell)}$, and $\lambda^{(m)}$. The approximations can be expressed by the following transformations:

$$\{\mathbf{Z}\} = [\Gamma]\,\{\psi\} \tag{8}$$

and

$$\left\{\frac{\partial \mathbf{Z}}{\partial \lambda}\right\} = [\bar{\Gamma}]\,\{\bar{\psi}\} \tag{9}$$

The columns of the matrices $[\Gamma]$ and $[\bar{\Gamma}]$ in Eqs. (8) and (9) are the global approximation vectors; and the elements of the vectors $\{\psi\}$ and $\{\bar{\psi}\}$ are the amplitudes of the approximation vectors that are, as yet, unknowns. Note that the number of basis vectors in Eqs. (8) and (9) is considerably smaller than the total number of degrees of freedom (components of each of the vectors $\{\mathbf{Z}\}$ and $\{\partial\mathbf{Z}/\partial\lambda\}$).

A Bubnov–Galerkin technique is now used to replace the governing equations of the panel postbuckling response and its first-order sensitivity coefficients, Eqs. (1) and (2), by the following reduced equations in the unknowns $\{\psi\}$ and $\{\bar{\psi}\}$

$$[\hat{K}]\{\psi\} + \{\hat{G}(\psi)\} - q_1\{\hat{Q}^{(1)}\} - q_2\{\hat{Q}^{(2)}\} = 0 \tag{10}$$

and

$$\left[[\bar{K}] + \left[\frac{\partial \bar{G}_{i'}}{\partial \psi_{j'}}\right]\right]\{\bar{\psi}\} = -\left[\frac{\partial \bar{K}}{\partial \lambda}\right]\{\psi\} + q_1\left\{\frac{\partial \bar{Q}^{(1)}}{\partial \lambda}\right\} \tag{11}$$

where

$$[\hat{K}] = [\Gamma]^t [K][\Gamma] \tag{12}$$

$$\{\hat{G}(\psi)\} = [\Gamma]^t \{G(\psi)\} \tag{13}$$

$$\{\hat{Q}^{(1)}\} = [\Gamma]^t \{Q^{(1)}\} \tag{14}$$

$$[\bar{K}] = [\bar{\Gamma}]^t [K][\bar{\Gamma}] \tag{15}$$

$$\left[\frac{\partial \bar{G}_{i'}}{\partial \psi_{j'}}\right] = [\Gamma]^t \left[\frac{\partial G_i}{\partial Z_j}\right][\Gamma] \tag{16}$$

$$\left[\frac{\partial \bar{K}}{\partial \lambda}\right] = [\bar{\Gamma}]^t \left[\frac{\partial K}{\partial \lambda}\right][\Gamma] \tag{17}$$

$$\left\{\frac{\partial \bar{Q}^{(1)}}{\partial \lambda}\right\} = [\bar{\Gamma}]^t \left\{\frac{\partial Q^{(1)}}{\partial \lambda}\right\} \tag{18}$$

Note that Eqs. (10) are nonlinear in the unknowns $\{\psi\}$ and Eqs. (11) are linear in $\{\bar{\psi}\}$. The vector $\{\hat{G}(\psi)\}$ in Eqs. (10) is obtained by replacing $\{Z\}$ in the vector $\{G(Z)\}$ by its expression in terms of $\{\psi\}$, Eqs. (8). The range of the indices i', j' is one to the number of components of $\{\bar{\psi}\}$.

Selection and Generation of Basis Vectors

The effectiveness of the proposed technique for calculating the sensitivity coefficients depends, to a great extent, on the proper choice of the basis vectors (the columns of the matrices $[\Gamma]$ and $[\bar{\Gamma}]$). An effective choice for the basis vectors used in approximating the response vector $\{Z\}$, Eqs. (8), was found to be the various-order path derivatives (derivatives with respect to the load parameters q_1 and q_2); that is, the columns of the matrix $[\Gamma]$ used in approximating $\{Z\}$, over a range of values of q_1 and q_2, consist of the response vector corresponding to a pair of values of q_1 and q_2 (viz., q_1^0, q_2^0) and its various-order derivatives with respect to q_1 and q_2, evaluated at the same pair of values (q_1^0, q_2^0), or

$$[\Gamma] = \left[\{Z\} \left\{\frac{\partial Z}{\partial q_1}\right\} \left\{\frac{\partial Z}{\partial q_2}\right\} \left\{\frac{\partial^2 Z}{\partial q_1^2}\right\} \left\{\frac{\partial^2 Z}{\partial q_1 \partial q_2}\right\} \left\{\frac{\partial^2 Z}{\partial q_2^2}\right\} \cdots\right]_{q_1^0,\, q_2^0} \tag{19}$$

The basis vectors used for approximating each of the sensitivity coefficients $\{\partial Z/\partial \lambda\}$ are selected to be combinations of: 1) the various-order path derivatives, that is, the columns of the matrix $[\Gamma]$; and 2) the first derivatives of $[\Gamma]$ with respect to λ, that is, $[\partial \Gamma/\partial \lambda]$. Therefore,

$$[\bar{\Gamma}] = \left[[\Gamma]\left[\frac{\partial \Gamma}{\partial \lambda}\right]\right] \tag{20}$$

The number of basis vectors in $[\bar{\Gamma}]$, for each λ, is twice the number of basis vectors in $[\Gamma]$.

The rationale for the particular choice of $[\bar{\Gamma}]$ is based on the following two facts:

1) Differentiating Eqs. (8) with respect to λ leads to

$$\left\{\frac{\partial \mathbf{Z}}{\partial \lambda}\right\} = [\Gamma]\left\{\frac{\partial \psi}{\partial \lambda}\right\} + \left[\frac{\partial \Gamma}{\partial \lambda}\right]\{\psi\} \tag{21}$$

that is, the expression for $\{\partial \mathbf{Z}/\partial \lambda\}$ includes both $[\Gamma]$ and $[\partial \Gamma/\partial \lambda]$.

2) The use of free parameters $\{\bar{\psi}\}$ instead of the fixed amplitudes $\{\psi\}$ and $\{\partial \psi/\partial \lambda\}$ in Eqs. (21) is expected to improve the accuracy of the approximation for $\{\partial \mathbf{Z}/\partial \lambda\}$ over a wide range of q_1, q_2. The free parameters $\{\bar{\psi}\}$ are obtained by applying the Bubnov–Galerkin technique to Eqs. (2), resulting in Eqs. (11).

The path derivatives (columns of the matrix $[\Gamma]$) are obtained by successive differentiation of the governing finite element equations of the panel, Eqs. (1), with respect to the parameters q_1 and q_2. The recursion relations for evaluating the path derivatives are given in Ref. 9. Note that only one matrix factorization is needed for generating all of the path derivatives.

The derivatives of $[\Gamma]$ with respect to λ and $[\partial \Gamma/\partial \lambda]$ are obtained by differentiating each of the recursion relations for evaluating the path derivatives with respect to λ. The resulting equations have the same left-hand sides as those of the original recursion relations and, therefore, no additional matrix factorizations are needed for generating $[\partial \Gamma/\partial \lambda]$.

Computational Procedure

The computational procedure for generating the nonlinear response vector $\{\mathbf{Z}\}$ and its sensitivity coefficients $\{\partial \mathbf{Z}/\partial \lambda\}$ can be conveniently divided into two distinct phases, namely, 1) evaluation of the basis vectors at a particular pair of values of q_1, q_2 (viz., q_1^0, q_2^0) and generation of the reduced equations, and 2) marching with the reduced equations in the solution space and generating the response and sensitivity coefficients for different combinations of q_1 and q_2. For each pair of values of q_1, q_2, the vector of reduced unknowns $\{\psi\}$ is obtained by solving the reduced nonlinear equations, Eqs. (10). Then the vectors $\{\bar{\psi}\}$ associated with the value of $\{\psi\}$ is evaluated by solving the reduced linear equations, Eqs. (11). The response vector and its sensitivity coefficients $\{\mathbf{Z}\}$ and $\{\partial \mathbf{Z}/\partial \lambda\}$ are obtained by using Eqs. (8) and (9). The process is repeated for different values of q_1 and q_2.

Computational Procedure for Evaluating the Sensitivity Coefficients of Detailed Response Characteristics

After the nodal displacements and stress-resultant parameters have been evaluated, a two-step computational procedure is used for evaluating the sensitivity coefficients of the stresses in the panel. In the first step, the in-plane strain components, the in-plane stresses, and their sensitivity coefficients in the different

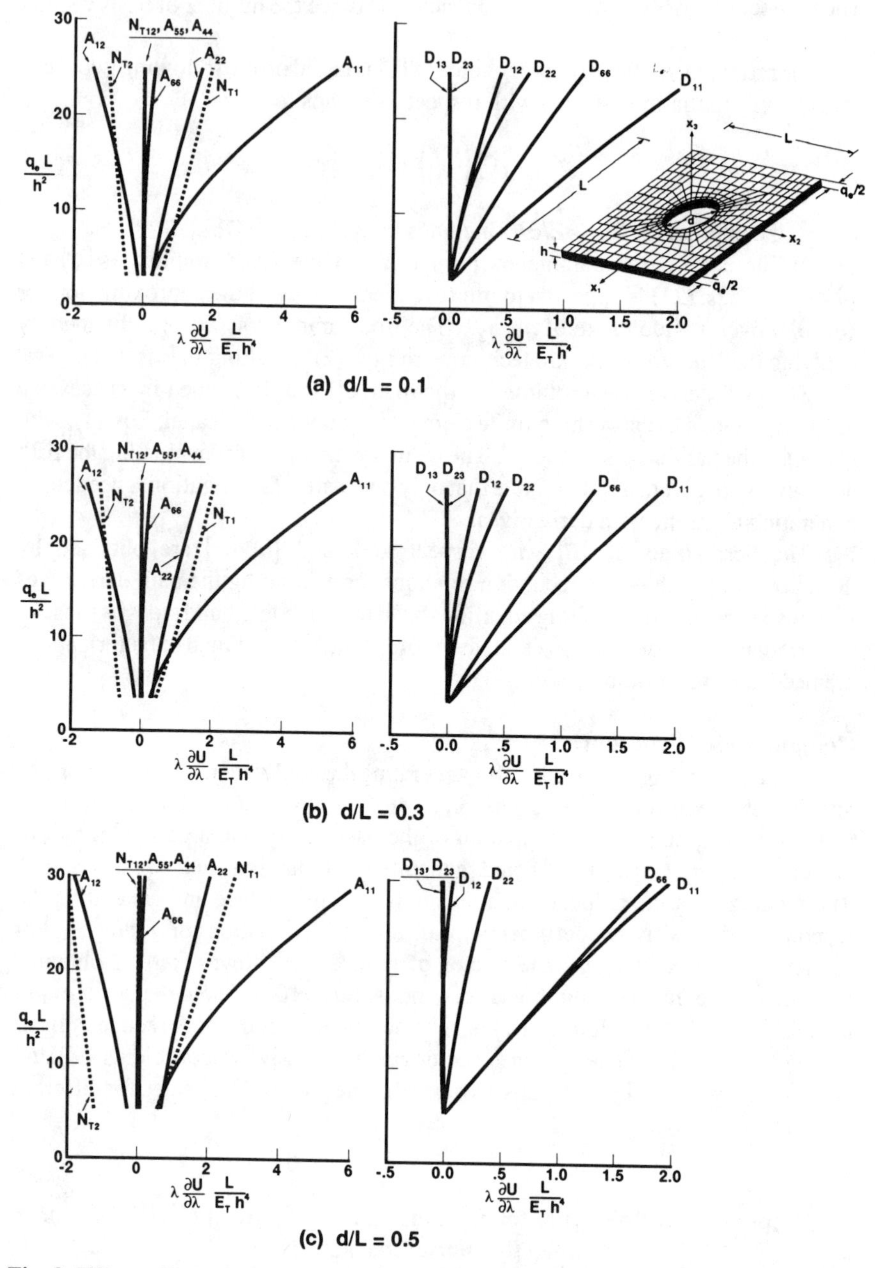

Fig. 3 Effect of hole diameter on the normalized sensitivity coefficients of the total strain energy U, with respect to panel parameters; thermally stressed, 16-layer quasi-isotropic panels with central circular cutout subjected to end shortening q_e, $T/T_{cr} = 0.5$ (see Fig. 1).

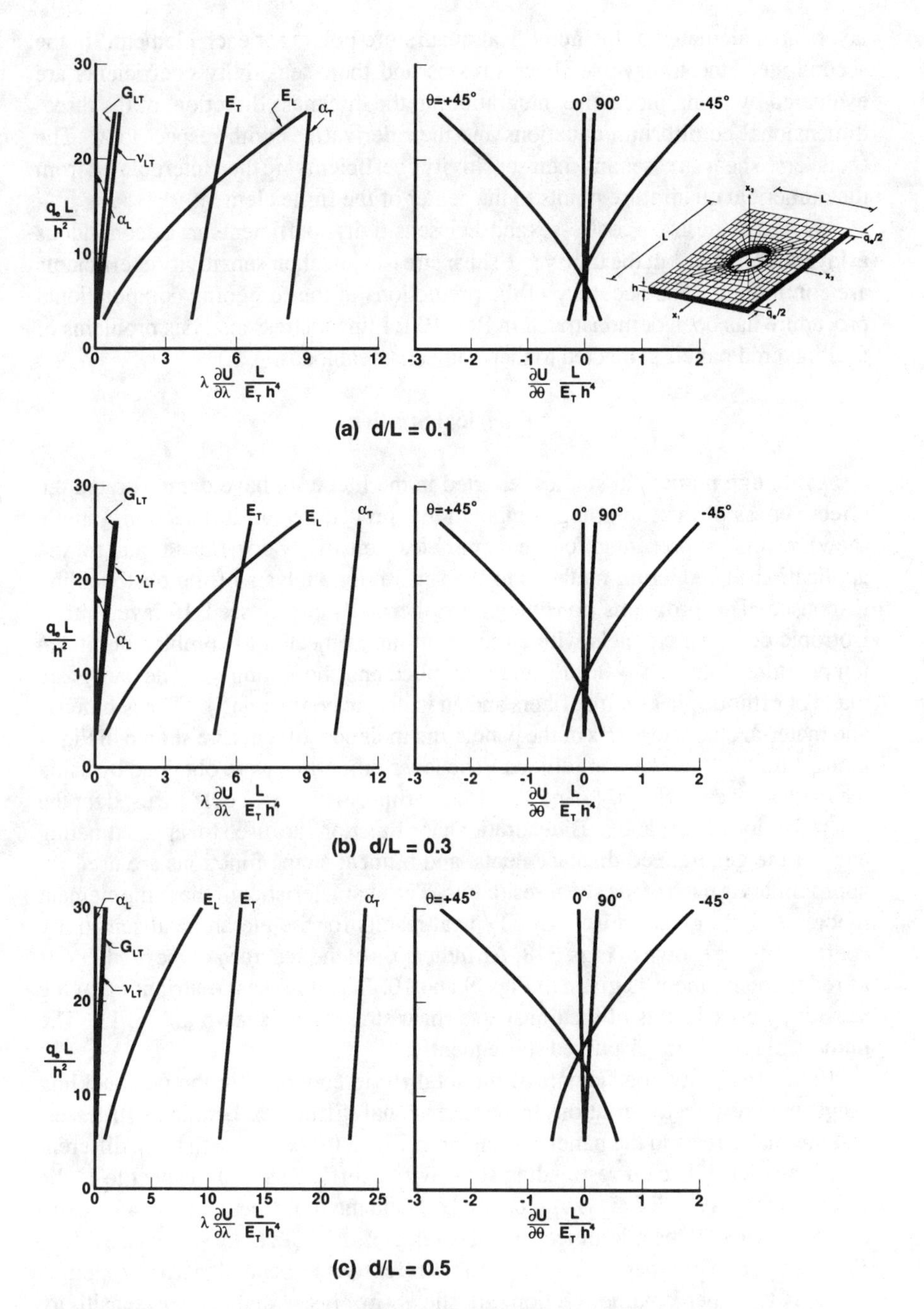

Fig. 4 Effect of hole diameter on the normalized sensitivity coefficients of the total strain energy U, with respect to layer parameters and fiber orientation angles; thermally stressed, 16-layer quasi-isotropic panels with central circular cutout subjected to end shortening q_e, $T/T_{cr} = 0.5$ (see Fig. 1).

layers are calculated at the numerical quadrature points for each element. In the second step, the transverse shear stresses and their sensitivity coefficients are evaluated by using piecewise integration, in the thickness direction, of the three-dimensional equilibrium equations and their derivatives with respect to λ. The transverse shear stresses and their sensitivity coefficients are then interpolated from the numerical quadrature points to the center of the finite element.

Note that the in-plane stresses and their sensitivity coefficients are discontinuous at layer interfaces, but the transverse shear stresses and their sensitivity coefficients are continuous. The accuracy of the predictions of the foregoing computational procedure has been demonstrated in Ref. 10 for linear stress analysis problems of multilayered panels subjected to thermomechanical loads.

Numerical Studies

Although numerical studies reported in the literature have demonstrated the effectiveness of the foregoing computational procedures, in this section, sample (new) results are presented for the hierarchical sensitivity coefficients and for the application of reduction methods to the sensitivity analysis of the postbuckling response. The problems considered are thermally prestressed 16-layer quasi-isotropic composite panels with circular cutouts subjected to combined uniform temperature change, $T = 0.5\ T_{cr}$ and an applied end shortening q_e. The panels are made of orthotropic graphite fibers and an isotropic epoxy matrix. The geometric and material characteristics of the panels and their constituents are shown in Fig. 1 along with the boundary conditions. The layer properties were obtained by using the method of cells (Refs. 11 and 12). Mixed finite element models are used for the discretization of the panels. Biquadratic shape functions are used for approximating each of the generalized displacements, and bilinear shape functions are used for approximating each of the stress resultants. The characteristics of the finite element model used are given in Ref. 13. Typical results for the hierarchical sensitivity coefficients are shown in Figs. 3–8. An indication of the accuracy of the predictions of reduction methods is given in Figs. 9 and 10. The thickness distribution of the sensitivity coefficients of the transverse shear stresses are shown in Fig. 11. The numerical results are discussed subsequently.

The sensitivity coefficients of the total strain energy U in the postbuckling range, with respect to variations in the extensional stiffnesses, bending stiffnesses, and thermal effects in the panel, are shown in Fig. 3 for panels with three different hole diameters. The corresponding sensitivity coefficients with respect to 1) the layer properties $E_L, E_T, G_{LT}, \nu_{LT}, \alpha_L$, and α_T, and the four fiber angles +45, -45, 0, and 90 deg, and 2) the micromechanical parameters $E_{f1}, E_{f2}, G_{f12}, \nu_{f12}, \nu_{f23}, \alpha_{f1}, \alpha_{f2}, E_m, \nu_m, \alpha_m$, and v_f, where subscripts f and m refer to the fiber and matrix properties and v_f is the fiber volume fraction, are shown in Figs. 4 and 5. The sensitivity coefficients with respect to the panel, layer, micromechanical properties, and fiber angles $\partial U/\partial\lambda$ and $\partial U/\partial\theta$ are normalized by multiplying each by $\lambda L/E_T h^4$ and $L/E_T h^4$, respectively, where λ refers to any of the material properties.

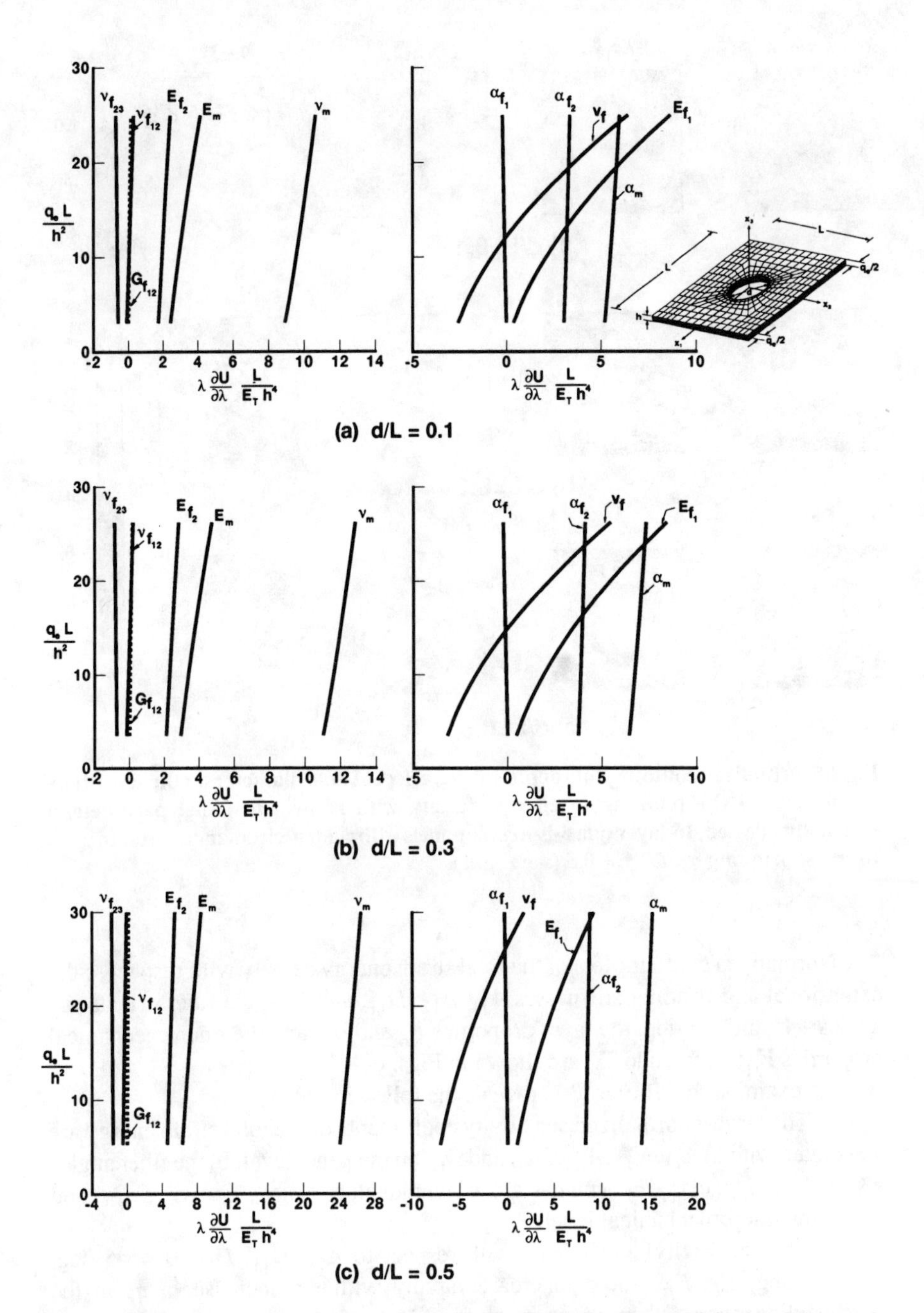

Fig. 5 Effect of hole diameter on the normalized sensitivity coefficients of the total strain energy U, with respect to micromechanical parameters; thermally stressed, 16-layer quasi-isotropic panels with central circular cutout subjected to end shortening q_e, $T/T_{cr} = 0.5$ (see Fig. 1).

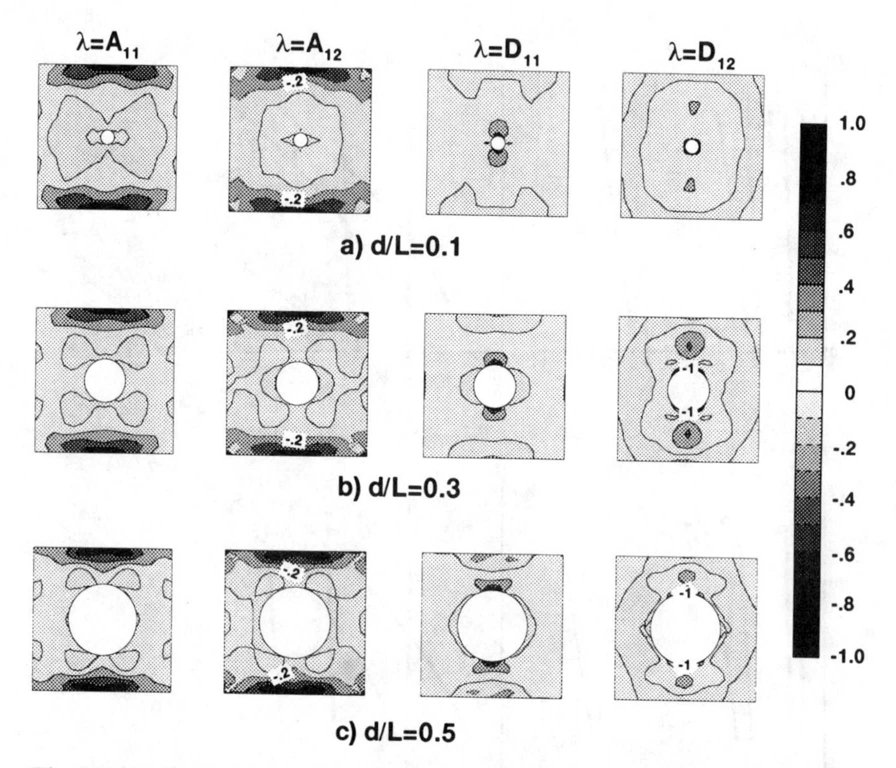

Fig. 6 Normalized contour plots depicting the effect of hole diameter on the sensitivity coefficients of the total strain energy density with respect to panel parameters; thermally stressed, 16-layer quasi-isotropic panels with central circular cutout subjected to end shortening q_e, $T/T_{cr} = 0.5$ (see Fig. 1).

Normalized contour plots of the total strain energy density with respect to the extensional and bending stiffnesses A_{11}, A_{12}, D_{11}, and D_{12}, the fiber orientation angles +45 and -45 deg, the layer properties E_L and α_T, and the micromechanical properties E_{f1}, v_f, ν_m, and E_m are shown in Figs. 6–8.

An examination of Figs. 3–8 reveals the following.

1) The largest normalized sensitivity coefficients of the total strain energy are associated with a) $A_{11}, N_{T1}, A_{22}, D_{11}$, and D_{66} on the panel level; b) the fiber angles +45 and -45 deg; c) α_T, E_L, and E_T on the layer level; and d) $\nu_m, E_{f1}, \alpha_m, E_m, v_f$ and α_{f2} on the micromechanical level.

2) The sensitivity coefficients with respect to A_{11}, D_{11}, D_{66}, $\theta = 45$ deg, $\theta = -45$ deg, E_L, E_{f1}, and v_f increase rapidly with the increase in q_e in the postbuckling range. Other sensitivity coefficients (such as those with respect to D_{12}, α_T, E_T, α_m, ν_m, and E_m) exhibit smaller variations with changes in q_e.

3) For a given applied end displacement q_e, the magnitudes of the sensitivity coefficients of the total strain energy with respect to D_{66}, α_T, E_T, ν_m, α_m, E_m, and E_{f2} exhibit a significant increase with the increase in the hole diameter.

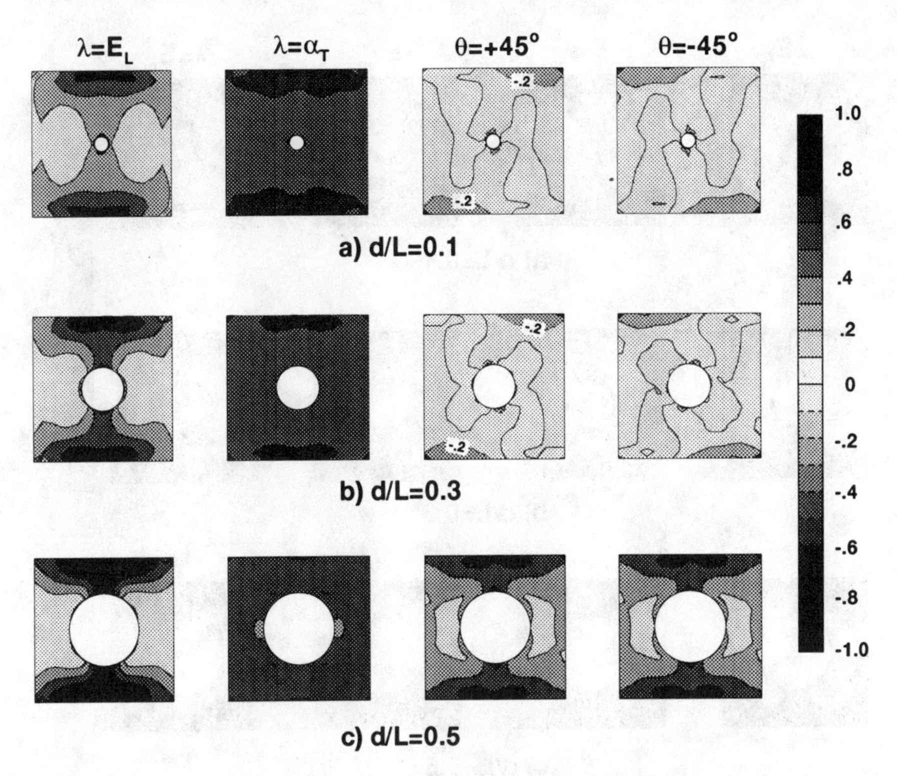

Fig. 7 Normalized contour plots depicting the effect of hole diameter on the sensitivity coefficients of the total strain energy density with respect to layer parameters; thermally stressed, 16-layer quasi-isotropic panels with central circular cutout subjected to end shortening q_e, T/T_{cr} = 0.5 (see Fig. 1).

4) The distributions of the sensitivity coefficients of the total strain energy density $\overline{U}$ differ from each other and change with the change in the hole diameter. However, for panels with $d/L \leq 0.3$, the distribution of the sensitivity coefficients of $\overline{U}$ with respect to E_L and E_{f1} are similar.

An indication of the accuracy of the sensitivity coefficients of the total strain energy U, with respect to the layer parameters and fiber angles, obtained by the reduction technique is given in Figs. 9 and 10. For each sensitivity coefficient λ, a single set of $2r$ global approximation vectors (where r is the number of global approximation vectors used in the response calculations) consisting of the response vector, its path derivatives with respect to q_e up to order r - 1, and the derivatives of these vectors with respect to λ was generated at $w_a/h = 1.4$, $T = 0.5\ T_{cr}$ and used throughout the analysis. As can be seen from Figs. 9 and 10, the sensitivity coefficients obtained by using $r = 6$ are almost indistinguishable from those obtained by the full system.

The thickness distributions of the transverse shear stresses σ_{31} and σ_{32} and their normalized sensitivity coefficients with respect to E_L and α_T are shown in Fig. 11

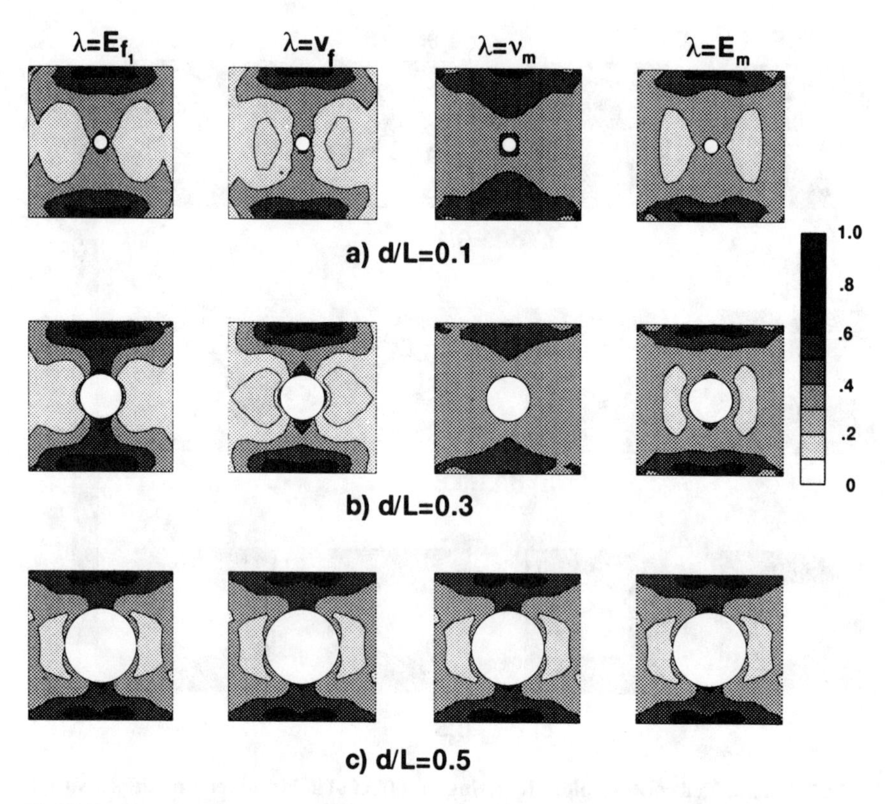

Fig. 8 Normalized contour plots depicting the effect of hole diameter on the sensitivity coefficients of the total strain energy density with respect to micromechanical parameters; thermally stressed, 16-layer quasi-isotropic panels with central circular cutout subjected to end shortening q_e, $T/T_{cr} = 0.5$ (see Fig. 1).

at the point of maximum transverse shear strain energy density $\bar{U}$ (the point marked by × in Fig. 11). As can be seen from Fig. 11, the thickness distributions of the sensitivity coefficients are different from those of the transverse shear stresses. This is particularly true for σ_{32}.

Future Directions for Research

Among the different aspects and applications of sensitivity analysis of thermomechanical postbuckling response which have high potential for research are the following.

1) Hierarchical sensitivity analysis can be extended to both temperature-sensitive and smart materials.

2) The sensitivity analysis can be used a) to study the effect of material imperfections at the constituent, layer, or laminate levels, on the postbuckling response; b) in conjunction with multiscale/multimodel computational strategies, for relating the local effects to the global response of the structure.

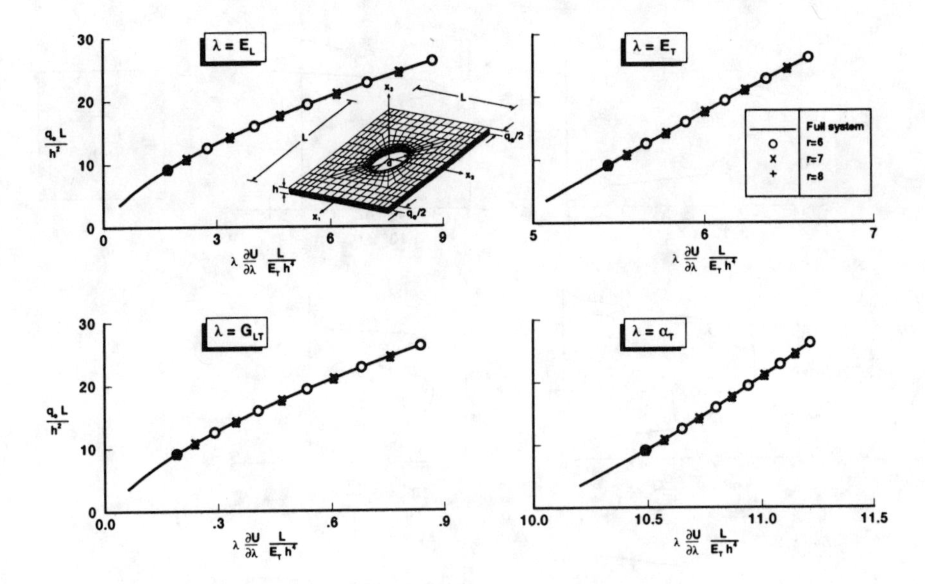

Fig. 9 Accuracy of sensitivity coefficients of the total strain energy, with respect to layer properties, obtained by reduction method; thermally stressed, 16-layer quasi-isotropic panels with central circular cutout subjected to end shortening q_e, $T/T_{cr} = 0.5$ (see Fig. 1).

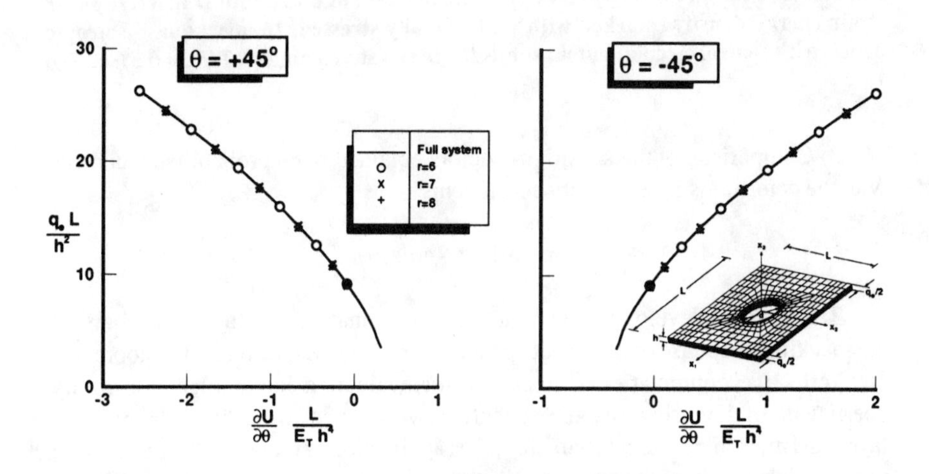

Fig. 10 Accuracy of sensitivity coefficients of the total strain energy, with respect to fiber orientation angles, obtained by reduction method; thermally stressed, 16-layer quasi-isotropic panels with central circular cutout subjected to end shortening q_e, $T/T_{cr} = 0.5$ (see Fig. 1).

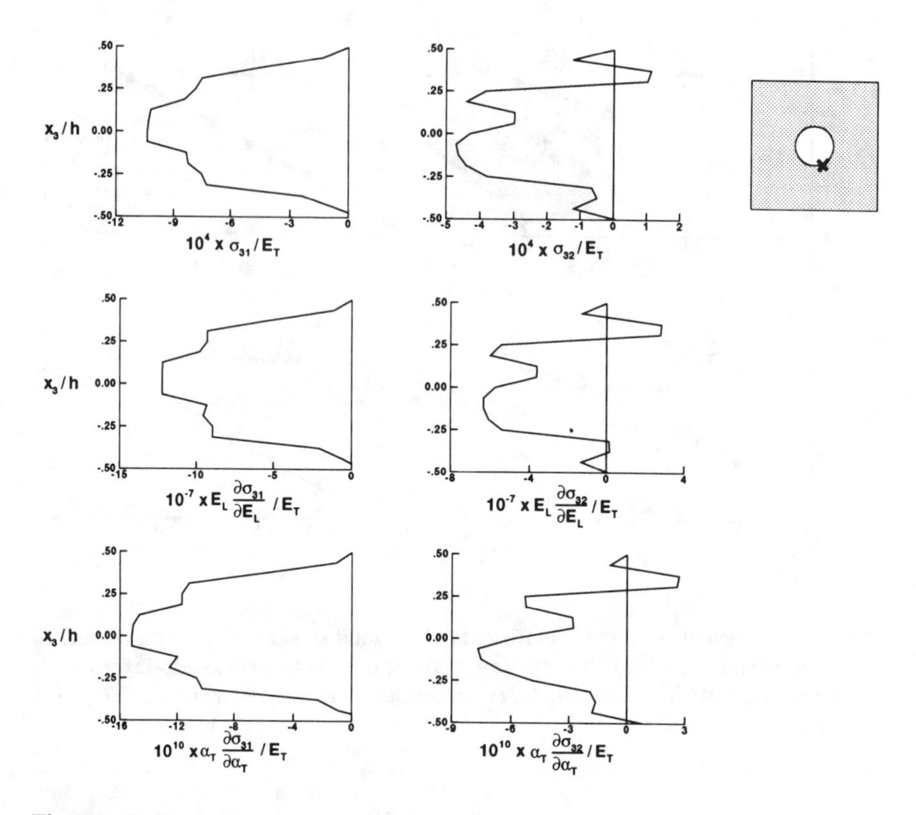

Fig. 11 Through-the-thickness distributions of the sensitivity coefficients of the transverse shear stresses σ_{32} and σ_{32} at the location of maximum transverse shear strain energy density (marked with $\times$); thermally stressed, 16-layer quasi-isotropic panels with central circular cutout subjected to end shortening q_e, $T/T_{cr} = 0.5$ (see Fig. 1).

3) Comparison of the sensitivity factors obtained from probabilistic analysis with the deterministic sensitivity coefficients.

Concluding Remarks

Three recent developments in the sensitivity analysis for thermomechanical postbuckling response of composite panels are reviewed. The three developments are: effective computational procedure for evaluating hierarchical sensitivity coefficients of the various response quantities with respect to the different laminate, layer, and micromechanical characteristics; application of reduction methods to the sensitivity analysis of the postbuckling response; and accurate evaluation of the sensitivity coefficients of transverse shear stresses.

The panels are modeled using two-dimensional, first-order shear deformation shallow shell theory with the effects of large displacements, moderate rotations, average transverse shear deformation through-the-thickness, and laminated anisotropic material behavior included. A linear Duhamel–Neumann type, constitutive model is used, and the material properties are assumed to be independent of temperature. A two-field mixed finite element model is used for the discretization, with the fundamental unknowns consisting of the generalized displacements at the nodes and stress-resultant parameters. The stress resultants are allowed to be discontinuous at interelement boundaries. The external loading consists of a uniform temperature change and an applied edge displacement.

The computational procedure for evaluating the hierarchical sensitivity coefficients consists of evaluating the sensitivity coefficients with respect to each of the panel (or laminate) stiffnesses and then generating the sensitivity coefficients with respect to the layer and micromechanical parameters as linear combinations of the sensitivity coefficients with respect to the panel parameters. Hierarchical sensitivity coefficients can be used to assess the effects of imperfections in the composite parameters on the postbuckling response. They can also help in bridging the gap between structural design and material development.

The success of application of reduction methods to the sensitivity analysis of the postbuckling response hinges on the proper selection of basis vectors. A good set of basis vectors consists of a combination of path derivatives and their first derivatives with respect to the panel, layer, or micromechanical parameters. The use of this set of basis vectors can make large-scale sensitivity calculations practical (particularly with parallel processing machines).

The accurate evaluation of the sensitivity coefficients of the transverse shear stresses in multilayered composite panels can be achieved by using first-order shear-deformation theory for evaluating the in-plane stress components and their sensitivity coefficients and then performing piecewise integration in the thickness direction of the three-dimensional equilibrium equations and their derivatives with respect to the composite parameters. The sensitivity coefficients of the transverse shear stresses are first evaluated at the numerical quadrature points and then interpolated to the center of the element.

Sample numerical results are presented for 16-layer quasi-isotropic panels with central circular cutout subjected to uniform temperature change and an applied edge shortening. The numerical results demonstrate the effectiveness of the computational procedures presented. Some of the future directions and applications of sensitivity analysis which have high potential for research are outlined.

Appendix A: Thermoelastic Constitutive Relations for the Laminate

The thermoelastic model used in the present study is based on the following assumptions.

1) The laminates are composed of a number of perfectly bonded layers.

2) Every point of the laminate is assumed to possess a single plane of thermoelastic symmetry parallel to the middle plane.

3) The material properties are independent of temperature.

4) The constitutive relations are described by lamination theory and can be written in the following compact form:

$$\begin{Bmatrix} N \\ M \\ Q \end{Bmatrix} = \begin{bmatrix} [A] & [B] & 0 \\ [B]^t & [D] & 0 \\ 0 & 0 & [A_s] \end{bmatrix} \begin{Bmatrix} \varepsilon \\ \kappa \\ \gamma \end{Bmatrix} - \begin{Bmatrix} N_T \\ M_T \\ 0 \end{Bmatrix} \tag{A1}$$

where $\{N\}$, $\{M\}$, $\{Q\}$, $\{\varepsilon\}$, $\{\kappa\}$, $\{\gamma\}$, and $\{N_T\}$ and $\{M_T\}$ are the vectors of extensional, bending, and transverse shear stress resultants, strain components and thermal effects of the laminate given by

$$\{N\}^t = [N_1 \;\; N_2 \;\; N_{12}] \tag{A2}$$

$$\{M\}^t = [M_1 \;\; M_2 \;\; M_{12}] \tag{A3}$$

$$\{Q\}^t = [Q_1 \;\; Q_2] \tag{A4}$$

$$\{\varepsilon\}^t = [\varepsilon_1 \;\; \varepsilon_2 \;\; 2\varepsilon_{12}] \tag{A5}$$

$$\{\kappa\}^t = [\kappa_1 \;\; \kappa_2 \;\; 2\kappa_{12}] \tag{A6}$$

$$\{\gamma\}^t = [2\varepsilon_{31} \;\; 2\varepsilon_{32}] \tag{A7}$$

$$\{N_T\}^t = [N_{T1} \;\; N_{T2} \;\; N_{T12}] \tag{A8}$$

and

$$\{M_T\}^t = [M_{T1} \;\; M_{T2} \;\; M_{T12}] \tag{A9}$$

The matrices $[A]$, $[B]$, $[D]$, and $[A_s]$ contain the extensional, coupling, bending, and transverse shear stiffnesses of the laminate which can be expressed in terms of the layer stiffnesses as follows:

$$([A] \;\; [B] \;\; [D]) = \sum_{k=1}^{NL} \int_{h_{k-1}}^{h_k} [\bar{Q}]^{(k)} \left([I] \;\; x_3[I] \;\; (x_3)^2[I]\right) \mathrm{d}x_3 \tag{A10}$$

$$[A_s] = \sum_{k=1}^{NL} \int_{h_{k-1}}^{h_k} [\bar{Q}_s]^{(k)} \mathrm{d}x_3 \tag{A11}$$

where $[\bar{Q}]^{(k)}$ and $[\bar{Q}_s]^{(k)}$ are the extensional and transverse shear stiffnesses of the kth layer (referred to the x_1, x_2, and x_3 coordinate system), $[I]$ is the identity matrix, h_k and h_{k-1} are the distances from the top and bottom surfaces of the kth layer to the middle surface, and NL is the total number of layers in the laminate. The expressions for the different coefficients of the matrices $[\bar{Q}]^{(k)}$ and $[\bar{Q}_s]^{(k)}$ in terms of the material and geometric properties of the constituents (fiber and matrix) are given in Refs. 14 and 15.

The vectors of thermal effects $\{N_T\}$ and $\{M_T\}$ are given by

$$[\{N_T\}\ \{M_T\}] = \sum_{k=1}^{NL} \int_{h_{k-1}}^{h_k} [\bar{Q}]^{(k)} \{\alpha\}^{(k)} [1\ \ x_3]\, T\, dx_3 \tag{A12}$$

where $\{\alpha\}$ is the vector of coefficients of thermal expansion (referred to the coordinates x_1, x_2, and x_3; see, for example, Refs. 16 and 17) and T is the temperature change.

Appendix B: Form of the Arrays in the Governing Discrete Equations of the Panel

The governing discrete equations of the panel, Eqs. (1), consist of both the constitutive relations and the equilibrium equations. The response vector $\{Z\}$ can be partitioned into subvectors of stress-resultant parameters $\{H\}$, and free (unconstrained) nodal displacements $\{X\}$, as follows:

$$\{Z\} = \begin{Bmatrix} H \\ X \end{Bmatrix} \tag{B1}$$

The different arrays in Eqs. (1) and (2) can be partitioned as follows:

$$[K] = \begin{bmatrix} -F & S_1 \\ S_1^t & 0 \end{bmatrix} \tag{B2}$$

$$\{G(Z)\} = \begin{Bmatrix} \bar{M}\ (X, \bar{X}_e) \\ \bar{N}\ (H, X, \bar{X}_e) \end{Bmatrix} \tag{B3}$$

$$\{Q^{(1)}\} = \begin{Bmatrix} \varepsilon_T \\ 0 \end{Bmatrix} \tag{B4}$$

$$\{\bar{Q}^{(2)}\} = \begin{Bmatrix} -[S_2]\{\bar{X}_e\} \\ 0 \end{Bmatrix} \tag{B5}$$

$$\left[\frac{\partial G_i}{\partial Z_j}\right] = \begin{bmatrix} 0 & \left[\dfrac{\partial \bar{M}_{I'}}{\partial X_J}\right] \\ \text{symm} & \left[\dfrac{\partial \bar{N}_I}{\partial X_J}\right] \end{bmatrix} \tag{B6}$$

$$\left[\frac{\partial K}{\partial \lambda}\right] = \begin{bmatrix} -\left[\frac{\partial F}{\partial \lambda}\right] & \mathbf{0} \\ \mathbf{0} & \mathbf{0} \end{bmatrix} \tag{B7}$$

$$\left\{\frac{\partial Q^{(1)}}{\partial \lambda}\right\} = \begin{Bmatrix} \frac{\partial \varepsilon_T}{\partial \lambda} \\ \mathbf{0} \end{Bmatrix} \tag{B8}$$

where $[\boldsymbol{F}]$ is the linear flexibility matrix; $[S_1]$ and $[S_2]$ are the linear strain-displacement matrices associated with the free nodal displacements $\{\boldsymbol{X}\}$ and the constrained (prescribed nonzero) edge displacements $q_2\{\bar{X}_e\}$; $\{\bar{\boldsymbol{M}}(X, \bar{X}_e)\}$ and $\{\bar{\boldsymbol{N}}(\boldsymbol{H}, \boldsymbol{X}, \bar{X}_e)\}$ are the subvectors of nonlinear terms; $\{\varepsilon_T\}$ is the subvector of normalized thermal strains; $\mathbf{0}$ is a null matrix or vector; and superscript t denotes transposition. The explicit form of $\{\varepsilon_T\}$ is given in Ref. 5.

For the purpose of obtaining analytic derivatives with respect to lamination parameters (e.g., fiber orientation angles of different layers), it is convenient to express $\partial[\boldsymbol{F}]/\partial\lambda$ in terms of $\partial[\boldsymbol{F}]^{-1}/\partial\lambda$ as follows:

$$\frac{\partial[F]}{\partial \lambda} = -[F]\frac{\partial[F]^{-1}}{\partial \lambda}[F] \tag{B9}$$

The explicit forms of $\partial[\boldsymbol{F}]^{-1}/\partial\lambda$ and $\{\partial\varepsilon_T/\partial\lambda\}$ are given in Ref. 5.

Analytic expressions are given in Ref. 18 for the laminate stiffnesses $[\boldsymbol{A}]$, $[\boldsymbol{B}]$, $[\boldsymbol{D}]$, and $[\boldsymbol{A}_s]$; the vectors of thermal effects $\{\boldsymbol{N}_T\}$ and $\{\boldsymbol{M}_T\}$; and their derivatives with respect to each of the material properties and fiber orientation angles.

Acknowledgments

This work was partially supported by NASA Cooperative Agreement NCCW-0011, by AFOSR Grant F49620-93-1-0184, and by NASA Grant NAG-1-1162. The numerical studies were performed on the CRAY Y-MP computer at NASA Ames Research Center. The author appreciates the help of Jeanne Peters and Catherine Richter, both of the University of Virginia, in generating the figures.

References

[1]Chen, L. W., Brunell, E. J., and Chen, L. Y., "Thermal Buckling of Initially Stressed Thick Plates," *Journal of Mechanical Design*, Vol. 104, No. 3, 1982, pp. 557–564.

[2]Snead, J. M., and Palazotto, A. N., "Moisture and Temperature Effects on the Instability of Cylindrical Composite Panels," *Journal of Aircraft*, Vol. 20, No. 9, 1983, pp. 777–783.

[3]Kossira, H., and Haupt, M., "Buckling of Laminated Plates and Cylindrical Shells Subjected to Combined Thermal and Mechanical Loads," *Buckling of Shell Structures, on*

Land, in the Sea and in the Air, edited by J. F. Jullien, Elsevier, London, England, UK, 1991, pp. 201–212.

[4]Noor, A. K., and Peters, J. M., "Thermomechanical Buckling of Multilayered Composite Plates," *Journal of Engineering Mechanics*, Vol. 118, No. 2, 1992, pp. 351–366.

[5]Noor, A. K., Starnes, J. H., Jr., and Peters, J. M., "Thermomechanical Buckling of Multilayered Composite Panels with Cutouts," *AIAA Journal*, Vol. 32, No. 7, 1994, pp. 1507–1519.

[6]Noor, A. K., and Burton, W. S., "Computational Models for High-Temperature Multilayered Composite Plates and Shells," *Applied Mechanics Reviews*, Vol. 45, No. 10, 1992, pp. 419–446.

[7]Noor, A. K., Starnes, J. H., Jr., and Peters, J. M., "Thermomechanical Postbuckling of Multilayered Composite Panels with Cutouts," *Proceedings of the AIAA/ASME/ ASCE/ AHS/ASC 35th Structures, Structural Dynamics and Materials Conference* (Hilton Head, SC), AIAA, Washington, DC, April 18–21, 1994, Pt. I, pp. 466–484.

[8]Noor, A. K., and Peters, J. M., "Reduced Basis Technique for Calculating Sensitivity Coefficients of Nonlinear Structural Response," *AIAA Journal*, Vol. 30, No. 7, 1992, pp. 1840–1847.

[9]Noor, A. K., and Peters, J. M., "Multiple-Parameter Reduced Basis Technique for Bifurcation and Postbuckling Analyses of Composite Plates," *International Journal for Numerical Methods in Engineering*, Vol. 19, No. 12, 1983, pp. 1783–1803.

[10]Noor, A. K., Kim, Y. H., and Peters, J. M., "Transverse Shear Stresses and Their Sensitivity Coefficients in Multilayered Composite Panels," *AIAA Journal*, Vol. 32, No. 6, 1994, pp. 1259–1269.

[11]Aboudi, J., "Micromechanical Analysis of Composites by the Method of Cells," *Applied Mechanics Reviews*, Vol. 42, No. 7, 1989, pp. 193–221.

[12]Aboudi, J., *Mechanics of Composite Materials - A Unified Micromechanical Approach*, Elsevier, Amsterdam, The Netherlands, 1991.

[13]Noor, A. K., and Andersen, C. M., "Mixed Models and Reduced/Selective Integration Displacement Models for Nonlinear Shell Analysis," *International Journal for Numerical Methods in Engineering*, Vol. 18, No. 10, 1982, pp. 1429–1454.

[14]Jones, R. M., *Mechanics of Composite Materials*, McGraw-Hill, New York, 1975.

[15]Tsai, S. W., and Hahn, H. T., *Introduction to Composite Materials*, Technomic, Westport, CT, 1980.

[16]Padovan, J., "Anisotropic Thermal Stress Analysis," *Thermal Stresses I*, edited by R. B. Hetnarski, Elsevier, Amsterdam, The Netherlands, 1986, pp. 143–262.

[17]Bert, C. W., "Analysis of Plates," *Composite Materials*, edited by C. C. Chamis, Vol. 7, Structural Design and Analysis, Pt. I, Academic, New York, 1975, pp. 149–206.

[18]Noor, A. K., and Tenek, L. H., "Stiffness and Thermal Coefficients for Composite Laminates," *Journal of Composite Structures*, Vol. 21, No. 1, 1992, pp. 57–66.

Laser Induced Thermal Stresses in Composite Materials

Jerry R. Couick*
Air Force Institute of Technology
Wright-Patterson Air Force Base, Ohio 45433

Abstract

A simple strength of materials solution to the bimetallic thermostat problem is extended to provide an estimate of thermal stresses in composite materials when subjected to high energy laser irradiation. In particular, a thin strip of a very long unidirectional fiber-reinforced composite lamina is modeled as a semi-infinite layered beam, with the relative thicknesses of the fiber and matrix layers determined by the volume fraction. The concept of interfacial shear compliance is used to formulate a second-order boundary-value problem for the interlaminar shearing force, allowing the axial stress-free boundary conditions on the ends of the beam to be enforced. The interlaminar shearing and normal (or "peeling") stresses are then derived from the shearing force by a simple equilibrium analysis. This elementary method is applicable for layered beams with any temperature distribution for which a particular solution to the governing differential equation may be found. The temperature must, however, be continuous and decay to zero at infinity. The insulated rod solution is of particular interest.

Nomenclature

A_1, A_2	=	ratio of thermal to mechanical properties
B	=	ratio of stiffnesses
b	=	thermal length of layer 1
$\hat{b}^2$	=	ratio of thermal diffusivities

* Aerospace Engineering, Structures Division, Wright Laboratory; also, Ph.D. Candidate.

E_1, E_2 = Young's modulus
h = total thickness
$\hat{h}$ = ratio of layer thicknesses
h_1, h_2 = layer thicknesses
I_1, I_2 = moment of inertia
k_1, k_2 = thermal conductivity
$\mathcal{M}_1, \mathcal{M}_2$ = internal bending moment
P_0 = concentrated corner force
Q = interlaminar shearing force
$\hat{Q}, \hat{P}_0, \hat{q}, \hat{p}$ = nondimensional forces and stresses
q, p = interlaminar shearing, normal stresses
V = internal shear force
v_f = composite volume fraction
x, ξ = longitudinal direction
y = thickness direction
α_1, α_2 = coefficient of linear thermal expansion
δ = free-edge boundary layer length
ϵ_1, ϵ_2 = interface strain
$\epsilon_{1_B}, \epsilon_{2_B}$ = interface strain due to bending
$\epsilon_{1_Q}, \epsilon_{2_Q}$ = interface strain due to Q
$\epsilon_{1_q}, \epsilon_{2_q}$ = interface strain due to q
$\epsilon_{1_T}, \epsilon_{2_T}$ = interface strain due to thermal expansion
$\kappa_{D_1}, \kappa_{D_2}$ = thermal diffusivity
κ_1, κ_2 = interfacial compliance coefficient
λ, k, μ = material property parameters
ρ = radius of curvature
$\sigma_{x_1}, \sigma_{x_2}$ = longitudinal stress
ϕ = nondimensional time

Background

While serving as Chief of the Laser Test Facility at the Air Force Weapons Laboratory (now the Phillips Laboratory), Kirtland Air Force Base, New Mexico, the author was privileged to work on several research programs involving laser-induced failure of filament wound graphite/epoxy (GE) and carbon/epoxy (CE) cylindrical pressure vessels. High-energy laser effects testing was conducted in order to determine damage thresholds as well as to verify analytical damage models. Extensive modeling efforts were conducted during and after the test series to allow interpretation of the results and to facilitate understanding the basic physics responsible for observed test results. The primary objective of modeling activity was, of course, to develop theoretical and/or computational models capable of accurately predicting the response of pressure vessels subjected to laser engagement scenarios not addressed by the experimental work.

The literature shows that at least six independent models evolved from the modeling activities. Although all of them are not reported in the open literature, they are all documented in some form (Refs. 1 and 2, for example). All six models use finite elements to address fracture mechanics issues. Linear elastic fracture mechanics and free-edge delamination appear to be the phenomena of most interest to the researchers. Whereas one of the models[2] accounts for through-the-thickness conduction, none of them address the problem of thermal transport along the fiber direction and what role, if any, thermal stesses play in the failure of composite cylinders. Although these factors may not be important in problems with rapid ablation, it seems possible that there exists a large number of engagement scenarios in which thermal issues are key to the outcome.

Problem Description

We now give a detailed description of the problem to be studied in the present work. The problem is to calculate the temperature and stresses in a semi-infinite layered Bernoulli–Euler beam consisting of two or more bonded dissimilar materials. Each individual layer is assumed to be isotropic, homogeneous, and linearly elastic. Each layer is assumed to be in perfect thermal and mechanical contact with neighboring layers. In general, each layer will have different thermal and mechanical material properites from adjacent layers. The beam is freely supported with no applied loads or tractions and with no residual stresses. The top and bottom of the beam are insulated, the temperature is held constant at infinity, and the exposed end is subjected to a known heat flux beginning at time zero. Initially, the beam is at a uniform temperature throughout. The problem is assumed to be quasisteady state which basically implies that whereas the temperature is a function of time, mechanical inertia terms in the equations of equilibrium are negligible. Therefore, time is merely a parameter in the equations for stress, strain, and displacement. Finally, another key assumption is that the volumetric expansion occurs on a long enough time scale that it does not cause the temperature to change. This final assumption allows the equilibrium and energy equations to be decoupled. The temperature and stress distributions may then be calculated separately with temperature changes preceding stresses.

Because of different temperature distributions and coefficients of thermal expansion in the two layers, interlaminar shear and normal stresses will occur along the interface of the two materials. Should these stresses exceed the strength of the interlaminar bond, failure will occur along the interface, initiating the process of delamination. Axial stresses due to bending and the interlaminar shearing force will also occur. We desire a simple engineering solution to this problem for temperature distributions representative of those caused when a laser beam strikes the exposed end ($x = 0$) of the beam. All temperatures are relative to the temperature at which the beam is stress-free, which is taken to be zero for convenience. Since the beam is of unit width, all forces and moments will be per unit width.

Earlier Work

Several elasticity solutions to the problem of isothermal bonded dissimilar materials have been obtained. Solutions of this type were obtained by Dundurs [3,4], Bogy [5,6], and others. They are typically derived from Airy stress function solutions to the problem of bonded dissimilar quarter-spaces. These solutions all indicate a singularity in the interlaminar stresses at the free-edge contact point, which, for brevity, will be referred to as the corner. None of the classical singularity solutions appear to be applicable to layered beams with nonuniform temperature distributions. Exact solutions to several approximations to the isothermal elasticity problem have been obtained. A classical solution of this type is the one due to Pipes and Pagano [7]; others include those of Puppo and Evensen [8], Goland and Reissner [9] and others. Various solutions have been obtained for finite layered beams, strips, and plates with uniform temperature change in all layers. The classical engineering solution of this type is Timoshenko's bimetallic thermostat solution [10]; others include those of Chen et al. [11], Williams [12], and Suhir [13]. Very few solutions have been obtained for problems with nonuniform temperature distributions. Ochoa and Marcano [14] considered beams with temperature being linear and sinusoidal in the longitudinal direction but with no transverse variation. They also considered temperature distributions given to be quadratic in both directions. Williams [15] and Yin [16] studied layered beams with through-the-thickness temperature variation only.

None of the referenced solutions appear to apply to the semi-infinite layered beam subjected to a constant laser heat flux at the exposed end. The temperature in this case is not uniform, linear, or even harmonic. In fact, under the current assumptions, the temperature is given by the insulated rod solution which contains error functions. It turns out, however, that the solution technique derived by Suhir [13] can be applied to the problem at hand provided certain assumptions are valid. These assumptions are discussed next.

Problem Solution

Following the work by Suhir [13], we seek an elementary strength of materials type of solution as opposed to an exact elasticity solution. A free body diagram of a section of the beam (cut at $\xi = x$) is shown in Fig. 1. Moment equilibrium of this section requires that

$$\mathcal{M}_1(x) + \mathcal{M}_2(x) = -h/2\, Q(x) \tag{1}$$

where $h = h_1 + h_2$. The radii of curvature of the two layers must be equal and are related to the bending moments by the standard formulas

$$\mathcal{M}_1(x) = E_1 I_1/\rho(x) \tag{2}$$
$$\mathcal{M}_2(x) = E_2 I_2/\rho(x) \tag{3}$$

where E is Young's modulus and I is the moment of inertia. By substituting Eqs. (2) and (3) into Eq. (1), the radius of curvature is found to be

$$\frac{1}{\rho(x)} = -\frac{h}{2}\left(\frac{1}{E_1 I_1 + E_2 I_2}\right) Q(x) \tag{4}$$

A system of interlaminar distributed forces must be present to hold the layers together, as shown in Fig. 1. Note that, although the figure appears to represent these forces as uniformly distributed, no assumption at all has been made regarding their distribution.

The section must be in force and moment equilibrium. Force equilibrium implies

$$Q(x) = \int_0^x q(\xi)\,\mathrm{d}\xi \tag{5}$$
$$V(x) = -P_0 + \int_0^x p(\xi)\,\mathrm{d}\xi \tag{6}$$

The traction-free boundary condition at $x = \infty$ requires that the cross-sectional shear force $V(x)$ be zero there. In the derivation of the interlaminar normal stress distribution $p(\xi)$ there is no requirement that it be self-equilibrating. In the event it is not, some other type of load system must be present to satisfy equilibrium. Since a great deal of research into the problem of thermal stresses in bonded dissimilar materials indicates a singularity in interlaminar normal stress at the corner, it seems appropriate to allow for the existence of a pair of concentrated loads P_0 acting at the origin of each strip. Since $V(\infty) \to 0$,

$$P_0 = \int_0^\infty p(\xi)\,\mathrm{d}\xi \tag{7}$$

Considering Eq. (5) and the traction-free requirement at ∞, it appears there is no similar load in the x direction. Moment equilibrium will be considered later, when the equations for the interlaminar normal stresses are developed.

To prevent dislocations from occurring along the interface the strain must be compatible. The interface strains of the two beams are given as follows

$$\epsilon_1 = \epsilon_{1_Q} + \epsilon_{1_q} + \epsilon_{1_B} + \epsilon_{1_T} \tag{8}$$

$$\epsilon_2 = \epsilon_{2_Q} + \epsilon_{2_q} + \epsilon_{2_B} + \epsilon_{2_T} \tag{9}$$

where ϵ_Q is the neutral axis strain due to the average force $Q(x)$, ϵ_q is a "correction term" due to the nonuniformity in the y direction of the distributed lateral load, ϵ_B is the strain due to bending, and ϵ_T is the neutral axis thermal strain. From simple beam theory we have

$$\begin{aligned} \epsilon_{1_Q} &= -\frac{Q(x)}{h_1 E_1}, & \epsilon_{2_Q} &= \frac{Q(x)}{h_2 E_2} \\ \epsilon_{1_B} &= \frac{h_1}{2\rho(x)}, & \epsilon_{2_B} &= -\frac{h_2}{2\rho(x)} \\ \epsilon_{1_T} &= \alpha_1 T_1(x), & \epsilon_{2_T} &= \alpha_2 T_2(x) \end{aligned} \tag{10}$$

where α is the coefficient of linear thermal expansion. Suffice it to say that the correction term ϵ_q is derived from the Ribière solution for a long and narrow strip and is related to the interlaminar shear stress by

$$\epsilon_{i_q}(x) = \kappa_i q'(x) \tag{11}$$

where the prime indicates differentiation and

$$\kappa_i = 2h_i/(3E_i) \tag{12}$$

The interested reader is referred to Suhir's paper [13] for the details of this development.

Suhir's solution is for finite strips with a uniform temperature increase. The concept of an interfacial shear compliance coefficient is critical because it allows the shear force $Q(x)$ to be defined by a differential equation, allowing the enforcement of zero normal stress (σ_{xx}) on the ends of the strip. Whereas it does not appear necessary to modify Suhir's technique for the problem at hand, it is important to defend the use of the same interfacial shear compliance coefficient. The interfacial shear compliance coefficient κ was shown earlier to apply to a finite bimaterial strip with a temperature distribution symmetric about the midlength of the strip. The problem of interest in the current study concerns a semi-infinite strip with a decaying temperature distribution. It is, therefore, appropriate to consider whether κ applies in this case. If both $T_1(x)$ and $T_2(x)$ decay to zero as $x \to \infty$, there is some value of x, say x_∞, for which all stresses are negligible for $x > x_\infty$. Note that x_∞ may be chosen such that the stresses are as negligible as desired. Consider now a mirror image of the same beam. If $T_1(-x) = T_1(x)$ and $T_2(-x) = T_2(x)$, the preceding comments hold for this beam as well. The two beams are identical to each other except they are semi-inifinite in opposite directions. By taking hypothetical cuts of each at the same distance from the free edge, we ensure continuity of stresses, temperature and their derivatives along the cut face. Cutting them at x_∞ and rejoining the cut surfaces produces a finite bimaterial beam with a symmetric temperature distribution. Therefore, Suhir's value of κ is applicable to the problem at hand.

Substituting Eqs. (10) into Eqs. (8) and (9), we obtain

$$\epsilon_1 = -Q(x)/(h_1E_1) + \kappa_1 q'(x) + \alpha_1 T_1(x) + h_1/2p(x) \quad (13)$$

$$\epsilon_2 = Q(x)/(h_2E_2) - \kappa_2 q'(x) + \alpha_2 T_2(x) - h_2/2p(x) \quad (14)$$

Equating Eq. (13) and (14) and substituting Eq. (4) into the result, we obtain

$$-\lambda Q(x) + \kappa q'(x) = \alpha_2 T_2(x) - \alpha_1 T_1(x) \quad (15)$$

where $\kappa = \kappa_1 + \kappa_2$ and

$$\lambda = \frac{1}{h_1E_1} + \frac{1}{h_2E_2} + \frac{h^2}{4}\left(\frac{1}{E_1I_1 + E_2I_2}\right) \quad (16)$$

By differentiating Eq. (5) and substituting the result into Eq. (15), the following ordinary differential equation in $Q(x)$ results.

$$Q''(x) - k^2Q(x) = \alpha_2/\kappa\, T_2(x) - \alpha_1/\kappa\, T_1(x) \quad (17)$$

where

$$k^2 = \frac{\lambda}{\kappa} = \frac{3}{2h_1^2}\left[\frac{B+1}{B+\hat{h}^2}\right] \quad (18)$$

with $B = \hat{E}\hat{h}$. Since the ends of the beam are specified to be traction free, $Q(x)$ must be zero there. Note that this condition will force only the axial stress, σ_{xx}, to be zero. The necessary boundary conditions are, therefore,

$$Q(0) = 0 \quad (19)$$

$$\lim_{x\to\infty} Q(x) = 0 \quad (20)$$

From Eq. (18) we see that k^2 is strictly positive. The solution to Eq. (17) is, therefore,

$$Q(x) = C_1e^{kx} + C_2e^{-kx} + Q_p(x) \quad (21)$$

where C_1 and C_2 are constants to be determined by applying the boundary conditions, and $Q_p(x)$ is a particular solution to Eq. (17). Applying Eqs. (19) and (20) to Eq. (21), we obtain

$$C_1 = -\lim_{x\to\infty}\left[Q_p(x)e^{-kx}\right] \quad (22)$$

$$C_2 = -Q_p(0) + \lim_{x\to\infty}\left[Q_p(x)e^{-kx}\right] \quad (23)$$

The interlaminar shear stress $q(x)$ is obtained by differentiating Eq. (5)

$$q(x) = Q'(x) = k\left(C_1e^{kx} - C_2e^{-kx}\right) + Q_p'(x) \quad (24)$$

The interlaminar normal stress, or "peeling" stress, may be obtained by considering the moment equilibrium of one of the beams shown in Fig. 1.

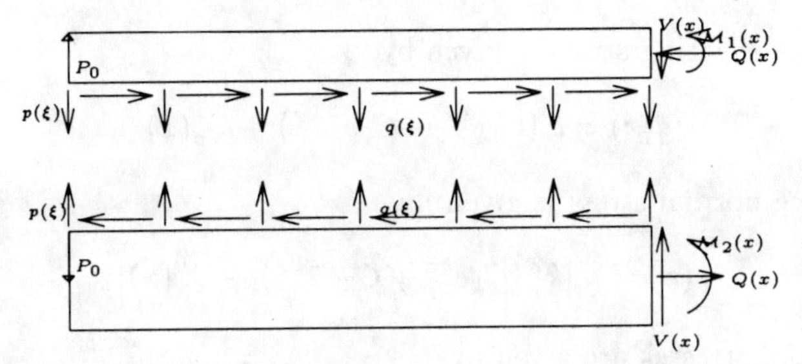

Fig. 1. Free-body diagram of the bimaterial beam.

Moment equilibrium of strip 1 requires that

$$\mathcal{M}_1(x) + \int_0^x \left[\int_0^\xi p(\xi')\, d\xi' \right] d\xi + \frac{h_1}{2} Q(x) - P_0 x = 0 \tag{25}$$

Substituting Eqs. (2) and (4) into Eq. (25) yields

$$\int_0^x \left[\int_0^\xi p(\xi')\, d\xi' \right] d\xi = \mu Q(x) + P_0 x \tag{26}$$

where

$$\mu = \frac{h}{2} \frac{E_1 I_1}{E_1 I_1 + E_2 I_2} - \frac{h_1}{2} \tag{27}$$

Differentiating Eq. (26) twice, we obtain the normal stress

$$p(x) = \mu Q''(x) = \mu \left[k^2 \left(C_1 e^{kx} + C_2 e^{-kx} \right) + Q_p''(x) \right] \tag{28}$$

We have yet to determine the "important" stress derived from simple beam theory, that is, the axial stress, which is given by

$$\sigma_x(x,y) = \begin{cases} -\dfrac{Q(x)}{h_1} + \dfrac{\mathcal{M}_1(x) y_1}{I_1}, & 0 < y \le h_1 \\ \dfrac{Q(x)}{h_2} + \dfrac{\mathcal{M}_2(x) y_2}{I_2}, & 0 > y \ge -h_2 \end{cases} \tag{29}$$

where

$$y_1 = h_1/2 - y \tag{30}$$

$$y_2 = -(h_2/2 + y) \tag{31}$$

The axial stresses are found by substituting Eqs. (2), (3), (30), and (31) into Eq. (29). All desired stresses have now been determined and the strength of materials solution to the problem is complete. The interface shearing force is given by

$$Q(x) = C_1 e^{kx} + C_2 e^{-kx} + Q_p(x) \tag{32}$$

the interface shearing stress is given by

$$q(x) = k\left(C_1 e^{kx} - C_2 e^{-kx}\right) + Q_p'(x) \tag{33}$$

the interface normal stress is given by

$$p(x) = \mu\left[k^2\left(C_1 e^{kx} + C_2 e^{-kx}\right) + Q_p''(x)\right] \tag{34}$$

and the axial stresses are given by

$$\sigma_{x_1} = -Q(x)\left\{\frac{1}{h_1} + \frac{hE_1(h_1/2 - y)}{2(E_1I_1 + E_2I_2)}\right\} \tag{35}$$

$$\sigma_{x_2} = Q(x)\left\{\frac{1}{h_2} + \frac{hE_2(h_2/2 + y)}{2(E_1I_1 + E_2I_2)}\right\} \tag{36}$$

The constants are given by

$$C_1 = -\lim_{x\to\infty}\left[Q_p(x)e^{-kx}\right] \tag{37}$$

$$C_2 = -Q_p(0) + \lim_{x\to\infty}\left[Q_p(x)e^{-kx}\right] \tag{38}$$

A concentrated force,

$$P_0 = \lim_{x\to\infty}\frac{C_1}{k}e^{kx} + \frac{1}{k}(C_2 - C_1) + \int_0^\infty Q_p(\xi)\,\mathrm{d}\xi \tag{39}$$

is found to act at the corner on the heated end of the beam. There is a tendency to require C_1 to be zero due the stress-free requirements at ∞. However, this cannot be properly addressed until the specific form of the particular solution $Q_p(x)$ is developed for a problem. It could very well be that the particular solution contains an exponential term, the presence of which cannot be eliminated at ∞ without the C_1 term.

Consider a semi-infinite bimaterial beam, initially at zero temperature everywhere, which is subjected at time zero to the thermal boundary conditions depicted in Fig. 2. The hash marks in the figure indicate insulated boundaries. Note that the interface is assumed to be an insulating layer, which causes T_1 and T_2 to be functions of x and t only, where t is time. The temperature distribution in this beam is given by the well-known insulated rod solution

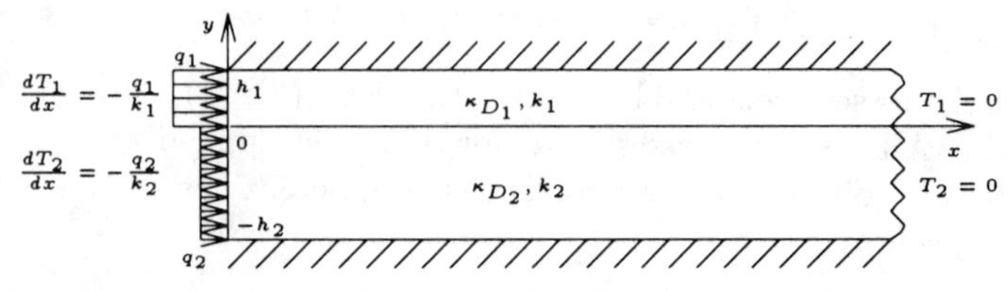

Fig. 2. Flux heated bimaterial semi-infinite beam.

$$T_1(x,t) = \frac{q_1 b}{k_1}\left\{\frac{1}{\sqrt{\pi}}e^{-\hat{x}^2} - \hat{x}(1-\operatorname{erf}\hat{x})\right\} \tag{40}$$

$$T_2(x,t) = \frac{q_2 b\hat{b}}{k_2}\left\{\frac{1}{\sqrt{\pi}}e^{-\tilde{x}^2} - \tilde{x}(1-\operatorname{erf}\tilde{x})\right\} \tag{41}$$

where

$$\hat{x} = \frac{x}{b}, \qquad \tilde{x} = \frac{\hat{x}}{\hat{b}}, \qquad b = 2\sqrt{\kappa_{D_1} t}, \qquad \hat{b} = \sqrt{\frac{\kappa_{D_2}}{\kappa_{D_1}}}$$

$$\operatorname{erf} x = \frac{2}{\sqrt{\pi}}\int_0^x e^{-\eta^2}\,d\eta$$

and k_i and κ_{D_i} are thermal conductivity and thermal diffusivity, respectively, of the ith strip. Substituting Eqs. (40) and (41) into Eq. (17), we obtain

$$Q''(x) - k^2 Q(x) = A_2\left\{\frac{1}{\sqrt{\pi}}e^{-\tilde{x}^2} - \tilde{x}(1-\operatorname{erf}\tilde{x})\right\} - A_1\left\{\frac{1}{\sqrt{\pi}}e^{-\hat{x}^2} - \hat{x}(1-\operatorname{erf}\hat{x})\right\} \tag{42}$$

where $A_1 = \alpha_1 q_1 b/(\kappa k_l)$ and $A_2 = \alpha_2 q_2 b\hat{b}/(\kappa k_2)$. The particular solution is obtained using the method of variation of parameters and is found to be

$$\begin{aligned}\hat{Q}_p(\hat{x},\tilde{x}) = &\left\{\frac{1}{\sqrt{\pi}}e^{-\hat{x}^2} - \hat{x}(1-\operatorname{erf}\hat{x})\right\} \\ &-\hat{A}\left\{\frac{1}{\sqrt{\pi}}e^{-\tilde{x}^2} - \tilde{x}(1-\operatorname{erf}\tilde{x})\right\} \\ &-\frac{1}{4\phi}e^{\phi^2}\left[e^{2\phi\hat{x}}\operatorname{erf}(\hat{x}+\phi) - e^{-2\phi\hat{x}}\operatorname{erf}(\hat{x}-\phi)\right] \\ &+\frac{1}{4\phi}\frac{\hat{A}}{\hat{b}}e^{\hat{b}^2\phi^2}\left[\begin{array}{l} e^{2\hat{b}\phi\tilde{x}}\operatorname{erf}(\tilde{x}+\hat{b}\phi) - \\ e^{-2\hat{b}\phi\tilde{x}}\operatorname{erf}(\tilde{x}-\hat{b}\phi)\end{array}\right]\end{aligned} \tag{43}$$

where $\phi = bk/2$ and $\hat{Q}_p(\hat{x},\tilde{x}) = Q_p(x)\left(k^2/A_1\right)$. Using Eqs. (22) and (23), the constants for the general solution are found to be

$$C_1 = \left(\frac{A_1 b^2}{16\phi^3}\right)\left[e^{\phi^2} - \frac{\hat{A}}{\hat{b}}e^{\hat{b}^2\phi^2}\right] \tag{44}$$

$$C_2 = \left(\frac{A_1 b^2}{16\phi^3}\right)\left[\begin{array}{l} -\frac{4\phi}{\sqrt{\pi}}(1-\hat{A}) - (1-2\operatorname{erf}\phi)e^{\phi^2} \\ +(1-2\operatorname{erf}\hat{b}\phi)\frac{\hat{A}}{\hat{b}}e^{\hat{b}^2}\phi^2\end{array}\right] \tag{45}$$

The general solution to the problem is then found to be

$$\hat{Q}(\hat{x},\tilde{x}) = \frac{1}{4\phi}e^{2\phi\hat{x}}\left\{e^{\phi^2}\operatorname{erfc}(\hat{x}+\phi) - \frac{\hat{A}}{\hat{b}}e^{\hat{b}^2\phi^2}\operatorname{erfc}(\tilde{x}+\hat{b}\phi)\right\}$$

$$+\frac{1}{4\phi}e^{-2\phi\hat{x}}\left\{\begin{array}{l} 2\left[\operatorname{erf}\phi\, e^{\phi^2} - \frac{\hat{A}}{\hat{b}}\operatorname{erf}\hat{b}\phi\, e^{\hat{b}^2\phi^2}\right] - \\ e^{\phi^2}\operatorname{erfc}(\hat{x}-\phi) + \frac{\hat{A}}{\hat{b}}e^{\hat{b}^2\phi^2}\operatorname{erfc}(\tilde{x}-\hat{b}\phi) \\ -\frac{4\phi}{\sqrt{\pi}}(1-\hat{A}) \end{array}\right\}$$

$$+\left\{\frac{1}{\sqrt{\pi}}\left[e^{-\hat{x}^2} - \hat{A}e^{-\tilde{x}^2}\right] - \left[\hat{x}\operatorname{erfc}\hat{x} - \hat{A}\tilde{x}\operatorname{erfc}\tilde{x}\right]\right\} \tag{46}$$

$$\hat{q}(\hat{x},\tilde{x}) = \frac{1}{4\phi}e^{2\phi\hat{x}}\left\{e^{\phi^2}\operatorname{erfc}(\hat{x}+\phi) - \frac{\hat{A}}{\hat{b}}e^{\hat{b}^2\phi^2}\operatorname{erfc}(\tilde{x}+\hat{b}\phi)\right\}$$

$$-\frac{1}{4\phi}e^{-2\phi\hat{x}}\left\{\begin{array}{l} 2\left[\operatorname{erf}\phi\, e^{\phi^2} - \frac{\hat{A}}{\hat{b}}\operatorname{erf}\hat{b}\phi\, e^{\hat{b}^2\phi^2}\right] - \\ e^{\phi^2}\operatorname{erfc}(\hat{x}-\phi) + \frac{\hat{A}}{\hat{b}}e^{\hat{b}^2\phi^2}\operatorname{erfc}(\tilde{x}-\hat{b}\phi) \\ -\frac{4\phi}{\sqrt{\pi}}(1-\hat{A}) \end{array}\right\}$$

$$-\frac{1}{2\phi}\left[\operatorname{erfc}\hat{x} - \frac{\hat{A}}{\hat{b}}\operatorname{erfc}\tilde{x})\right] \tag{47}$$

$$\hat{p}(\hat{x},\tilde{x}) = \frac{1}{4\phi}e^{2\phi\hat{x}}\left\{e^{\phi^2}\operatorname{erfc}(\hat{x}+\phi) - \frac{\hat{A}}{\hat{b}}e^{\hat{b}^2\phi^2}\operatorname{erfc}(\tilde{x}+\hat{b}\phi)\right\}$$

$$+\frac{1}{4\phi}e^{-2\phi\hat{x}}\left\{\begin{array}{l} 2\left[\operatorname{erf}\phi\, e^{\phi^2} - \frac{\hat{A}}{\hat{b}}\operatorname{erf}\hat{b}\phi\, e^{\hat{b}^2\phi^2}\right] - \\ e^{\phi^2}\operatorname{erfc}(\hat{x}-\phi) + \frac{\hat{A}}{\hat{b}}e^{\hat{b}^2\phi^2}\operatorname{erfc}(\tilde{x}-\hat{b}\phi) \\ -\frac{4\phi}{\sqrt{\pi}}(1-\hat{A}) \end{array}\right\} \tag{48}$$

$$\hat{P}_0 = \frac{-1}{\sqrt{\pi}}(1-\hat{A}) + \frac{1}{2\phi}\left\{\begin{array}{l} 1 - e^{\phi^2}\operatorname{erfc}\phi - \\ \frac{\hat{A}}{\hat{b}}(1 - e^{\hat{b}^2\phi^2}\operatorname{erfc}\hat{b}\phi) \end{array}\right\} \tag{49}$$

where

$$\hat{Q}(\hat{x},\tilde{x}) = Q(x)\left(k^2/A_1\right)$$

$$\hat{q}(\hat{x},\tilde{x}) = q(x)\left(k/A_1\right)$$

$$\hat{p}(\hat{x},\tilde{x}) = p(x)\left(1/A_1\mu\right)$$

$$\hat{P}_0 = k\int_0^\infty \hat{p}(\xi)\,\mathrm{d}\xi$$

$$\operatorname{erfc} x = 1 - \operatorname{erf} x$$

It is generally accepted that beam theory is not applicable very near bounding surfaces. There is a boundary layer near the free edge where beam theory solutions are to be used with great caution. Kuo [17] showed the length of this region to be less than three times the thickness of the thinner layer for bimaterial beams. This is on the order of the total beam thickness. For the purpose of evaluating the length of the boundary-layer region as a function of time, let the boundary layer be defined as that region of the beam within one total thickness of the free edge. If δ denotes the boundary-layer length, we have

$$\delta = h/b \tag{50}$$

Using the relation $\phi = bk/2$ and Eq. (18), this becomes

$$\delta = \frac{1}{2\phi}(1+\hat{h})\sqrt{\frac{3}{2}}\sqrt{\frac{B+1}{B+\hat{h}^2}} \tag{51}$$

Results

Equations (46) through (49) are plotted in Figs. 3 through 6 for various values of the parameter ϕ with $\hat{A} = 0$. The ratio of the axial stresses $\hat{\sigma}_1$ and $\hat{\sigma}_2$ to $\hat{Q}$ are shown in Figs. 7 and 8. Also, since the maximum values of $\hat{q}$ and $\hat{p}$ occur at $\hat{x} = \tilde{x} = 0$(See Suhir [13] for a discussion of the maximum stresses), they are of special interest and are found from Eqs. (47) and (48)

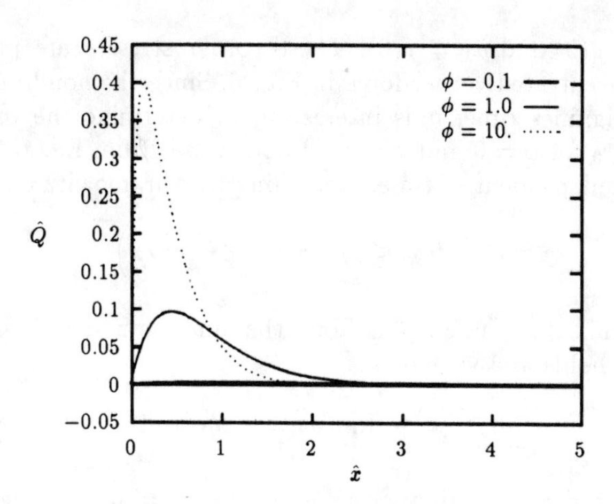

Fig. 3. Nondimensional resultant axial force, Eq. (46), with T_1 given by Eq. (40) and $T_2 = 0$; plotted for various values of the nondimensional time $\phi = k\sqrt{\kappa_{D_1} t}$.

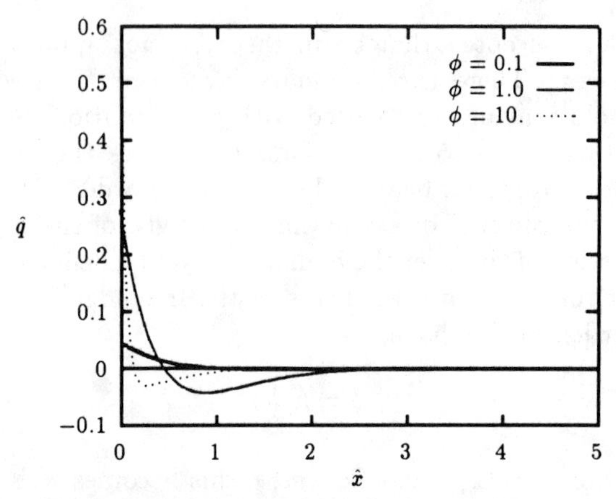

Fig. 4. Nondimensional interlaminar shearing stress, Eq. (47), with T_1 given by Eq. (40) and $T_2 = 0$; plotted for various values of the nondimensional time $\phi = k\sqrt{\kappa_{D_1} t}$.

to be

$$\hat{q}_0 = \frac{1}{2\phi}\left\{e^{\phi^2}\operatorname{erfc}\phi - \frac{\hat{A}}{\hat{b}}e^{\hat{b}^2\phi^2}\operatorname{erfc}\hat{b}\phi - \left(1 - \frac{\hat{A}}{\hat{b}}\right)\right\} + \frac{1}{\sqrt{\pi}}(1 - \hat{A}) \tag{52}$$

$$\hat{p}_0 = \frac{-1}{\sqrt{\pi}}(1 - \hat{A}) \tag{53}$$

where $\hat{q}_0 = \hat{q}(0)$ and $\hat{p}_0 = \hat{p}(0)$. These corner stresses are plotted along with the concentrated corner force in Fig. 6. Since the nondimensional parameter ϕ signifies time, it is interesting to determine the magnitude of $\hat{P}_0, \hat{q}_0$, and $\hat{p}_0$ for $\phi = 0$ and $\phi = \infty$. First, we see from Eq. (53) that $\hat{p}_0$ is apparently independent of time. According to Abramowitz and Stegun [18],

$$e^{\phi^2}\operatorname{erfc}\phi \le 1/\left(\phi + \sqrt{\phi^2 + 4/\pi}\right)$$

Therefore, $\lim_{\phi\to\infty} e^{\phi^2}\operatorname{erfc}\phi \le 0$. Since the function is strictly nonnegative, the equality holds and we have

$$\lim_{\phi\to\infty}\hat{P}_0 = -\lim_{\phi\to\infty}\hat{q}_0 = -1/\sqrt{\pi}\,(1 - \hat{A}) \tag{54}$$

Now, in to order to evaluate the limits at $x = 0$, we must resort to L'Hôpital's rule. It is found that both $\hat{q}_0$ and $\hat{P}_0$ approach 0 as $\phi \to 0$.

Equation (51) is plotted in Figs. 9 and 10 as a function of the nondimensional time ϕ for various values of the ratio of stiffnesses B. If $\hat{h} =$

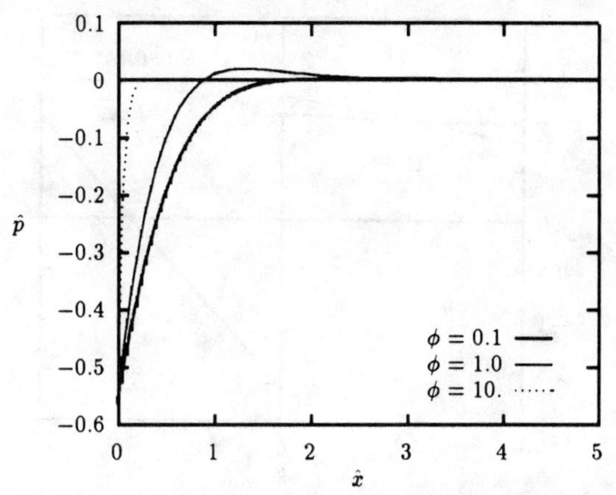

Fig. 5. Nondimensional interlaminar peeling stress, Eq. (48), with T_1 given by Eq. (40) and $T_2 = 0$; plotted for various values of the nondimensional time $\phi = k\sqrt{\kappa_{D_1} t}$.

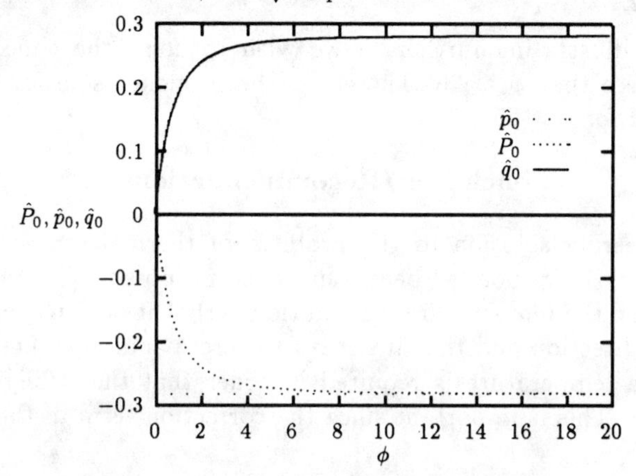

Fig. 6. Concentrated corner force $\hat{P}_0$ [Eq. (49)], corner stresses, $\hat{p}_0$ and $\hat{q}_0$ [Eqs. (53) and (52)]; plotted as a function of ϕ, with $\hat{A} = \hat{b} = 1/2$.

1, then $\delta \neq \delta(B)$. The resulting value of δ is then a lower bound. Also, referring to Fig. 9, it appears that δ is not a strong function of B for $\hat{h} \neq 1$. Choosing $B = 0$, we obtain an upper bound on δ. Therefore, δ must be bounded as follows.

$$\frac{1}{\phi}\sqrt{\frac{3}{2}} \quad \leq \delta \leq \quad \frac{1}{2\phi}\left(1 + \frac{1}{\hat{h}}\right)\sqrt{\frac{3}{2}} \tag{55}$$

To illustrate the utility of the bounds on δ, suppose $\phi = 1$ and $\hat{h} = 1/3$. Then $\delta \approx 2.5$. Referring to Fig. 4, we see that all of the significant changes in the interlaminar shearing stress occurs in the region $\hat{x} \leq \delta$. And this is within the boundary-layer region where the proposed solution may not

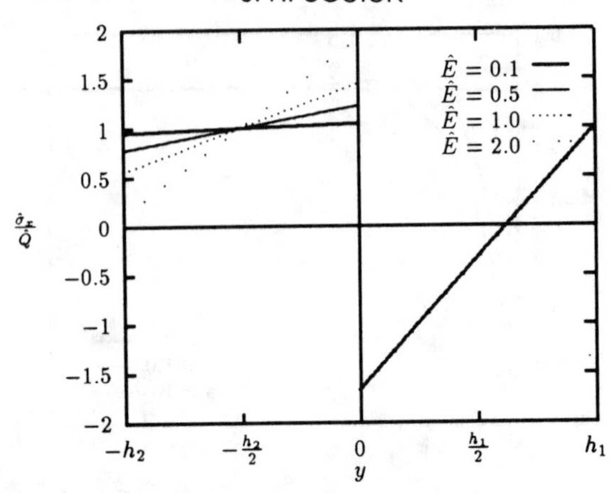

Fig. 7. Nondimensional axial stress, Eqs. (35) and (36), with $\hat{h} = 1/3, v_f = 0.75$; plotted through the thickness for various values of $\hat{E}$.

be reliable. Illustrating further, if we wish to know the time at which $\delta = 0.1$, we see that $\phi \approx 25$. Therefore, the proposed solution would be recommended for $\phi \geq 25$.

Conclusion/Recommendations

An engineering solution to the problem of thermal stresses in semi-infinite (or very long) bonded beams has been proposed. The only restrictions placed on the temperature distribution is that it be continuous in the longitudinal direction and that it decays to zero at infinity. The decaying nature of the temperature is required in order that the Ribière solution be applicable. This is important since the correction term in the interface

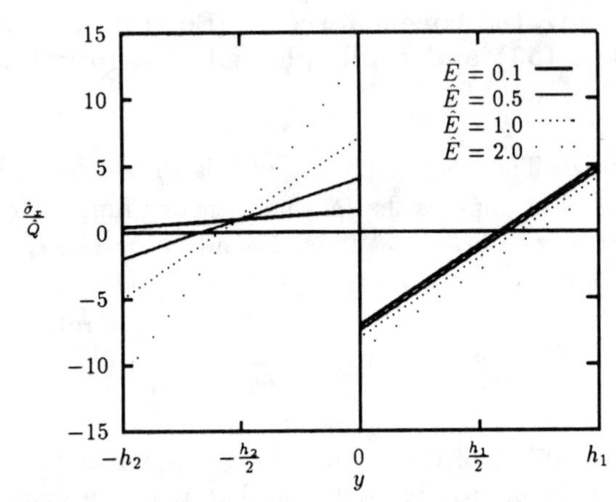

Fig. 8. Nondimensional axial stress, Eqs. (35) and (36), with $\hat{h} = 1, v_f = 0.5$; plotted through the thickness for various values of $\hat{E}$.

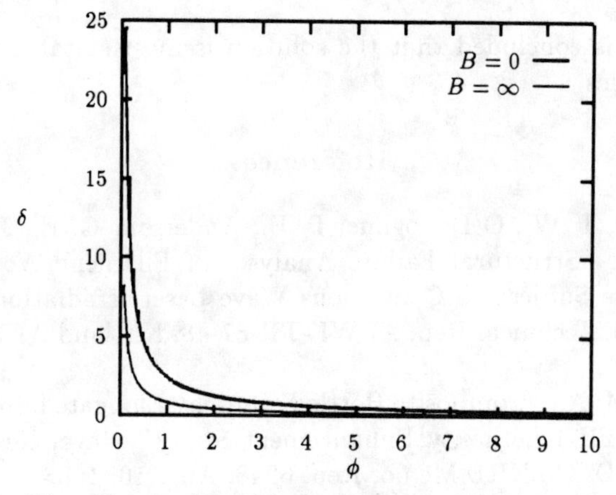

Fig. 9. Nondimensional boundary-layer length δ, with $\hat{h} = 1/3, v_f = 0.75$; plotted as a function of ϕ.

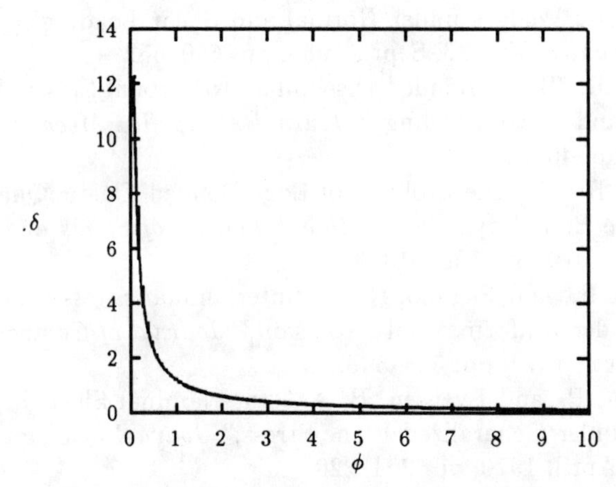

Fig. 10. Nondimensional boundary-layer length δ, with $\hat{h} = 1, v_f = 0.5$; plotted as a function of ϕ.

strain equations is derived from the Ribière solution, and it is the presence of this term which allows formulation of the boundary-value problem in such a manner as to permit enforcement of the stress-free end conditions.

The primary significance of the present work is that is may be used to find closed-form expressions for thermal stresses in layered beams with quite complicated temperature distributions. This is in contrast to most earlier work which was done for problems with uniform or very simple temperature distributions. The solution should be accepted with caution within one boundary-layer length of the free edge. The boundary-layer length was shown to be inversely proportional to the nondimensional time.

Therefore, it is concluded that the solution is most suitable for relatively long time scales.

References

[1]Cardinal, J. W., O'Donoghue, P. E., Anderson, C. E., Jr., and Kanninen, M. F., "Structural Failure Analyses of Filament Wound Pressurized Cylinders Subject to Continuous Wave Laser Irradiation," Air Force Weapons Lab. Technical Rep. AFWL-TR-87-48, Kirtland AFB, NM, June, 1988.

[2]Tamm, M. A., "Composite Bottle Failure by Elongated Spot CW Laser Irradiation LTH-1 Lethality Enhancement Study," Naval Research Lab. , Washington, D. C., NRL Memo. Rep. 6248, Aug. 10, 1988.

[3]Dundurs, J., "Effect of Elastic Constants on Stress in a Composite Under Plane Deformation," *Journal of Composite Materials,* Vol. 1, July, 1967, pp. 310–322.

[4]Dundurs, J., "Discussion of D. B. Bogy's 'Edge-Bonded Dissimilar Orthogonal Elastic Wedges under Normal and Shear Loading', " *Journal of Applied Mechanics,* Vol. 36, Sept., 1969, pp. 650–652.

[5]Bogy, D. B., "Edge-Bonded Dissimilar Orthogonal Elastic Wedges Under Normal and Shear Loading," *Journal of Applied Mechanics,* Vol. 35, Sept., 1968, pp. 460–466.

[6]Bogy, D. B., "On the Problem of Edge-Bonded Elastic Quarter-Planes Loaded at the Boundary," *International Journal of Solids and Structures,* Vol. 6, No. 9, 1970, pp. 1287–1313.

[7]Pipes, R. B., and Pagano, N. J., "Interlaminar Stresses in Composite Laminates under Uniform Axial Extension," *Journal of Composite Materials,* Vol. 4, Oct., 1970, pp. 538–548.

[8]Puppo, A. P., and Evensen, H. A., "Interlaminar Shear in Laminated Composites under Generalized Plane Stress," *Journal of Composite Materials,* Vol. 4, April, 1970, pp. 204–220.

[9]Goland, M., and Reissner, E., "The Stresses in Cemented Joints," *Journal of Applied Mechanics,* Vol. 11, March, 1944, pp. A17–A27.

[10]Timoshenko, S. P., "Analysis of Bi-Metal Thermostats," *Journal of the Optical Society of America,* Vol. 11, Sept., 1925, pp. 233–255.

[11]Chen, D., Cheng, S., and Gerhardt T. D., "Thermal Stresses in Laminated Beams," *Journal of Thermal Stresses,* Vol. 5, No. 1, 1982, pp. 67–84.

[12]Williams, H. E., "Asymptotic Analysis of the Thermal Stresses in a Two-Layer Composite with an Adhesive Layer," *Journal of Thermal Stresses,* Vol. 8, No. 2, 1985, pp. 183–203.

[13]Suhir, E., "Stresses in Bi-Metal Thermostats," *Journal of Applied Mechanics,* Vol. 53, Sept. 1986, pp. 657–660.

[14]Ochoa, O. O., and Marcano, V. M., "Thermal Stresses in Laminated Beams," *International Journal of Solids Structures,* Vol. 20, No. 6, 1984, pp. 579–587.

[15]Williams, H. E., "Thermal Stresses in Bonded Solar Cells-the Effect of the Adhesive Layer," *Journal of Thermal Stresses,* Vol. 6, No. 3, 1983, pp. 231–252.

[16]Yin,W. L., "Effects of Inclined Free Edges on the Thermal Stresses in a Layered Beam," *Journal of Electronic Packaging,* Vol. 115, June, 1993, pp. 208–213.

[17]Kuo, A. Y., "Thermal Stresses at the Edge of a Bimetallic Thermostat," *Journal of Applied Mechanics,* Vol. 56, Sept., 1989, pp. 585–589.

[18]Abramowitz M., and Stegun I. A., *Handbook of Mathematical Functions,* Dover, New York, 1972, p. 298.

Quantification of Uncertainties of Hot-Wet Composite Long-Term Behavior

Christos C. Chamis*
NASA Lewis Research Center
Cleveland, Ohio 44135

Surendra N. Singhal†
NYMA, Inc.
Brook Park, Ohio 44142

Abstract

Formal methods are presented and described which can be used to quantify the uncertainties of hot-wet composite long term behavior. These methods are embedded in a computer code which combines probabilistic composite mechanics with probabilistic finite element structural analysis and with a multifactor interaction equation which represents the combined effects of all factors that influence composite behavior. The quantification of uncertainties is then performed via computational simulation for cumulative probability of failure and dominant sensitivity factors. Simulation results compare well with limited available hot-wet data. Also, typical results show that the cyclic stress and sustained stress can be evaluated for specified probabilities.

Introduction

An important consideration to assure the successful application of polymer matrix composites (PMCs) in aircraft structures is their hot-wet long-term behavior and their respective uncertainties. The traditional approach is long-term testing which requires a large number of tests to capture the effect of all pertinent variables, hence, it is time consuming and costly. The large number of experiments are necessary because of the many factors that influence composite behavior (Fig. 1). An alternative approach is computational simulation supplemented with a few strategically selected experiments. A computer code integrated probabilistic assessment of composite structures (IPACS) has been under development at

---

* Senior Aerospace Scientist, Structures Division.

† Director of Structures, Materials, and Instrumentation.

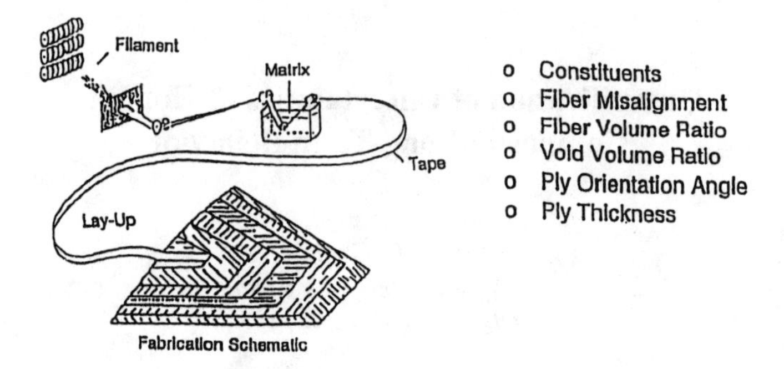

Fig. 1 Source of scatter, fabrication process.

NASA Lewis for the past several years. IPACS consists mainly of two major moduli: probabilistic composite mechanics and probabilistic finite element structural analysis. A block diagram of IPACS is shown in Fig. 2, and a more detailed description is provided in Ref. 1. IPACS integrates the effects of changes in the constituent materials in the structural response progressively through the various scales inherent in composite structures (Fig. 3).

IPACS has been modified to quantify the uncertainties of the hot-wet long-term behavior of PMCs using a modified multifactor interaction model (MFIM). The MFIM represents the stress/time and temperature/moisture nonlinear behavior of PMCs. The objective of this article is to demonstrate IPACS simulation of hot-wet

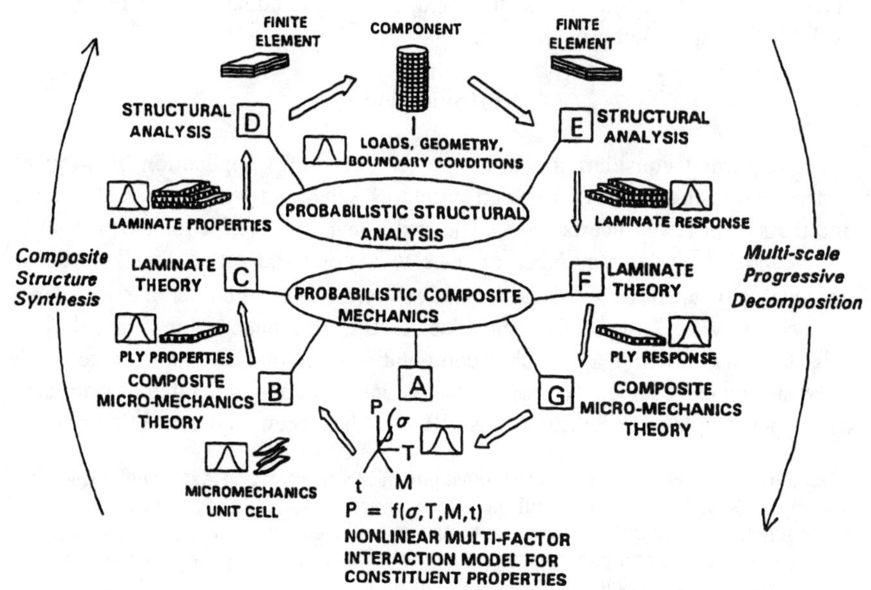

Fig. 2 Integrated probabilistic assessment of composite structures (IPACS).

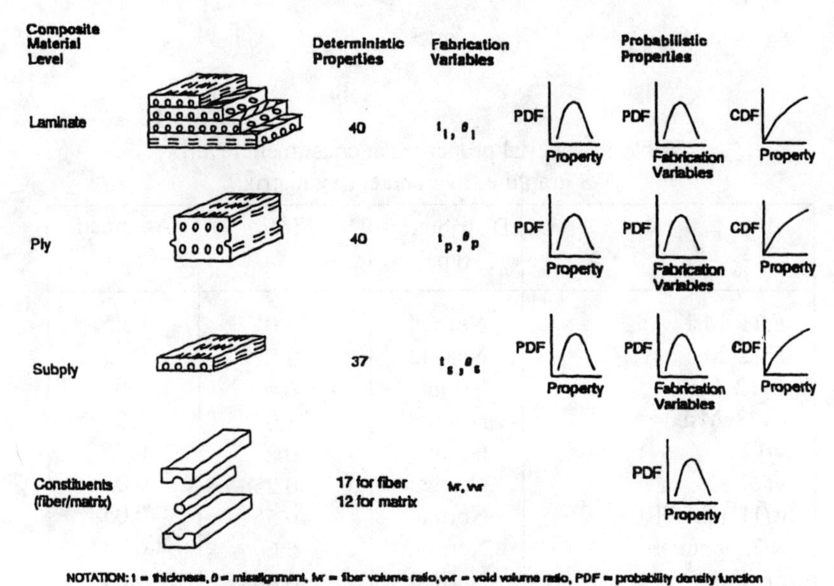

Fig. 3 Multilevel simulation of probabilistic properties of composites.

long-term behavior of composites. The demonstration cases provide results for material properties at all composite scales (fiber/matrix, ply, and laminate) and strain response for long-time exposure of PMCs to sustained stress levels.

The specific cases include moduli and strengths. The results are presented in terms of cumulative probability density functions and their respective sensitivity factors. Most of the results included are for unidirectional composites because experimental data was available for comparisons. Limited room temperature and combined cyclic loadings with sustained stress results are also shown for some laminates to demonstrate the effectiveness of the probabilistic simulation in IPACS.

Probabilistic Simulation of Select Room Temperature Composite Properties

A few comparisons are in order to show that the probabilistic simulation bounds experimental data. The comparisons are for AS-graphite fiber/epoxy matrix laminate. The constituent material properties and other information required for the probabilistic simulation are listed in Tables 1–3. The constituent material properties are listed in Table 1. Those for fabrication ply or laminate variables are listed in Table 2. The boundary conditions variables for the structure are listed in Table 3. The values in these tables are nominal means and assumed scatter.

Comparisons with moduli of an angle ply laminate are shown in Fig. 4. The experimental data is within the predicted scatter for all three moduli. Probabilistic

Table 1 Material properties at constituent level, A-S graphite fiber and epoxy matrix

	Distribution type	Mean	Assumed scatter
*Ef*11 Msi	Normal	31.0	0.05
*Ef*22 Msi	Normal	2.0	0.05
*Gf*12 Msi	Normal	2.0	0.05
*Gf*23 Msi	Normal	1.0	0.05
νf12	Normal	0.2	0.05
νf23	Normal	0.25	0.05
αf11 ppm/°F	Normal	–0.55	0.05
αf22 ppm/°F	Normal	5.6	0.05
ρf lb/in.^{3}	Normal	0.063	0.05
Nf	Constant	10,000	0.00
df in.	Normal	0.0003	0.05
Cf Btu/in./°F	Normal	0.17	0.05
*Kf*11 Btu in./hr/in.^{2}/°F	Normal	580	0.05
*Kf*22 Btu in./hr/in.^{2}/°F	Normal	58	0.05
*Kf*33 Btu in./hr/in.^{2}/°F	Normal	58	0.05
SfT Ksi	Weibull	400	0.05
SfC Ksi	Weibull	400	0.05
Em Msi	Normal	0.5	0.05
Gm Msi	Normal	0.185	0.05
νm	Normal	0.35	0.05
αm ppm/°F	Normal	42.8	0.05
ρm lb/in.^{3}	Normal	0.0443	0.05
Cm Btu/in./°F	Normal	0.25	0.05
Km Btu in./hr/in.^{2}/°F	Normal	1.25	0.05
SmT Ksi	Weibull	15	0.05
SmC Ksi	Weibull	35	0.05
SmS Ksi	Weibull	13	0.05
βm in./in./1% moist	Normal	0.004	0.05
Dm in.^{3}/s	Normal	0.002	0.05

Table 2 Fabrication variables at ply level

	Distribution type	Mean	Coefficient of variation
fvr	Normal	0.60	0.05
vvr	Normal	0.02	0.05
θp deg	Normal	0.00	0.9 (stdv)
tpsk in.	Normal	0.005	0.05
tpst in.	Normal	0.02	0.05

Table 3 Uncertainties in the structural level

	Distribution type	Mean	Coefficient of variation
KcTR lb/in.	Normal	30E+06	0.20
KcTO lb-in./rad	Normal	12E+02	0.20
FX kips	Normal	288	0.05
Fy kips	Normal	5.76	0.05
Mxx kips-ft	Normal	576	0.05

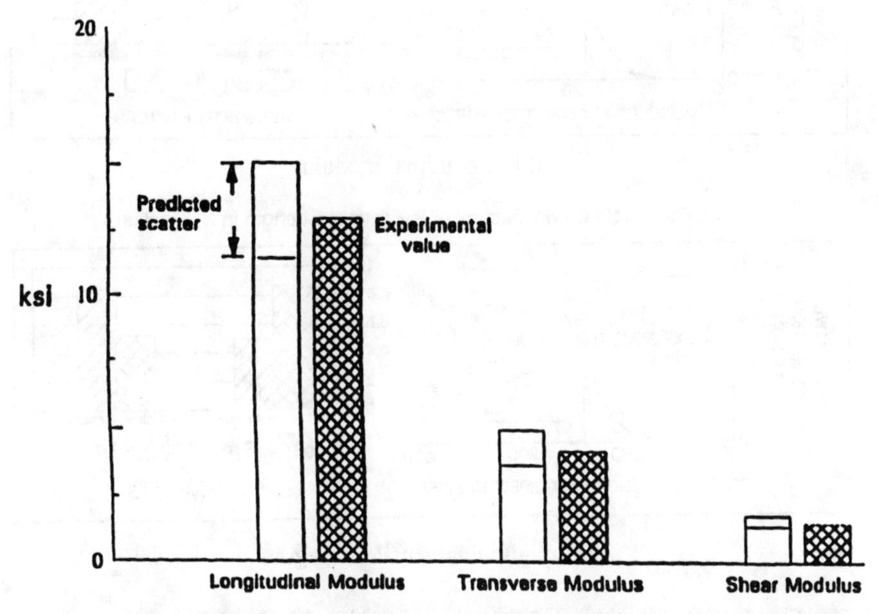

Fig. 4 Probabilistic composite mechanics verified with experimental data $[0_2 \pm 45/0_2/90/01]$.

simulation results are more useful when presented in cumulative distribution function form together with the sensitivity factors of dominant variables. These types of results are shown in Fig. 5a, for room temperature longitudinal modulus and in Fig. 5b, for longitudinal tensile strength. The experimental data (mean) is close to the simulated mean. Of all of the factors listed in Tables 1 and 2, only the longitudinal fiber modulus and the fiber volume ratio influence the ply longitudinal modulus with equal weight at both probability levels. The fiber tensile strength and the fiber volume ratio influence the ply longitudinal strength. However, these factors have different weights and flip-flop for the two different probability levels.

The results described in this section constitute an attestment that the probabilistic simulation in IPACS adequately represent the physics and the mechanics required to account for uncertainties in composite properties.

Probabilistic Simulation of Hot-Wet Composite Properties

The hot-wet properties of unidirectional composites are probabilistically simulated and compared with experimental data from Ref. 2.

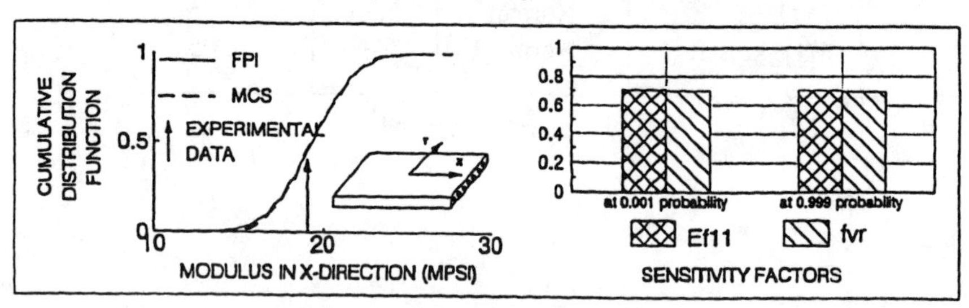

a) Longitudinal modulus

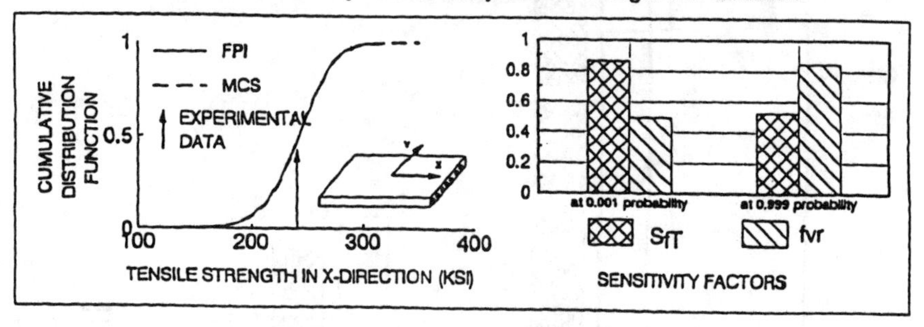

b) Longitudinal tensile strength

Fig. 5 Probabilistic simulation of unidirectional composite properties; As-Graphite fiber/Epoxy; 0.6 mean fiber volume ratio.

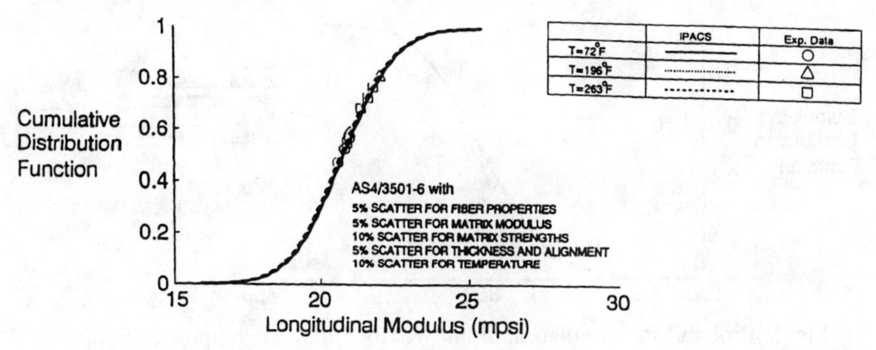

Fig. 6 Probabilistic simulation of temperature effects on composite longitudinal modulus; A-S/E; mean fiber volume ratio.

Longitudinal Modulus

The simulated results and comparisons for this modulus are shown in Fig. 6. The simulated results for the three different temperatures are on the same distribution. Temperature and moisture have no effect on this composite property. The corresponding data fall on the same distribution. This is to be expected since the ply longitudinal modulus is not sensitive to hot-wet environmental conditions. The only scatter in the longitudinal modulus is that due to the scatter in the constituent properties.

Transverse Modulus

The temperature simulated results for the corresponding transverse modulus are shown in Fig. 7. Note that the comparisons for the three different ranges of temperature are in very good agreement. Comparisons for shear modulus and Poisson's ratios show similar agreement, although results are not shown here.

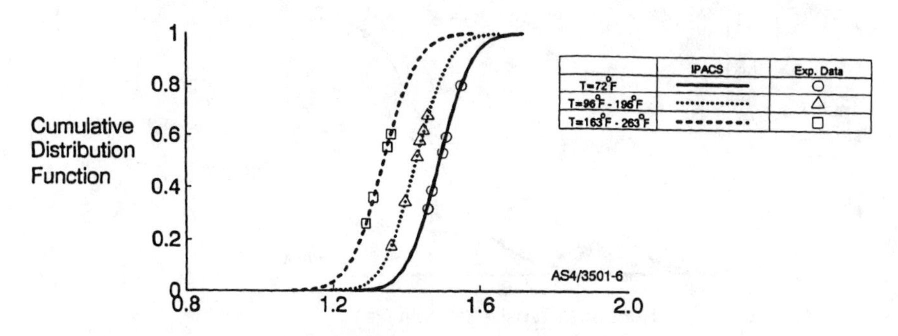

Fig. 7 Probabilistic simulation of hot-wet transverse ply modulus; A-S/E; 0.6 mean fiber volume ratio.

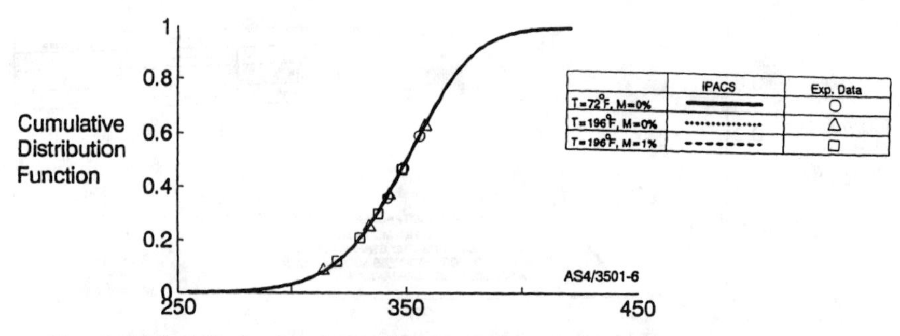

Fig. 8 Probabilistic simulation of hot-wet longitudinal composite strength; A-S/E; 0.6 mean fiber volume ratio.

Longitudinal Tensile Strength

The hot-wet comparison results for this strength are shown in Fig. 8 for three conditions. Both the experimental data and the simulated results fall on the same curve. This is an expected result since the longitudinal tensile strength is insensitive to environmental conditions.

Transverse and Intralaminar Shear Strengths

The hot-wet simulated transverse strength for three environmental conditions is shown in Fig. 9. The experimental data for the two thermal conditions fall on the simulated curves. However, that with the 1% moisture is at the low end of its corresponding probability curve. It is not clear why those data congregate at the very low-probability level. Some conjectures are 1) there may be eccentricities present during testing since the specimen is relatively soft 2) incorrectly recorded data or 3) additional factors which are not accounted for in the probabilistic simulation.

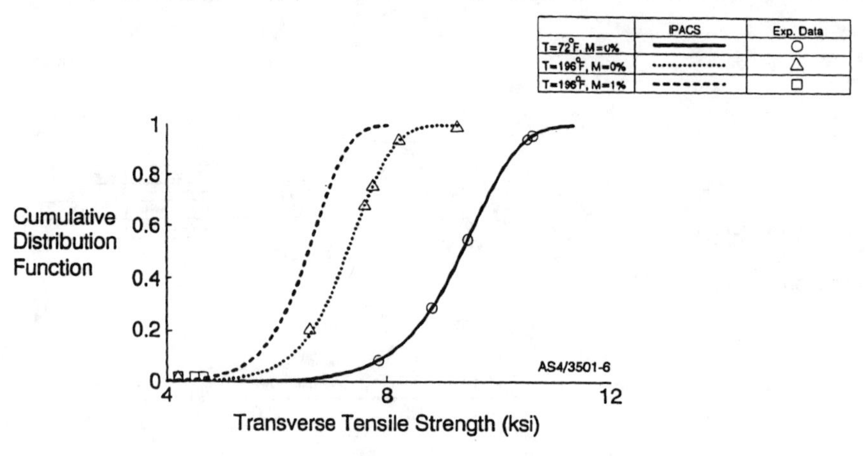

Fig. 9 Probabilistic simulation of hot-wet ply transverse strength; A-S/E; 0.6 mean fiber volume ratio.

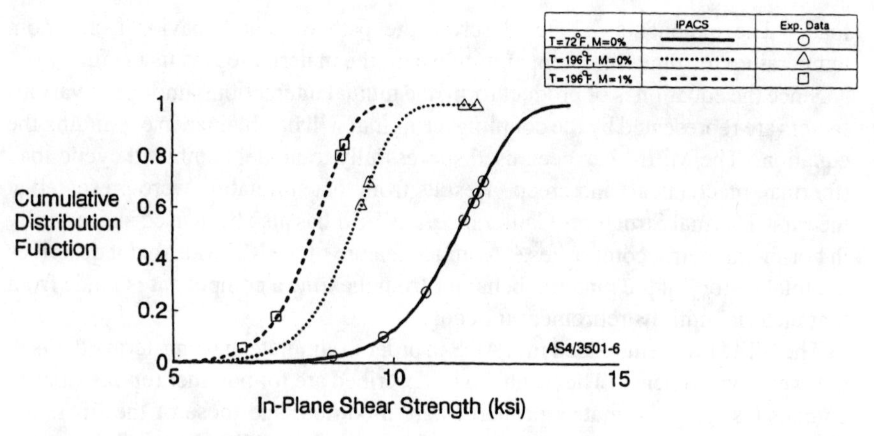

Fig. 10 Probabilistic simulation of hot-wet ply intralaminar shear strength; A-S/E; 0.6 mean fiber volume ratio.

The hot-wet simulated intralaminar shear strength is shown in Fig. 10 for three environmental conditions. The experimental data fall on the respective simulated curves. Because these data coincide with the simulated curves, the first two conjectures, mentioned earlier, may be responsible for the deviations in the transverse tensile strength.

The conclusion from the results and comparisons presented thus far is that probabilistic simulation adequately describes the ambient and hot-wet behavior of composites.

Long-Term Behavior

Long-term behavior includes both cyclic and time dependent as well as combinations. The results described in the preceding section were obtained from the composite mechanics which accounted for hot-wet (hygrothermal) effects. The effects for cyclic load and time dependence need to be modeled. A convenient way to model those effects is through the use of a multifactor interaction model MFIM. In equation from this model is

$$\frac{P}{P_0} = \underbrace{\left(\frac{T_{gw} - T}{T_{gd} - T_0}\right)^{1/2} \left(1 - \frac{\sigma}{S_f}\right)^m}_{\substack{\text{empirical} \\ \text{(verified)}}} \underbrace{\left(1 - \frac{\sigma t}{S_f t_f}\right)^n \left(1 - \frac{\sigma_m N_m}{S_f N_{fm}}\right)^p \left(1 - \frac{\sigma_t N_t}{S_f N_{ft}}\right)^q}_{\substack{\text{assumed} \\ \text{(unverified)}}}$$

where P is the property; T temperature, S strength, σ stress, N number of cycles, t time; the subscripts are gw wet glass transition temperature, gd dry glass transition temperature; 0 reference condition, f final, m mechanical load, and t thermal load. The exponents m, etc. describe the path of that behavior from room temperature to some final temperature where the material looses its usefulness.

Since the equation is of product form, the mutual interactions among the various factors are represented by the coupling terms that will be obtained in expanding the equation. The MFIM has been used successfully to model combined cyclic load (thermal, mechanical) and creep. Results from that simulation were presented at the First Thermal Structures Conference.[3] MFIM has also been used successfully in hot-metal matrix composites.[4] A unique feature of the MFIM is that it describes the total history of the material behavior from the time a component is made from that material until its retirement for cause.

The MFIM was embedded in IPACS in order to simulate the long-term effects in hot-wet environments. The results to be described are for polymer (epoxy) matrix composites, only the matrix properties are modified, and those of the fibers are assumed to be unchanged. Those of the fibers can be modified as well, however, as would be the case for high-temperature metal matrix and ceramic matrix composites.

Note that the cyclic load terms and the time term in the MFIM have the same form. It is instructive to examine some typical deterministic results to see how the composite properties and stresses change prior to presenting probabilistic simulation results. The time-dependent strains for a ±45 graphite/epoxy angle-ply laminate are shown in Fig. 11, for combined cyclic loads and sustained stress. These figures show bilinear behavior of the laminate strains along the load and transverse to it. Note the behavior shown in the figures is exaggerated because of the relatively expanded scale. Note also that the laminate has not failed yet nor have the stresses stabilized as would be expected if a secondary creep stage was present. The corresponding microstresses in the matrix and in the fiber are shown in Fig. 12. It is worth noting that the longitudinal stress in the matrix decreases ("relaxes") whereas that in the fiber increases. It is also worth noting that combinations of combined cyclic loading and time exist in which the matrix would completely shed its stress. The corresponding ply transverse tensile strength is shown in Fig. 13. This figure shows that the ply transverse tensile strength decreases with time.

The curves in Fig. 11–13 can be used to establish ply microcracking, ply failure, and laminate failure. For example, ply microcracking will occur when microstresses in the matrix exceed those of its corresponding strength. (The matrix strength exhibits similar behavior as the ply transverse strength.) Ply failure will occur when the ply stresses exceed their corresponding strengths, either individually or combined. Laminate failure will occur where all of the plies fail in that section.

The important point and observation is that a simple model is available to predict the combined cyclic loading and time behavior of polymer composites.

It was mentioned earlier that long-term behavior is represented by cyclic loading or sustained stress. IPACS with MFIM was used to simulate the probabilistic long term behavior of a pseudo-isotropic laminate. Typical results obtained from such simulation for mechanical cyclic loading (fatigue) are shown in Fig. 14, where both the cumulative curve and the dominant sensitivity factors are plotted.[5] Five factors influence the fatigue life when the cyclic stress is 90% of the

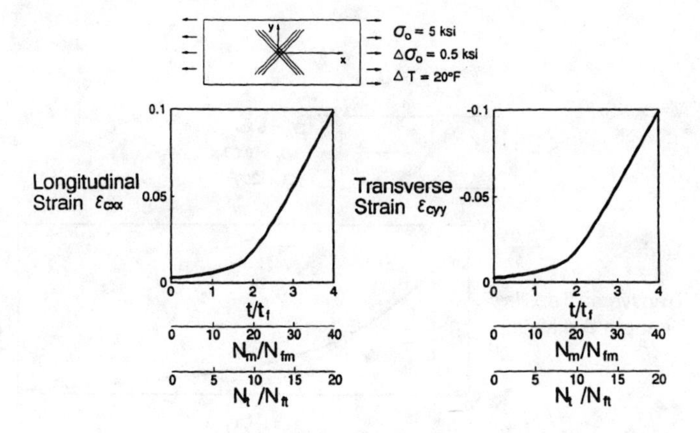

Fig. 11 Combined effects on the laminate strain of a graphite/epoxy (+45/-45) composite.

corresponding strength. Note it was mentioned earlier that the hot-wet conditions have rather negligible effects on fiber dominated properties which is the case when the cyclic load is applied along the 0-deg ply of pseudoisotropic laminate. Therefore, the results shown in Fig. 14 are not expected to be about the same for hot-wet environments typical of polymer-epoxy composites.

Similar results to those in Fig. 14 can be generated for other laminate cyclic stress to 0-deg ply longitudinal strength ratios. Those results can then be plotted to obtain the well known S/N (stress/cycle) curves as is shown in Fig. 15, for specified probability levels. The important point to be noted is that formal methods can and have been developed to simulate the hot-wet long term behavior of polymer-epoxy matrix composites. These methods are generic and are applicable to all types of composites and all types of structures made therefrom.

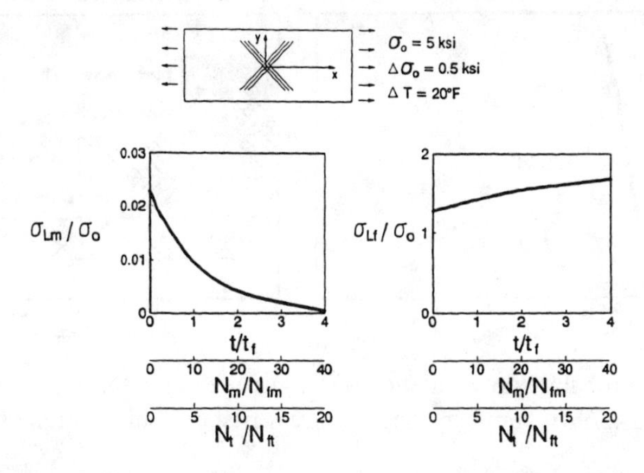

Fig. 12 Combined effects on the micro-stresses of a graphite/epoxy (+45/-45) composite.

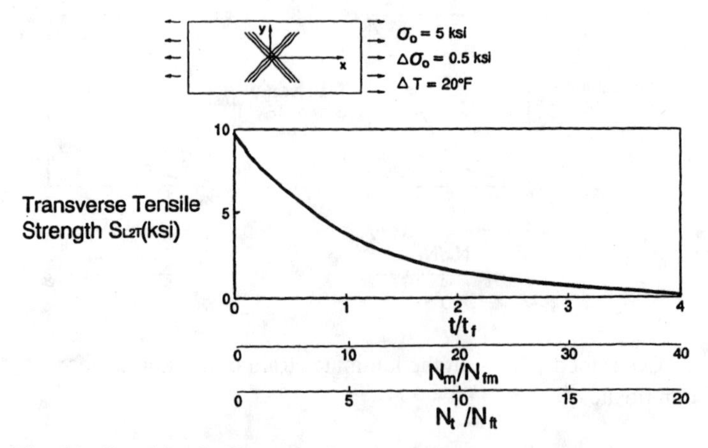

Fig. 13 Combined effects on the ply strength of a graphite/epoxy (+45/-45) composite.

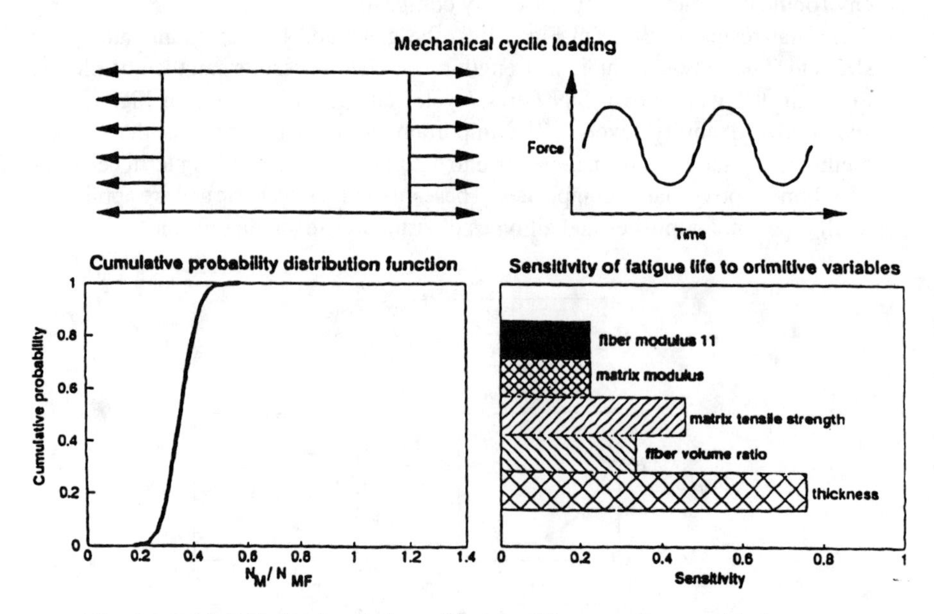

Fig. 14 Probabilistic simulation of fatigue life curve for graphite/epoxy $(0/\pm45/90)_s$ laminate. Applied stress/first ply failure strength = 0.9.

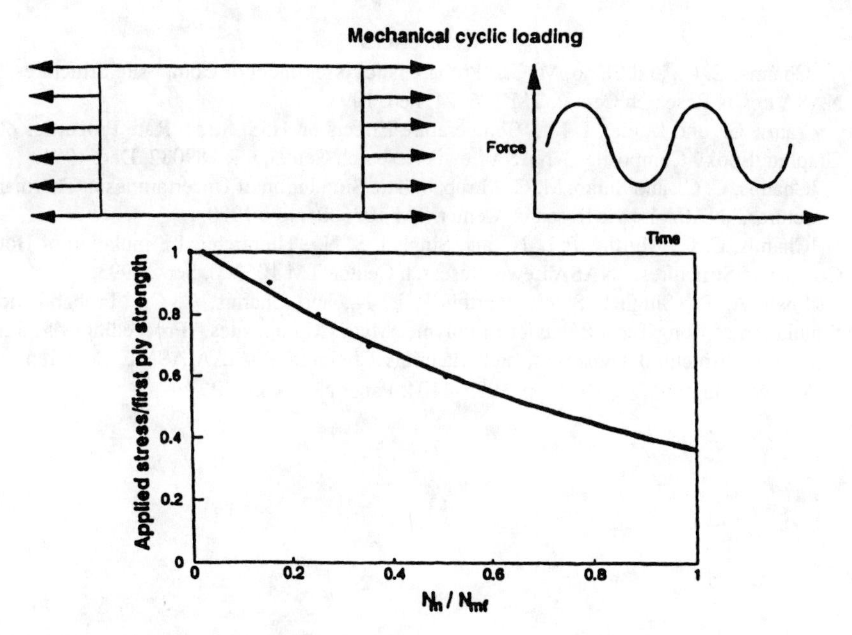

Fig. 15 Reliability based fatigue life curve reliability = 0.999 for graphite/epoxy $(0/\pm45/90)_s$ laminate.

Concluding Remarks

Formal method were presented and described which can be used to quantify the uncertainties of hot-wet composite long-term behavior. These methods are embedded in a computer code which combines probabilistic composite mechanics with probabilistic finite element structural analysis and with a multifactor interaction equation which represents the combined effects of all factors that influence composite behavior. The quantification of uncertainties is then performed via computational simulation for cumulative probability of failure and dominant sensitivity factors. Important results from these simulations for polymer-epoxy fiber composites are summarized as follows.

1) Probabilistic simulation bounds the experimental results for longitudinal, transverse and shear moduli and fc. longitudinal tensile strength.
2) Available hot-wet experimental data fall on the same simulated curves for fiber dominated properties (longitudinal modulus and longitudinal tensile strength).
3) Available hot-wet data fall on respective simulated curves for matrix dominated properties (transverse and shear moduli and transverse and intralaminar shear strengths).
4) Normalized plots for long-term behavior (sustained stress, mechanical and thermal cyclic loadings) are similar.
5) A pseudoisotropic laminate will survive 50% of its life with 0.999 probability when subjected to a cyclic stress which is 60% of its static strength.

References

[1]Chamis, C. C., and Shiao, M. C., "Probabilistic Assessment of Composite Structures," NASA Lewis Research Center, TM 106024, Feb. 1993.

[2]Yaniv, G., and Daniel, I. M., "Temperature Effects on High Strain Rate Properties of Graphite/Epoxy Composites," NASA Lewis Research Center, CR 189082, Dec. 1991.

[3]Chamis, C. C., and Shiao, M. C., "Probabilistic Simulation of Uncertainties in Thermal Structures," NASA Lewis Research Center, TM 103680, Nov. 1990.

[4]Chamis, C. C., Murthy, P. L. N., and Singhal, S. N., "Hierarchical Simulation of Hot Composite Structures," NASA Lewis Research Center, TM 106200, Sept. 1993.

[5]Shah, A. R., Singhal, S. N., Murthy, P. L. N., and Chamis, C. C., "Probabilistic Simulation of Long Term Behavior in Polymer Matrix Composites," Proceedings of 35th Structures, Structural Dynamics, and Materials Conference, AIAA/ASME Conference, AIAA, Washington, DC, 1994, pp. 1095–1104. Paper No. 1445-CP.

Minimizing Thermal Deformation by Using Layered Structures

Robert C. Wetherhold* and Jianzhong Wang†

State University of New York, Buffalo, New York 14260-4400

Abstract

Many thin structural components such as beams, plates, and shells experience a through-thickness temperature variation. This temperature variation can produce both an out-of-plane (bending) curvature as well as an in-plane expansion. Given that these thin components interact with or connect to other components, we often wish to minimize the thermal deformation or to match the thermal deformation of another component. This is accomplished by combining layers of material with a positive thermal expansion coefficient with layers possessing a negative thermal expansion coefficient. A three-layer beam is demonstrated which can eliminate thermal curvature while lowering in-plane expansion or can match a desired in-plane expansion while lowering thermal curvature. A five-layer beam is demonstrated which can eliminate thermal curvature as well as match (within limits) a desired in-plane expansion. The beam results are independent of the actual temperature values, within the limitations of steady-state heat transfer and constant material properties. If the temperature excursion is large enough that the properties vary, a zero thermal curvature beam may still be constructed. This temperature-dependent case is pre ented and discussed.

Introduction

When a structure is subjected to a uniform or nonuniform temperature field, it normally reacts by producing deformations composed of (in-plane) expansion and (out-of-plane) bending. These deformations are usually undesirable since they distort the structure and can cause stresses when the structure's parts expand unequally. With the availability of materials such as graphite and Kevlar fibers, which possess a negative axial thermal expansion

---

* Associate Professor, Department of Mechanical and Aerospace Engineering.

† Graduate Student, Department of Mechanical and Aerospace Engineering.

coefficient and high stiffness, a composite lamina may be made which exhibits a negative axial coefficient of thermal expansion (CTE). When these composite laminae are suitably layered with other composite (or metal or ceramic) laminae, we can create a laminated structure with a near-zero CTE in the plane.[1] Zero CTE ensures dimensional stability and minimal thermal mismatch stresses amongst similar parts of a structure subjected to a uniform temperature field. If we join zero CTE components with nonzero CTE components, however, there will be mismatch stresses and distortion even for a uniform temperature.

For the case of a nonuniform temperature field, the problem becomes more complex. In actual application, the structure usually sees a temperature variation through its thickness as one side is exposed to sunlight, an impinging gas or liquid, or a heat source. The zero CTE components do not provide protection against this temperature field; they will bend, and the bending deformation can be considerable.

In this chapter, we investigate ways to control thermal deformation and mismatches in thermal deformation. A three-layer symmetric beam [composite/metal/composite] offers one independent variable - the relative thickness of the composite and metal layers - and can be used to satisfy one condition. This condition can be either to eliminate thermal bending or to match in-plane expansion. A five-layer symmetric beam offers two variables (two relative thicknesses) and can be used to satisfy two conditions. Thus, we can eliminate thermal bending while matching an in-plane expansion, including zero CTE.

Beam Theory

Three-Layer Beam

In this section, we develop the equations required to describe the deformation of the symmetric beam shown in Fig. 1. To do so, we evaluate the thermal and mechanical moment and force resultants and relate them, through laminate stiffnesses, to the bending curvature and in-plane strain. This also allows us to calculate the CTE of the beam. The center layer could be any isotropic material (e.g., metal or ceramic); in the following, we assume it is

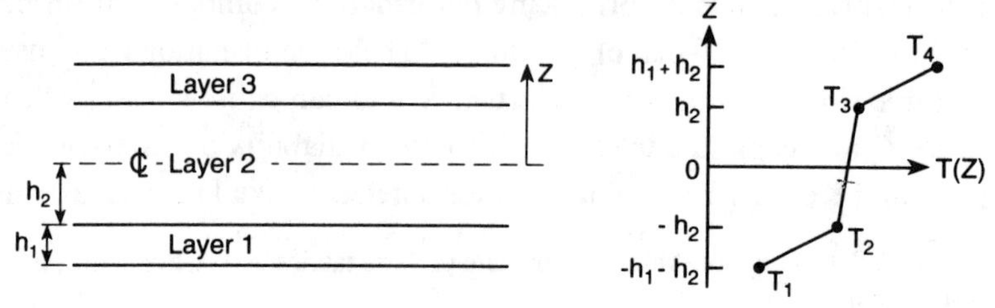

Fig. 1 Three-layer beam.

metal. We follow an analysis and notation similar to that for a laminated, symmetric plate,[2] reducing the analysis to one dimension for a shear-indeformable beam. The following is a summary; further details may be found in Refs. 3 and 4. The midplane strain ε_x^0 and the bending curvature κ_x are related to applied force and moment resultants by

$$N_x = A\varepsilon_x^o - N_x^T \tag{1}$$

$$M_x = D\kappa_x - M_x^T \tag{2}$$

The plate has stiffnesses in-plane (A) and out-of-plane or bending (D) given by

$$A = \int_{-h_1-h_2}^{h_1+h_2} E\,dz = 2[E_1 h_1 + E_2 h_2] > 0 \tag{3a}$$

$$D = \int_{-h_1-h_2}^{h_1+h_2} Ez^2\,dz = \frac{2}{3}\{E_1[(h_1+h_2)^3 - h_2^3] + E_2 h_2^3\} > 0 \tag{3b}$$

with the subscripts 1 and 2 corresponding to properties of composite and metal. The usual definitions for the mechanical and thermal force and moment resultants apply.

Consider the case of zero applied mechanical loading ($N_x = M_x = 0$) to obtain the effective in-plane thermal expansion coefficient α_x and the thermal curvature κ_x. The effective laminate CTE can be found for the case of a uniform temperature field.

$$\alpha_x = \frac{1}{A}\int_{-h_1-h_2}^{h_1+h_2} E\alpha\,dz \tag{4}$$

A complete representation of the temperature field is given by

$$T(z) = \begin{cases} T_1 + \dfrac{(T_2 - T_1)}{h_1}(z + h_1 + h_2) & z\ \varepsilon\ (-h_1 - h_2, -h_2) \\[2ex] T_2 + \dfrac{(T_3 - T_2)}{2h_2}(z + h_2) & z\ \varepsilon\ (-h_2, h_2) \\[2ex] T_3 + \dfrac{(T_4 - T_3)}{h_1}(z - h_2) & z\ \varepsilon\ (h_2, h_1 + h_2) \end{cases} \tag{5}$$

where k_1 and k_2 are thermal conductivities of composite and metal, h_1 and $2h_2$ are thicknesses of composite and metal layers, and T is measured from the stress-free temperature. The constant heat flux through each layer provides a relation between the interface temperatures.

$$\frac{k_1}{h_1}(T_2 - T_1) = \frac{k_2}{2h_2}(T_3 - T_2) = \frac{k_1}{h_1}(T_4 - T_3) \tag{6}$$

For the temperature field of Eq. (5) and using Eq. (6), the thermal moment may be given in normalized form as[3,4]

$$\bar{M}^T = \frac{M_x^T}{(T_3 - T_2)h_2^2 E_2 \alpha_2} = \phi^3\left(\frac{ab}{3c}\right) + \phi^2\left(\frac{ab}{2c} + \frac{ab}{2}\right) + \phi(ab) + \frac{1}{3} \tag{7}$$

where a, b and c are ratios of lamina properties E_1/E_2, α_1/α_2, and k_1/k_2 and ϕ is h_1/h_2. The subscripts 1 and 2 again refer to the composite and metal. The curvature may be given in a normalized form as

$$\bar{\kappa} = \frac{\kappa_x h_2}{\alpha_2(T_3 - T_2)} = \frac{3}{2}\frac{\bar{M}^T}{\{a[(\phi + 1)^3 - 1] + 1\}} \tag{8}$$

The effective laminate (in-plane) thermal expansion coefficient α_x is given in normalized form by

$$\bar{\alpha} = \frac{\alpha_x}{\alpha_2} = \frac{ab\phi + 1}{a\phi + 1} \tag{9}$$

Note that the limits on $\bar{\alpha}$ are expressed by $\bar{\alpha}\, \varepsilon\ [b, 1]$. This demonstrates a reduction of the effective thermal expansion.

If we wish to eliminate thermal bending, we set $\bar{M}^T = 0$; this is equivalent to setting $\kappa_x = 0$ (or $\bar{\kappa} = 0$). The cubic equation in ϕ which arises is independent of the actual temperature difference.[3,4]

$$\phi^3\left(\frac{ab}{3c}\right) + \phi^2\left(\frac{ab}{2c} + \frac{ab}{2}\right) + \phi(ab) + \frac{1}{3} = 0 \tag{10}$$

Since the property ratios a, $c > 0$ and the ratio $b < 0$, Descartes' rule of signs ensures a single real positive root for ϕ. This ϕ value is substituted into Eq. (9) to determine the reduced in-plane expansion. If, on the other hand, we wish to specify that $\bar{\alpha} = \eta$ (a desired value), we solve for ϕ as

$$\phi = \frac{1 - \eta}{a(\eta - b)} \tag{11}$$

This ϕ value would be substituted into Eq. (8) to determine the (reduced) bending curvature.

As a practical engineering concern, we may wish to replace an existing all-metal beam with a laminate as shown in Fig. 1. Assume that the center layer of the three-layer beam is of identical material as the all-metal beam, and that we wish the beams to have equal flexural stiffness (D). If the all-metal beam is of height $2h_0$ and the ϕ value is fixed by a solution for either zero curvature [Eq. (10)] or to match a required expansion coefficient [Eq. (11)], then the ratio of thicknesses is given by

$$\frac{h_1 + h_2}{h_o} = \frac{(\phi + 1)}{\{a[(\phi + 1)^3 - 1] + 1\}^{1/3}} \tag{12}$$

Five-Layer Beam

In considering a symmetric five-layer beam (see Fig. 2), there are many similarities but some distinct differences when compared with the three-layer system. The temperature field in the steady state and the constant heat flux conditions have representations similar to Eqs. (5) and (6). The evaluations for the thermal curvature and thermal expansion coefficient follow as in Eq. (2) with $M_x = 0$ and in Eq. (4). The differences between five-layer and three-layer beams are both operational (more layers and properties) and conceptual (two independent relative layer thicknesses). It is the latter difference which has significance; two independent layer thickness ratios permit us to eliminate thermal bending and satisfy an in-plane thermal expansion requirement (within limits).

The details of the through-thickness temperature field and heat flux conditions allow the evaluation of the thermal moment and thermal force

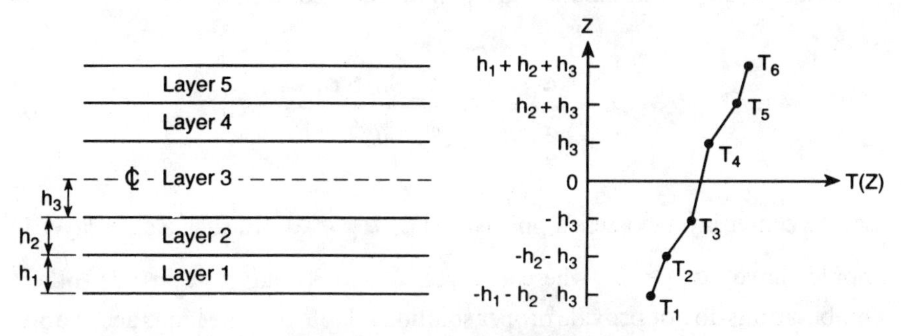

Fig. 2 Five-layer beam.

resultants. This leads, in turn, to the thermal curvature κ_x and the in-plane expansion coefficient α_x; for the full details, see Ref. 4. The curvature equation is given by

$$\kappa_x = \left(\frac{3}{2}\right) \frac{\alpha_3(T_4-T_3)}{h_3} \bar{M}^T \{a_1[(\phi_1+\phi_2+1)^3 - (\phi_2+1)^3] + a_2[(\phi_2+1)^3-1] + 1\}^{-1} \tag{13}$$

where $\bar{M}^T$ is the normalized thermal moment resultant

$$\bar{M}^T = \frac{M_x^T}{(T_4-T_3)h_2^3E_3\alpha_3} \tag{14}$$

and M_x^T can be schematically expressed as

$$M_x^T = (T_4-T_3)F(\phi_1,\phi_2,a_1,a_2,b_1,b_2,c_1,c_2)$$

the property ratios a_i, b_i, and c_i are given by

$$a_i = E_i/E_3 \quad b_i = \alpha_i/\alpha_3 \quad c_i = k_i/k_3$$

and the thickness ratios by

$$\phi_i = h_i/h_3$$

for layer $i = 1,2$. The curvature can thus be schematically expressed as

$$\kappa_x = (T_4-T_3)f(\phi_1,\phi_2,a_1,a_2,b_1,b_2,c_1,c_2) \tag{15}$$

The CTE α_x for the laminate is given in normalized form as

$$\bar{\alpha} = \frac{\alpha_x}{\alpha_3} = \frac{a_2b_2\phi_2 + a_1b_1\phi_1 + 1}{a_2\phi_2 + a_1\phi_1 + 1} \tag{16}$$

Let the center layer possess a positive CTE, $\alpha_3 > 0$. In such a case, layer 1 should have $\alpha_1 > 0$ whereas layer 2 should have $\alpha_2 < 0$; other combinations do not provide proper solutions. Note that the limits achievable by the normalized CTE are

$$\bar{\alpha} \ \varepsilon \left[\frac{\alpha_2}{\alpha_3}, \ \frac{\max(\alpha_1, \alpha_3)}{\alpha_3} \geq 1 \right] \tag{17a}$$

If the center layer possesses a negative CTE, $\alpha_3 < 0$, then layers 1 and 2 must exhibit $\alpha_1 < 0$ and $\alpha_2 > 0$ to obtain proper solutions. The achievable limits for the normalized CTE become

$$\bar{\alpha} \ \varepsilon \left[\frac{\alpha_2}{\alpha_3}, \ \frac{\min(\alpha_1, \alpha_3)}{\alpha_3} \geq 1 \right] \tag{17b}$$

The two equations in ϕ_1, ϕ_2 to be solved simultaneously are

$$\kappa_x = 0 \quad (\overline{M}^T = 0) \tag{18}$$

and

$$\bar{\alpha} = \eta = \text{desired constant}$$

Note that Eq. (18) is again independent of the value of the temperature difference $(T_4 - T_3)$. Since Eq. (19) is linear, it is reasonable to solve it for ϕ_1 and reduce Eq. (18) to a single cubic equation in ϕ_2 only. The signs of the coefficients depend on the values of a_i, b_i, c_i and η, and the application of Descartes' rule of signs is not obvious. It is clear that we cannot set η outside the limits given by Eq. (17). What must be determined by specific application is the practical range of η which will return real, positive roots for ϕ_1 and ϕ_2.

We may wish to replace a monolithic (one material) beam with the laminate shown in Fig. 2. Assume that the center layer of the laminate is of the same material as the monolithic beam and that we wish the beams to have equal flexural stiffness (D). If the monolithic beam is of height $2h_0$ and the ϕ_1 and ϕ_2 values are fixed by the solution for Eqs. (18) and (19), the ratio of thicknesses is given by

$$(h_1 + h_2 + h_3)/h_o = (\phi_1 + \phi_2 + 1)\{a_1 + [(\phi_1 + \phi_2 + 1)^3 - (\phi_2 + 1)^3] \tag{20}$$

$$+ \ a_2[(\phi_2 + 1)^3 - 1] \ + \ 1\}^{1/3}$$

Applications

Three-Layer Beam

Consider the metal and composite material properties given in Table 1. The composites are unidirectional and are aligned with the x axis. Let the center layer be aluminum, with outer layers of composite with CTE $\alpha < 0$. We enforce the condition of zero thermal curvature, viz. Eq. (10). The thickness ratio ϕ, normalized CTE (α_x / α_2), and thickness relative to an all-aluminum beam of equal stiffness [$(h_1 + h_2)/h_0$, eq. (12)] are given in Table 2. Note that required composite layer to aluminum layer thickness ratio ϕ is small (all less than 0.3), indicating a reasonable solution. All of the normalized thermal expansion coefficients are reduced (all less than 1.0), indicating decreased in-plane thermal strain. Because of the all-metal beam with the same bending stiffness: $(h_1 + h_2)/h_0 < 1$.

Five-Layer Beam

Consider the same properties given in Table 1. As a first step, we choose to eliminate thermal bending ($\kappa_x = 0$) and eliminate in-plane expansion ($\alpha_x = 0$). The resulting layer thicknesses to satisfy these conditions are found in Table 3 for a beam with a metal center layer ($\alpha_3 > 0$). Matching the flexural stiffness of the five-layer beam to that of a homogeneous beam of the same material as the center layer of the five-layer beam gives the thickness ratio [$(h_1 + h_2 + h_3)/h_o$], see Eq. (20). For a beam whose center layer displays a negative CTE ($\alpha_3 < 0$), the results are given in Table 4.

Table 1 Properties for layer thickness calculation

	AS Gr/Ep	Kevlar/Ep	P100 Gr/Ep	E-Gl/Ep	Al[6]	Cu[9]
E, 10^9Pa	138[5]	76[5]	480[1]	38.6[5]	69	124
α, 10^{-6} C^{-1}	-0.3[5]	-4.0[5]	-1.22[1]	8.6[5]	24	17.3
k, W/m K	0.71[7]	0.16[7]	2.0[8]	0.34[7]	180	358

Table 2 Values resulting from using various composite outer layers

	AS Gr/Ep	Kevlar/Ep	P100 Gr/Ep
$\phi = (h_1/h_2)$	0.293	0.0549	0.128
(α_x/α_2)	0.626	0.933	0.504
$(h_1 + h_2)/h_0$	0.866	0.995	0.709

We may wish to minimize or eliminate the mismatch of this beam to another beam. In this case, we may set the bending curvature κ_x to zero, while matching a desired normalized CTE value of η. The results for an example beam are shown below in Table 5. Although the value η = - 0.04 is within the attainable CTE limits for $\bar{\alpha}$ [Eq. (17a)), it is not possible to satisfy both $\kappa_x = 0$ and $\bar{\alpha} = -0.04$. This can be visualized as follows. Reviewing Eq. (13) with fixed material properties, we form an implicit equation $f_1(\phi_1, \phi_2) = 0$ to satisfy zero curvature. The curve of this equation must intercept the line given by Eq. (16), $\bar{\alpha} = \eta$, in quadrant I of (ϕ_1, ϕ_2) space for the results to be physically meaningful. (Negative values of ϕ_i are irrelevant.) In Fig. 3, the line for $\eta = 0$ intersects the curve for $\bar{M}^T = 0$

Table 3 Results for zero thermal deformation with positive center layer CTE

Layer 1	Layer 2	Layer 3	ϕ_1	ϕ_2	$\frac{h_1 + h_2 + h_3}{h_0}$
Cu	P100 Gr/Ep	Cu	0.527	5.59	0.674
Aℓ	P100 Gr/Ep	Cu	0.671	5.56	0.692
E-Gℓ/Ep	P100 Gr/Ep	Cu	1.32	4.41	0.775
Cu	P100 Gr/EP	Aℓ	0.418	4.36	0.554
Aℓ	P100 Gr/Ep	Aℓ	0.532	4.33	0.568
E-Gℓ/Ep	P100 Gr/Ep	Aℓ	1.04	3.42	0.635

Table 4 Results for zero thermal deformation with negative center layer CTE

Layer 1	Layer 2	Layer 3	ϕ_1	ϕ_2	$\frac{h_1+h_2+h_3}{h_0}$
P100 Gr/Ep	E-Gℓ/Ep	P100 Gr/Ep	0.554	2.74	1.35
P100 Gr/Ep	Aℓ	P100 Gr/Ep	0.702	0.602	1.09
P100 Gr/Ep	Cu	P100 Gr/Ep	0.707	0.466	1.06
Kevlar/Ep	E-Gℓ/Ep	Kevlar/Ep	0.643	1.50	1.09
Kevlar/Ep	Aℓ	Kevlar/Ep	0.714	0.315	1.00
Kevlar/Ep	Cu	Kevlar/Ep	0.718	0.243	0.976

(also $\bar{\kappa} = 0$) at the unique point (0.532, 4.33). The line for $\eta = -0.04$ diverges from the curve $\bar{M}^T = 0$ as we proceed into quadrant I. The only intersection of the line and the curve occurs outside of quadrant I and, thus, no physically meaningful solution exists.

Temperature-Dependent Properties

Material properties such as Young's modulus, CTE, and thermal conductivity (E, α, k) are functions of temperature. For moderate temperature excursions, this is usually not important. The influence of the temperature field on properties must be taken into account if there is a large-temperature excursion. In this section, we consider the material properties (E, α, k) to be functions of temperature one-at-a-time, and re-evaluate the thermal curvature and laminate effective CTE for a three-layer laminated beam. For the sake of simplicity, we assume all properties are linear functions of the form

$$P(T) = P^0[1+\beta(T-T_0)] \tag{21}$$

where $P^0 = P(T_0)$, β is the coefficient (constant over a certain temperature range), and T is the temperature. Even though Eq. (21) is only an approximation, it will improve the degree of accuracy compared to use of a constant value.

Table 5 [Al/P100 Gr-Ep/Al]$_s$

η	ϕ_1	ϕ_2
-0.04	-1, -1.03	0, -0.439
-0.02	1.25	10.7
-0.01	0.743	6.20
0	0.532	4.33
0.02	0.342	2.67
0.04	0.252	1.90
0.06	0.199	1.46
0.20	0.0721	0.491

The definitions for laminate stiffnesses, thermal force and moment, and thermal expansion and thermal curvature referred to in the three-layer beam section are still valid; we only need to substitute the property in the form of Eq. (21). First we consider E as a linear function of temperature (α and k constant) and let $T_0 = 0$. The thermal moment becomes

$$M_x^T = \int_{(-h_1-h_2)}^{-h_2} E_1^o(1+\beta_1 T)T\alpha_1 z\, dz + \int_{-h_2}^{h_2} E_2^o(1+\beta_2 T)T\alpha_2 z\, dz + \int_{h_2}^{(h_1+h_2)} E_1^o(1+\beta_1 T)T\alpha_1 z\, dz \tag{22a}$$

which moment can be separated into two contributions

$$M_x^T = \left(M_x^T\right)_o + \left(M_x^T\right)_a \tag{22b}$$

where $\left(M_x^T\right)_o$ is the nominal thermal moment at $\beta_1 = \beta_2 = 0$, and $\left(M_x^T\right)_a$ is the additional thermal moment from considering $E = E(T)$. We may normalize the bending moment as in Eq. (7),

$$\bar{M}_x^T = \left(\bar{M}_x^T\right)_o + \left[\left(\bar{M}_x^T\right)_o \beta_1 + \frac{\beta_2 - \beta_1}{3}\right](T_3 + T_2) \tag{22c}$$

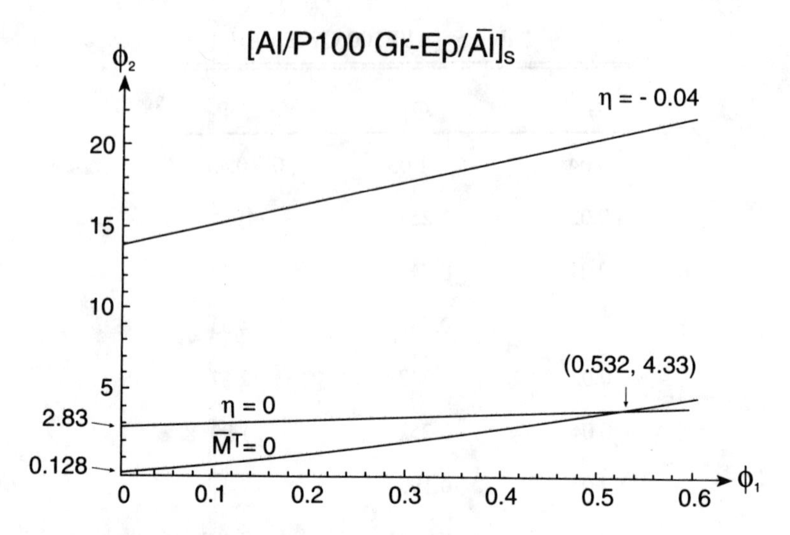

Fig. 3 Intersecting and nonintersecting curves in quadrant I.

where $\left(\bar{M}_x^T\right)_o$ is given by Eq. (7).

The bending stiffness D must be evaluated for $E = E(T)$.

$$D = \int_{(-h_1-h_2)}^{(h_1+h_2)} E(T)z^2 \, dz = (D_o + D_a) \tag{23}$$

where D_0 is the nominal stiffness given in Eq. (3b), and D_a is the additional term which arises when $E = E(T)$,

$$D_a = \left[\frac{1}{3}\left(\beta_1 E_1^o h_1^3 + \beta_2 E_2^o h_2^3\right) + \beta_1 E_1^o h_1 h_2 (h_1 + h_2)\right](T_3 + T_2) \tag{24}$$

Substituting Eqs. (22b) and (23) into Eq. (2) with $M_x = 0$ gives the thermal curvature

$$\kappa_x = \frac{\left(M_x^T\right)_o + \left(M_x^T\right)_a}{D_o + D_a} \tag{25}$$

The re-evaluation of the thermal force resultant (N_x^T) can be done similar to that of the thermal moment (M_x^T). To get the effective laminate CTE, the in-plane stiffness A also must be re-evaluated

$$A = \int_{(-h_1-h_2)}^{(h_1+h_2)} E(T)\, dz = A_o + A_a \tag{26}$$

where A_0 is the nominal in-plane stiffness given in Eq. (3a), and A_a is the additional term

$$A_a = \left(\beta_1 E_1^o h_1 + \beta_2 E_2^o h_2\right)(T_3 + T_2) \tag{27}$$

The normalized effective in-plane thermal expansion coefficient can be evaluated.

$$\bar{\alpha} = \frac{\alpha_x}{\alpha_2} = \frac{2N_x^T}{\alpha_2(A_o + A_a)(T_3 + T_2)} \tag{28}$$

Note that the laminate stiffnesses A and D are now functions of temperature.

If we wish to eliminate bending curvature, we can set $\bar{M}_x^T = 0$; this is equivalent to setting $\kappa_x = 0$. The cubic equation in ϕ now is dependent on the temperature field and can be expressed as

$$\left(\bar{M}_x^T\right)_o [1 + \beta_1(T_3 + T_2)] + \frac{\beta_2 - \beta_1}{3}(T_3 + T_2) = 0 \tag{29}$$

Whether we can have a meaningful solution [a positive root of Eq. (29)] depends on the midplane temperature ($T_3 + T_2$) and the temperature coefficients β_1 and β_2. Descartes' rule of signs ensures a single positive real root when

either $\quad \begin{matrix} \beta_1(T_3+T_2) > -1 \\ \beta_2(T_3+T_2) > -1 \end{matrix} \quad$ *or* $\quad \begin{matrix} \beta_1(T_3+T_2) < -1 \\ \beta_2(T_3+T_2) < -1 \end{matrix}$

is satisfied. On the other hand, if we wish to set $\bar{\alpha} = \eta$ (a desired value), we need to solve Eq. (28) for ϕ.

Considering $\alpha = \alpha(T)$ (E and k constant) is functionally equivalent to considering $E = E(T)$ as far as the thermal moment and force resultant are concerned; additional terms arise. To evaluate the curvature and the effective

CTE, the temperature independent stiffnesses (D,A) use the forms of Eqs. (3a) and (3b).

Considering $k=k(T)$ (α and E constant) is more problematic as the piecewise linear form of the temperature field [Eq. (5)] no longer holds. Consider the simplest case when we have a laminate in the form of [metal/composite/metal] and only the composite's thermal conductivity varies with temperature; the temperature field within the top and bottom layer is linear in z, resembling Eq. (5). Notice that the temperatures at the interface T_2 and T_3 need to be re-evaluated as the form of k_2 is now changed in Eq. (6). The temperature field of the middle layer is one of the roots of the equation

$$\left(\frac{\beta_2}{2}k_2^o\right)T^2 + k_2^o T - \left[k_2^o\left(T_2 + \frac{\beta_2}{2}T^2\right) + (k_m)_2\xi(T_3 - T_2)\right] = 0 \qquad (30)$$

where

$$\xi = \frac{z + h_2}{2h_2} \in (0,1) \quad (k_m)_2 = k_2\left(1 + \beta_2\frac{T_3 + T_2}{2}\right)$$

is the midplane or average thermal conductivity of the composite layer according to the boundary temperatures and β_2 (Refs. 10 and 11). For example, when $T_3 > T_2 > 0$ and $\beta_2 > -1/T_3$, the root

$$T = \frac{-q_1 + \sqrt{q_1^2 - 4q_2q_0}}{2q_2}$$

must be picked to ensure $T \in (T_2, T_3)$, where q_2, q_1, and q_0 are the second, first and zeroth order coefficients of the quadratic equation [Eq. (30)]. Using the temperature field we re-evaluate (N_x^T, M_x^T). Using the stiffnesses D and A defined in Eqs. (3a) and (3b), we can get the thermal curvature and effective CTE by using Eqs. (2) and (1).

A numerical example was set up to show the influence of temperature-dependent properties $E = E(T)$ and $k = k(T)$. Using the nominal material properties of Table 1, assume

$$\beta_{Al} = -0.0009\ 1/K \qquad T_1 = 0\ K$$

$$\beta_{comp} = -0.0005\ 1/K \qquad T_4 = 100\ K$$

[guarantees a single positive real root for Eq. (29) for $E=E(T)$ where α and k are constant]. For $k=k(T)$ (α and E constant), assume $\beta_2 = -0.001$ Wm/K^2 and

Table 6 Thickness ratio and normalized CTE for zero thermal curvature when $E=E(T)$

		ASGr-Ep /Al/ ASGr-Ep	Kev-Ep /Al/ Kev-Ep	P100Gr-Ep /Al/ P100Gr-Ep	Al/ ASGr-Ep /Al	Al/ Kev-Ep /Al	Al/ P100Gr-Ep /Al
$E=E(T)$[a]	ϕ	0.287	0.0544	0.126	0.00866	0.0620	0.116
	(α_x/α_2)	0.626	0.933	0.505	0.665	0.641	0.674
E constant	ϕ^b	0.293	0.0556	0.128	0.00830	0.0594	0.112
	$(\alpha_x/\alpha_2)^b$	0.626	0.933	0.504	0.665	0.642	0.674
Penalty	$(\kappa_x/\kappa_x^o)^c$	-2.51e-4	-6.04e-4	-1.10e-3	-6.39e-4	-6.43e-3	-3.82e-3
	$(\alpha_x/\alpha_2)^d$	0.621	0.931	0.500	0.679	0.655	0.686

[a]Relative thickness and normalized CTE with temperature-dependent property.
[b]Relative thickness and normalized CTE with temperature-independent property.
[c]Normalized thermal curvature by substituting ϕ^b into the actual curvature equation.
[d]Normalized CTE by substituting ϕ^b into the actual effective laminate CTE equation.

Table 7 Thickness ratio and normalized CTE for zero in-plane expansion when $E=E(T)$

		ASGr-Ep /Al/ ASGr-Ep	Kev-Ep /Al/ Kev-Ep	P100Gr-Ep /Al/ P100Gr-Ep	Al/ ASGr-Ep /Al	Al/ Kev-Ep /Al	Al/ P100Gr-Ep /Al
E=E(T)[a]	ϕ	39.5	5.38	2.79	0.0266	0.195	0.376
	(κ_x/κ_x^o)	-0.0152	-0.153	-0.0798	0.0313	0.286	0.217
E constant	ϕ^b	40	5.45	2.83	0.0250	0.184	0.354
	$(\kappa_x/\kappa_x^o)^b$	-0.0155	-0.157	-0.0818	0.0313	0.287	0.215
Penalty	$(\kappa_x/\kappa_x^o)^c$	-0.0152	-0.153	-0.0799	0.0286	0.264	0.199
	$(\alpha_x/\alpha_2)^d$	**-1.48e-4**	**-1.70e-3**	**-5.80e-4**	**0.0574**	**0.0498**	**0.0553**

[a]Relative thickness and normalized thermal curvature with temperature-dependent property.
[b]Relative thickness and normalized thermal curvature with temperature-independent property.
[c]Normalized thermal curvature by substituting ϕ^b into the actual curvature equation.
[d]Normalized CTE by substituting ϕ^b into the actual effective laminate CTE equation.

Table 8 Thickness ratio and normalized CTE for zero thermal curvature when $k=k(T)$

		Al/ ASGr-Ep /Al	Al/ Kev-Ep /Al	Al/ P100Gr-Ep /Al
k=k(T)[a]	ϕ	0.00829	0.0594	0.112
	(α_x/α_2)	0.648	0.625	0.657
k constant	ϕ^b	0.00830	0.0594	0.112
	$(\alpha_x/\alpha_2)^b$	0.665	0.642	0.674
Penalty	$(\kappa_x/\kappa_x^o)^c$	**1.02e-5**	**1.76e-5**	**6.34e-5**
	$(\alpha_x/\alpha_2)^d$	0.648	0.625	0.656

[a]Relative thickness and normalized CTE with temperature-dependent property.
[b]Relative thickness and normalized CTE with temperature-independent property.
[c]Normalized thermal curvature by substituting ϕ^b into the actual curvature equation.
[d]Normalized CTE by substituting ϕ^b into the actual effective laminate CTE equation.

Table 9 Thickness ratio and normalized CTE for zero in-plane expansion when $k=k(T)$

		Al/ ASGr-Ep /Al	Al/ Kev-Ep /Al	Al/ P100Gr-Ep /Al
k=k(T)[a]	ϕ	0.0246	0.180	0.347
	(κ_x/κ_x^o)	0.0305	0.280	0.210
k constant	ϕ^b	0.0250	0.184	0.354
	$(\kappa_x/\kappa_x^o)^b$	0.0313	0.287	0.215
Penalty	$(\kappa_x/\kappa_x^o)^c$	0.0313	0.287	0.216
	$(\alpha_x/\alpha_2)^d$	**-0.0173**	**-0.0151**	**-0.0165**

[a]Relative thickness and normalized thermal curvature with temperature-dependent property.
[b]Relative thickness and normalized thermal curvature with temperature-independent property.
[c]Normalized thermal curvature by substituting ϕ^b into the actual curvature equation.
[d]Normalized CTE by substituting ϕ^b into the actual effective laminate CTE equation.

the same boundary temperatures T_1 and T_4. In the following tables, (κ_x/κ_x^0) is the normalized thermal curvature compared to an all-aluminum beam having identical bending stiffness and identical boundary temperatures T_1 and T_4. The values in bold face indicate the penality or error from ignoring the dependency of the property on temperature.

It can be seen from Table 6 that with the temperature coefficients assumed, the change for ϕ when $E=E(T)$ and the goal is to eliminate the thermal curvature is insignificant. The solution obtained from the temperature-independent case can still be used without introducing large error.

Table 7 shows that when using the laminate configuration of [composite/metal/composite], the error from the consideration of $E=E(T)$ is minimal when we want to eliminate the in-plane expansion. When the laminates are in the form of [metal/composite/metal], however, the effect can not be ignored.

It can be seen from Tables 8 and 9 that the results are similar with $k=k(T)$ or $E = E(T)$. When the goal is to eliminate thermal curvature, the relative thickness ratios solved with temperature-independent property are still valid; on the other hand, when we want to eliminate the in-plane expansion, we must solve the problem using the actual temperatures.

Closure

This paper demonstrates that symmetric laminated beams can be constructed with useful thermal deformation properties. By combining laminae with positive CTE with laminae with negative CTE, we can control or tailor the laminate CTE and/or the laminate thermal curvature. The design of the beam can be naturally expressed in terms of property ratios (given) and lamina thickness ratios (to be found). A three-layer beam offers one thickness ratio, and thus can be used either to control laminate CTE or eliminate thermal curvature. A five-layer beam offers two thickness ratios, and can be used to control CTE and eliminate curvature.

If the temperature excursion is large enough that the properties have to be considered as functions of both temperature and space. A zero thermal curvature laminated beam may still be constructed. In this paper, we demonstrate how to calculate the penalty for assuming that the properties are constant.

References

[1]Tompkins, S.S., and Funk, J.G., "Sensitivity of the Coefficient of Thermal Expansion of Selected Graphite Reinforced Composite Laminates to

Lamina Thermoelastic Properties," Society for the Advancement of Material and Processing Engineering Quarterly, Vol. 23, 1992, p. 55.

[2]Whitney, J.M., Structural Analysis of Laminated Anisotropic Plates, Technomic, Lancaster, PA, 1987.

[3]Wetherhold, R.C., and Wang, J., "A Self-Correcting, Thermal-Curvature-Stable Bending Element," Journal of Composite Materials. Vol. 28, 1994, p. 1588.

[4]Wetherhold, R.C., and Wang, J., "Tailoring Thermal Deformation by using Layered Beams", to be published in Composites Science and Technology.

[5]Tsai, S.W., Composites Design, Think Composites, Dayton, OH, 1987, p. B-2.

[6]Anon., Aluminum Standards and Data, Aluminum Association, Washington, DC, 1986, p. 30.

[7]Anon., Engineered Materials Handbook, Vol. 1, Composites, ASM International, Metals Park, OH, 1987, p. 399.

[8]Stonier, R.A., "Spacecraft Structures, Development", International Encyclopedia of Composites, Vol. 5, Editor Lee, S.M., VCH, New York, 1991, p. 241.

[9]Ross, R.B., Metallic Materials Specification Handbook, 4th ed, Chapman and Hall, New York, 1992, p. 96.

[10]Chapman, A.J., Heat Transfer, 4th ed. MacMillan, New York, 1984, p. 44.

[11]Boley, B.A., and Weiner, J.H., Theory of Thermal Stresses, Wiley, New York, 1960, p. 141.

Harmonic Generalized Thermoelastic Waves in Anisotropic Laminated Composites

Muhammad A. Hawwa* and Adnan H. Nayfeh†

May 25, 1995

Abstract

The propagation of harmonic waves in a laminated composite consisting of an arbitrary number of layered anisotropic plates is studied. In the context of the generalized theory of thermoelasticity, each layer is allowed to have a monoclinic degree of symmetry in the thermoelastic sense. Waves are allowed to propagate along any angle from the normal to the plates as well as along any azimuthal angle. The problem is treated analytically using a combination of linear orthogonal transformations and the transfer matrix method to reach the characteristic equation of the composite. Numerical illustrations are given in the form of dispersion and attenuation curves of a representative layering.

*Research Scientist, Department of Mechanical Engineering, Virginia Polytechnic Institute and State University, Blacksburg, Virginia 24061.

†Professor, Department of Aerospace Engineering and Engineering Mechanics, University of Cincinnati, Cincinnati, Ohio 45221.

I. Introduction

Motivated by the trend toward employing composite structures in environments with varying temperatures, this paper is concerned with the interaction of generalized thermoelastic waves with composite media. Generalized thermoelastic waves imply that, besides the mechanical waves, heat transfers in the form of waves propagating with finite speeds. This wave-like heat flow is known in the literature as the second sound. Second sound becomes of critical importance in low-temperature environments as well as under suddenly changing temperature conditions.

To summarize the literature on the modeling of generalized thermoelastic waves in solid structures, we mention the phenomenological models which have been studied by Lord and Shulman[1] who formulated the dynamical theory of generalized thermoelasticity for isotropic solids and gave some physical justifications to support their model; Achenbach[2] who employed the theory of propagating surfaces of discontinuity to study the effect of thermal relaxation time on the speeds of the wave fronts, and Nayfeh and Nemat-Nasser[3] who presented a detailed analysis of plane waves in isotropic unbounded media. In addition to the cited models, other generalized theories were developed on the entropy production inequality by Muller,[4] Green and Lindsay,[5] and Suhubi.[6]

Free generalized thermoelastic waves have been considered in isotropic media by Puri,[7,8] Agarwal,[9] and Tao and Prevost.[10] The interaction of generalized thermoelastic waves with anisotropic media was studied by Banerjee and Pao,[11] Dhaliwal and Sherief,[12] Singh and Sharma,[13] Sharma and Sidhu,[14] and Sharma and Singh.[15]

Guided waves were considered by several researchers. Nayfeh and Nemat-Nasser,[3] Sinha and Sinha,[16] and Agarwal[17] studied thermoelastic Rayleigh waves. Harinath,[18] and Harinath and Muthuswamy[19] studied thermoelastic Stonely type waves. Mondal,[20] Mondal and Jana,[21] Massalas,[22] Massalas and Kalpakidis,[23] and Massalas and Tsolakidis[24] considered the propagation of generalized thermoelastic waves in thin plates. Erby and Suhubi[25] studied the problem of longitudinal waves in a circular rod. Massalas and Kalpakidis[26] considered a waveguide of a rectangular cross section.

In contrast, little work has been conducted on thermoelastic waves in layered and/or composite media. Sve[27] used the classical theory of thermoelasticity to treat obliquely traveling waves in a bilaminated isotropic medium. His analysis was conducted by using potential functions. The results were given in the form of an exact dispersion relation resulting from the vanishing of a 12th-order determinant. Numerical results were given for the quasielastic modes in the form of dispersion and attenuation curves.

In the present paper, we consider the propagation of harmonic thermoelastic waves in laminated composites. The study is carried out in the context of the generalized theory of thermoelasticity. This choice allows the results corresponding to the classical theory to be obtained by setting the thermal relaxation time equal to zero. Each layer is allowed to have up to a monoclinic degree of anisotropy. The problem is formulated in the format of the transfer matrix method. The solutions can be used for the cases of higher symmetry materials such as orthotropic, transversely isotropic, and cubic by specializing the constitutive constants for these materials. Numerical illustrations are given in the form of dispersion and attenuation (damping) characteristics. Both the quasielastic and quasithermal modes are considered. The effect of the inclusion of thermal relaxation time is investigated.

II. Problem Formulation

Consider the layered medium shown in Fig. 1. Each layer is allowed to possess up to the monoclinic thermoelastic symmetry. In our analysis, the planes of elastic symmetry are assumed to coincide with those of thermal

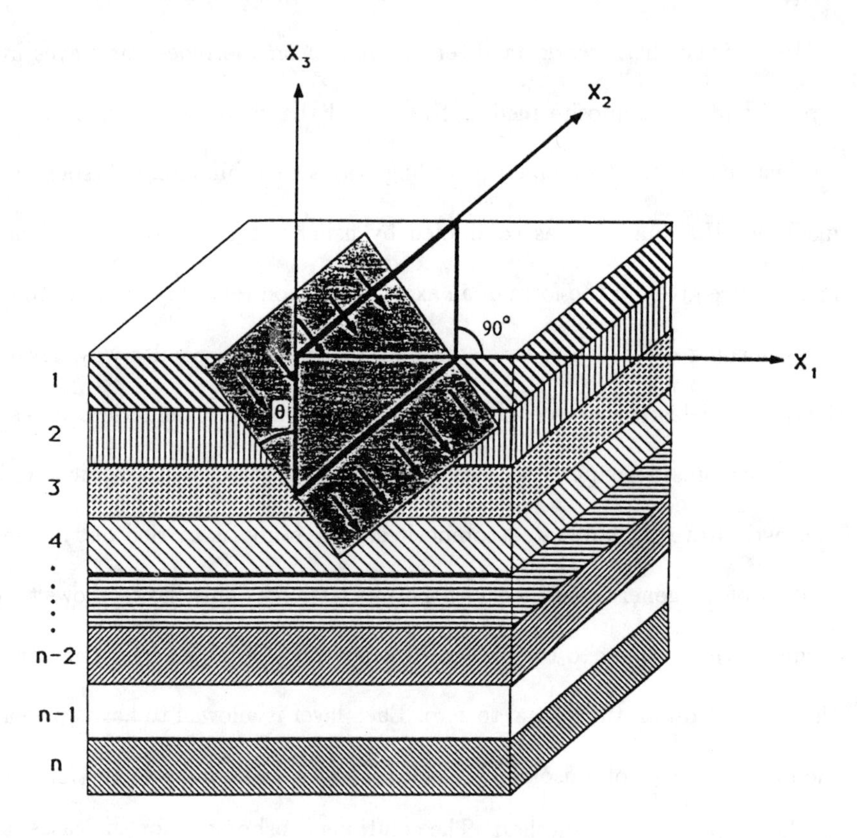

Fig. 1 Layered medium.

symmetry. This assumption is an idealistic one for natural materials, but it holds well for manufactured structures, such as fiber reinforced composites. Indicial notations are used to facilitate the treatment with tensor quantities. Thus, the frame of reference x_i stands for (x_1, x_2, x_3). With this notation, consider the array, which consists of an arbitrary number n of monoclinic layers perfectly bonded at their interfaces and stacked normal to the x_3 axis. Hence, the plane of each layer is parallel to the $x_1 - x_2$ plane which is also chosen to coincide with the bottom surface of the layered array. To maintain generality, it is assumed that each layer is arbitrarily oriented in the $x_1 - x_2$ plane. In order to be able to describe the relative orientation of the layers, a local Cartesian coordinate $(x'_i)_m$ is assigned for each layer $m, m = 1, 2, ..., n$, such that its origin is located in the bottom plane of the layer with $(x'_3)_m$ normal to it as shown in Fig. 2. Thus, layer m extends from $0 < (x'_3)_m < d_m$ where d_m is its thickness. According to this notation the total thickness of the layered array d equals the sum of the thicknesses of each layer and, hence, the plate occupies the region

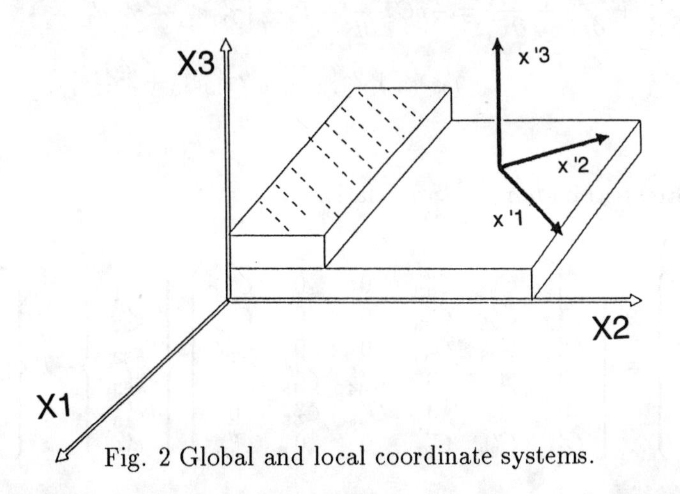

Fig. 2 Global and local coordinate systems.

$0 < (x_3) < d$. Equivalently, the orientation of the mth layer in the x_i space can be described by a rotation of an angle ϕ_m between $(x_1')_m$ and x_i. Hence, once all orientation angles ϕ_m are specified, the geometry of the plate is defined.

The generalized thermoelastic field equations for the anisotropic case were derived by Dhaliwal and Sherief.[12] With respect to the primed coordinate system (x_i') of layer m, they are given by the following:

1. Equations of motion:

$$\frac{\partial \sigma_{11}'}{\partial x_1'} + \frac{\partial \sigma_{12}'}{\partial x_2'} + \frac{\partial \sigma_{13}'}{\partial x_3'} = \rho \frac{\partial^2 u_1'}{\partial t^2} \tag{1}$$

$$\frac{\partial \sigma_{12}'}{\partial x_1'} + \frac{\partial \sigma_{22}'}{\partial x_2'} + \frac{\partial \sigma_{23}'}{\partial x_3'} = \rho \frac{\partial^2 u_2'}{\partial t^2} \tag{2}$$

$$\frac{\partial \sigma_{13}'}{\partial x_1'} + \frac{\partial \sigma_{23}'}{\partial x_2'} + \frac{\partial \sigma_{33}'}{\partial x_3'} = \rho \frac{\partial^2 u_3'}{\partial t^2} \tag{3}$$

2. Energy equation:

$$\frac{\partial q_1'}{\partial x_1'} + \frac{\partial q_2'}{\partial x_2'} + \frac{\partial q_3'}{\partial x_3'} = -\rho C_e \frac{\partial T}{\partial t} - T_0 \frac{\partial}{\partial t}\{\gamma_{11}' e_{11}' + \gamma_{22}' e_{22}' + \gamma_{33}' e_{33}' + 2\gamma_{23}' e_{23}' + 2\gamma_{13}' e_{13}' + 2\gamma_{12}' e_{12}'\} \tag{4}$$

3. Stress-strain-temperature relations:

$$\begin{Bmatrix} \sigma_{11}' \\ \sigma_{22}' \\ \sigma_{33}' \\ \sigma_{23}' \\ \sigma_{13}' \\ \sigma_{12}' \end{Bmatrix} = \begin{bmatrix} C_{11}' & C_{12}' & C_{13}' & 0 & 0 & C_{16}' \\ C_{12}' & C_{22}' & C_{23}' & 0 & 0 & C_{26}' \\ C_{13}' & C_{23}' & C_{33}' & 0 & 0 & C_{36}' \\ 0 & 0 & 0 & C_{44}' & C_{45}' & 0 \\ 0 & 0 & 0 & C_{45}' & C_{55}' & 0 \\ C_{16}' & C_{26}' & C_{36}' & 0 & 0 & C_{66}' \end{bmatrix} \begin{Bmatrix} e_{11}' \\ e_{22}' \\ e_{33}' \\ e_{23}' \\ e_{13}' \\ e_{12}' \end{Bmatrix} - T \begin{Bmatrix} \gamma_{11}' \\ \gamma_{22}' \\ \gamma_{33}' \\ \gamma_{23}' \\ \gamma_{13}' \\ \gamma_{12}' \end{Bmatrix} \tag{5}$$

where the contracting subscript notations are used, i.e: $1 \rightarrow 11$, $2 \rightarrow 22$, $3 \rightarrow 33$, $4 \rightarrow 23$, $5 \rightarrow 13$, $6 \rightarrow 12$ to relate $c'_{ijk\ell}$ to C'_{pq} , where $(i, j, k, \ell = 1, 2, 3, \text{and} \quad p, q = 1, 2, \ldots, 6)$.

4. Heat conduction constitutive relations:

$$(1+\tau_0\frac{\partial}{\partial t})\left\{\begin{array}{c} q'_1 \\ q'_2 \\ q'_3 \end{array}\right\} = \begin{bmatrix} k'_{11} & k'_{12} & 0 \\ k'_{12} & k'_{22} & 0 \\ 0 & 0 & k'_{33} \end{bmatrix} \left\{\begin{array}{c} \frac{\partial}{\partial x'_1} \\ \frac{\partial}{\partial x'_2} \\ \frac{\partial}{\partial x'_3} \end{array}\right\} T \tag{6}$$

where σ'_{ij} are the components of the stress tensor, e'_{ij} are the components of the strain tensor, u'_i are the displacement components in x'_i directions, ρ is the material density, $c'_{ijk\ell}$ (C'_{pq}) are the elastic constants of the material, q'_i are the components of heat flux in x'_i direction, T is the temperature change from the absolute basic temperature T_0, $\gamma'_{ij}(= c'_{ijk\ell}\alpha'_{kl})$ are the thermoelastic coupling constants, where α'_{kl} are the elements of the linear thermal expansion tensor, k'_{ij} are the components of the thermal conductivity tensor, C_e is the specific heat at constant strain, and τ_0 is the thermal relaxation time.

It is assumed that plane waves propagate in the $x_1 - x_3$ plane, at an arbitrary angle θ measured from the normal x_3. This assumption makes it suitable to conduct the analysis in terms of the global coordinate system (see Nayfeh et al.[28]).

Since σ'_{ij}, $e'_{k\ell}$, $c'_{ijk\ell}$, q'_i, γ'_{ij}, and k'_{ij} are tensors, they are transformed and written in terms of the global coordinate system according to

$$\sigma_{mn} = \beta_{mi}\beta_{nj} \quad \sigma'_{ij} \tag{7a}$$

$$e_{op} = \beta_{ok}\beta_{p\ell} \quad e'_{k\ell} \tag{7b}$$

$$c_{mnop} = \beta_{mi}\beta_{nj}\beta_{ok}\beta_{p\ell}\ \ c'_{ijk\ell} \tag{7c}$$

$$q_r = \beta_{ri}\ \ q'_i \tag{7d}$$

$$k_{rs} = \beta_{ri}\beta_{sj}\ \ k'_{ij} \tag{7e}$$

$$\gamma_{st} = \beta_{si}\beta_{tj}\ \ \gamma'_{ij} \tag{7f}$$

where the β are the direction cosines of the primed system x'_i with respect to the unprimed system x_j. For a rotation of angle ϕ in the $x'_1 - x'_2$ plane, the transformation tensor reduces to

$$\begin{bmatrix} \cos\phi & \sin\phi & 0 \\ -\sin\phi & \cos\phi & 0 \\ 0 & 0 & 1 \end{bmatrix} \tag{8}$$

The equations of motion, Eqs. (1-3) are expressed in terms of the rotated system x_i as

$$\frac{\partial \sigma_{11}}{\partial x_1} + \frac{\partial \sigma_{12}}{\partial x_2} + \frac{\partial \sigma_{13}}{\partial x_3} = \rho\frac{\partial^2 u_1}{\partial t^2} \tag{9}$$

$$\frac{\partial \sigma_{12}}{\partial x_1} + \frac{\partial \sigma_{22}}{\partial x_2} + \frac{\partial \sigma_{23}}{\partial x_3} = \rho\frac{\partial^2 u_2}{\partial t^2} \tag{10}$$

$$\frac{\partial \sigma_{13}}{\partial x_1} + \frac{\partial \sigma_{23}}{\partial x_2} + \frac{\partial \sigma_{33}}{\partial x_3} = \rho\frac{\partial^2 u_3}{\partial t^2} \tag{11}$$

and the transformed energy equation reads:

$$\frac{\partial q_1}{\partial x_1} + \frac{\partial q_2}{\partial x_2} + \frac{\partial q_3}{\partial x_3} = -\rho C_e\frac{\partial T}{\partial t} - T_0\frac{\partial}{\partial t}\{\gamma_{11}e_{11} + \gamma_{22}e_{22} + \gamma_{33}e_{33} + 2\gamma_{23}e_{23} + 2\gamma_{13}e_{13} + 2\gamma_{12}e_{12}\} \tag{12}$$

If Eqs. (7) are applied to the relations (5) and (6), the following constitutive relations are yielded:

$$\begin{Bmatrix} \sigma_{11} \\ \sigma_{22} \\ \sigma_{33} \\ \sigma_{23} \\ \sigma_{13} \\ \sigma_{12} \end{Bmatrix} = \begin{bmatrix} C_{11} & C_{12} & C_{13} & 0 & 0 & C_{16} \\ C_{12} & C_{22} & C_{23} & 0 & 0 & C_{26} \\ C_{13} & C_{23} & C_{33} & 0 & 0 & C_{36} \\ 0 & 0 & 0 & C_{44} & C_{45} & 0 \\ 0 & 0 & 0 & C_{45} & C_{55} & 0 \\ C_{16} & C_{26} & C_{36} & 0 & 0 & C_{66} \end{bmatrix} \begin{Bmatrix} e_{11} \\ e_{22} \\ e_{33} \\ e_{23} \\ e_{13} \\ e_{12} \end{Bmatrix} - T \begin{Bmatrix} \gamma_{11} \\ \gamma_{22} \\ \gamma_{33} \\ \gamma_{23} \\ \gamma_{13} \\ \gamma_{12} \end{Bmatrix} \tag{13}$$

and

$$\left(1 + \tau_0 \frac{\partial}{\partial t}\right) \begin{Bmatrix} q_1 \\ q_2 \\ q_3 \end{Bmatrix} = \begin{bmatrix} k_{11} & k_{12} & 0 \\ k_{12} & k_{22} & 0 \\ 0 & 0 & k_{33} \end{bmatrix} \begin{Bmatrix} \frac{\partial}{\partial x_1} \\ \frac{\partial}{\partial x_2} \\ \frac{\partial}{\partial x_3} \end{Bmatrix} T \tag{14}$$

where the transformation relations between the C_{pq}, and those between k_{ij} and γ_{ij} are in the Appendix. Notice that regardless of the rotational angle ϕ, the zero entries in Eqs. (5) and (6) will remain zero in Eqs. (13) and (14). This is a property of monoclinic materials.

Substituting from relations (13) and (14) into Eqs. (9-11) and (12) results in the following system of equations for the displacements u_1, u_2, and u_3, and the temperature change T:

$$\begin{aligned} &\left[C_{11}\frac{\partial^2}{\partial x_1^2} + 2C_{16}\frac{\partial^2}{\partial x_1 \partial x_2} + C_{66}\frac{\partial^2}{\partial x_2^2} + C_{55}\frac{\partial^2}{\partial x_3^2} - \rho\frac{\partial^2}{\partial t^2}\right] u_1 + \left[C_{61}\frac{\partial^2}{\partial x_1^2}\right. \\ &\left. + (C_{12} + C_{66})\frac{\partial^2}{\partial x_1 \partial x_2} + C_{26}\frac{\partial^2}{\partial x_2^2} + C_{45}\frac{\partial^2}{\partial x_3^2}\right] u_2 + \left[(C_{36} + C_{45})\frac{\partial^2}{\partial x_2 \partial x_3}\right. \\ &\left. + (C_{13} + C_{55})\frac{\partial^2}{\partial x_1 \partial x_3}\right] u_3 + \left(\gamma_{11}\frac{\partial}{\partial x_1} + \gamma_{12}\frac{\partial}{\partial x_2}\right) T = 0 \end{aligned} \tag{15}$$

$$\begin{aligned} &\left[C_{16}\frac{\partial^2}{\partial x_1^2} + (C_{12} + C_{66})\frac{\partial^2}{\partial x_1 \partial x_2} + C_{26}\frac{\partial^2}{\partial x_2^2} + C_{45}\frac{\partial^2}{\partial x_3^2}\right] u_1 + \left[C_{66}\frac{\partial^2}{\partial x_1^2}\right. \\ &\left. + 2C_{26}\frac{\partial^2}{\partial x_1 \partial x_2} + C_{22}\frac{\partial^2}{\partial x_2^2} + C_{44}\frac{\partial^2}{\partial x_3^2} - \rho\frac{\partial^2}{\partial t^2}\right] u_2 + \left[(C_{23} + C_{44})\frac{\partial^2}{\partial x_2 \partial x_3}\right. \\ &\left. + (C_{36} + C_{45})\frac{\partial^2}{\partial x_1 \partial x_3}\right] u_3 + \left(\gamma_{12}\frac{\partial}{\partial x_1} + \gamma_{22}\frac{\partial}{\partial x_2}\right) T = 0 \end{aligned} \tag{16}$$

$$\left[(C_{36}+C_{45})\frac{\partial^2}{\partial x_2\partial x_3}+(C_{13}+C_{55})\frac{\partial^2}{\partial x_1\partial x_3}\right]u_1+\left[(C_{23}+C_{44})\frac{\partial^2}{\partial x_2\partial x_3}\right.$$
$$\left.+(C_{36}+C_{45})\frac{\partial^2}{\partial x_1\partial x_3}\right]u_2+\left[C_{55}\frac{\partial^2}{\partial x_1^2}+C_{45}\frac{\partial^2}{\partial x_1\partial x_2}+C_{44}\frac{\partial^2}{\partial x_2^2}+C_{33}\frac{\partial^2}{\partial x_3^2}\right.$$
$$\left.-\rho\frac{\partial^2}{\partial t^2}\right]u_3+\left(\gamma_{33}\frac{\partial}{\partial x_3}\right)T=0 \tag{17}$$

$$-T_0\left(\frac{\partial}{\partial t}+\tau_0\frac{\partial^2}{\partial t^2}\right)\left\{\left[\gamma_{11}\frac{\partial}{\partial x_1}+\gamma_{12}\frac{\partial}{\partial x_2}\right]u_1+\left[\gamma_{12}\frac{\partial}{\partial x_1}+\gamma_{22}\frac{\partial}{\partial x_2}\right]u_2\right.$$
$$\left.+\left[\gamma_{33}\frac{\partial}{\partial x_3}\right]u_3\right\}+\left[k_{11}\frac{\partial^2}{\partial x_1^2}+2k_{12}\frac{\partial^2}{\partial x_1\partial x_2}+k_{22}\frac{\partial^2}{\partial x_2^2}+k_{33}\frac{\partial^2}{\partial x_3^2}\right.$$
$$\left.-\rho C_e\left(\frac{\partial}{\partial t}+\tau_0\frac{\partial^2}{\partial t^2}\right)\right]T=0 \tag{18}$$

III. Formal Solutions

Having identified the plane of incidence to be the $x_1 - x_3$ plane, then for an angle of incidence θ, the following solution for the displacements and the temperature change is proposed:

$$(u_1, u_2, u_3, T) = (\tilde{u}, \tilde{v}, \tilde{w}, \tilde{h})e^{i\xi(x_1 \sin\theta + px_3 - ct)} \tag{19}$$

where ξ is the wave number, $c = \omega/\xi$, ω is the circular frequency, p is still an unknown parameter, and $\tilde{u}, \tilde{v}, \tilde{w}$, and $\tilde{h}$ are the amplitudes of u_1, u_2, u_3, and T, respectively. Notice that although solutions (19) are explicitly independent of x_2, an implicit dependence is contained in the transformation. Furthermore, notice the nonvanishing of the transverse displacement component u_2 in Eq. (19). The assumption of this solution is a consequence to the work by Henneke[29] that the variables are independent of the direction normal to the propagation direction.

Substituting the solution (19) into the field Eqs. (15-18) leads to the four coupled equations

$$[B(p)]\mathbf{U} = 0 \tag{20}$$

where $\mathbf{U} = \{\tilde{u}, \tilde{v}, \tilde{w}, \tilde{h}\}^T$, and

$$B_{11} = C_{11}\sin^2\theta - \rho c^2 + C_{55}p^2$$

$$B_{12} = C_{16}\sin^2\theta + C_{45}p^2$$

$$B_{13} = (C_{13} + C_{55})p\sin\theta$$

$$B_{14} = i\gamma_{11}\sin\theta c/\omega$$

$$B_{21} = B_{12}$$

$$B_{22} = C_{66}\sin^2\theta - \rho c^2 + C_{44}p^2$$

$$B_{23} = (C_{36} + C_{45})p\sin\theta$$

$$B_{24} = i\gamma_{12}\sin\theta c/\omega$$

$$B_{31} = B_{13}$$

$$B_{32} = B_{23}$$

$$B_{33} = C_{55}\sin^2\theta - \rho c^2 + C_{33}p^2$$

$$B_{34} = i\gamma_{33}pc/\omega$$

$$B_{41} = i\gamma_{11}T_0\sin\theta c(i + \tau_0\omega)$$

$$B_{42} = i\gamma_{12}T_0\sin\theta c(i + \tau_0\omega)$$

$$B_{43} = i\gamma_{33}T_0 pc(i + \tau_0\omega)$$

$$B_{44} = \rho C_e c^2(i/\omega + \tau_0) - k_{11}\sin^2\theta - p^2 k_{33} \tag{21}$$

The existence of a nontrivial solutions for $\tilde{u}, \tilde{v}, \tilde{w}$, and $\tilde{h}$ requires the vanishing of the determinant in Eq. (20), and yields the eighth-degree polynomial equation

$$p^8 + A_1 p^6 + A_2 p^4 + A_3 p^2 + A_4 = 0 \tag{22}$$

where the coefficients A_1, A_2, A_3, and A_4 are given in the Appendix. Equation (22) admits eight solutions for p. Equation (20) can be written in the following form:

$$\begin{bmatrix} B_{11} & B_{13} & B_{14} \\ B_{22} & B_{23} & B_{24} \\ B_{23} & B_{33} & B_{34} \end{bmatrix} \begin{Bmatrix} V_j \\ W_j \\ H_j \end{Bmatrix} = \begin{Bmatrix} -B_{11} \\ -B_{22} \\ -B_{13} \end{Bmatrix} \tag{23}$$

This allows the use of Cramer's rule to find the ratios $V_j = \tilde{v_j}/\tilde{u_j}$, $W_j = \tilde{w_j}/\tilde{u_j}$, and $H_j = \tilde{h_j}/\tilde{u_j}$ corresponding to each p_j, $j = 1, ..., 8$. Combining these ratios with the elastic and thermal constitutive relations (13) and (14), and using superposition, the formal solutions for the displacements, stresses, temperature change, and heat flux can be written in the expanded matrix form:

$$\begin{Bmatrix} u_1 \\ u_2 \\ u_3 \\ T \\ \sigma_{33} \\ \sigma_{13} \\ \sigma_{23} \\ q_3 \end{Bmatrix} = \begin{bmatrix} 1 & 1 & 1 & 1 & 1 & 1 & 1 & 1 \\ V_1 & V_2 & V_3 & V_4 & V_5 & V_6 & V_7 & V_8 \\ W_1 & W_2 & W_3 & W_4 & W_5 & W_6 & W_7 & W_8 \\ H_1 & H_2 & H_3 & H_4 & H_5 & H_6 & H_7 & H_8 \\ D_{11} & D_{12} & D_{13} & D_{14} & D_{15} & D_{16} & D_{17} & D_{18} \\ D_{21} & D_{22} & D_{23} & D_{24} & D_{25} & D_{26} & D_{27} & D_{28} \\ D_{31} & D_{32} & D_{33} & D_{34} & D_{35} & D_{36} & D_{37} & D_{38} \\ Q_1 & Q_2 & Q_3 & Q_4 & Q_5 & Q_6 & Q_7 & Q_8 \end{bmatrix} \begin{Bmatrix} \tilde{u}_1 E_1 \\ \tilde{u}_2 E_2 \\ \tilde{u}_3 E_3 \\ \tilde{u}_4 E_4 \\ \tilde{u}_5 E_5 \\ \tilde{u}_6 E_6 \\ \tilde{u}_7 E_7 \\ \tilde{u}_8 E_8 \end{Bmatrix} \tag{24}$$

where

$$E_j = e^{i\omega p_j x_3 / c}$$

$$D_{1j} = i\omega[C_{13}\sin\theta + C_{36}\sin\theta V_j + C_{33}p_j W_j]/c - \gamma_{33}H_j$$

$$D_{2j} = i\omega[C_{13}(p_j + W_j \sin\theta) + C_{44}p_j V_j]/c$$

$$D_{3j} = i\omega[C_{55}(p_j + W_j \sin\theta) + C_{45}p_j V_j]/c$$

$$Q_j = -i\omega p_j k_{33} H_j/[c(1 - i\omega\tau_0)], \quad j = 1, 2, \ldots, 8 \tag{25}$$

IV. The Transfer Matrix Method

Equation (24) can be used to relate the displacements, stresses, temperature change, and heat flux at $(x_3')_m = d_m$ to those at $(x_3')_m = 0$. This is done by specializing Eq. (24) to these two locations, eliminating the common amplitudes $u_1, \ldots, u_8$ and getting

$$\mathbf{F}_m^+ = [A_m]\mathbf{F}_m^-, \qquad m = 1, 2, \ldots, n \tag{26}$$

here

$$\mathbf{F}_m^{\mp} = \{u_1, u_2, u_3, T, \sigma_{33}, \sigma_{23}, \sigma_{13}, q_3\}_{m\mp}^T \tag{27}$$

are the column vectors of the field variables specialized to the upper and lower surfaces of the layer, m, respectively, and

$$[A_m] = [\chi_m][E_m][\chi_m]^{-1} \tag{28}$$

where $[\chi_m]$ is the 8×8 square matrix of Eq. (24) and $[E_m]$ is an 8×8 diagonal matrix whose entries are $e^{i\omega p_j d_m/c}$.

Matrix $[A_m]$ constitutes the transfer matrix for the thermoelastic monoclinic layer m. By applying the above given procedure for each layer followed by invoking continuity of displacements, stress components, temperature change, and heat flux at the layer interfaces, the field variables at the top of the layered array, $x_3 = d$, can be related to those at its bottom, $x_3 = 0$, via the transfer matrix multiplication

$$[A] = [A_n][A_{n-1}]\dots[A_1] \tag{29}$$

which results in

$$\mathbf{F}^+ = [A]\mathbf{F}^- \tag{30}$$

where now $\mathbf{F}^+$ and $\mathbf{F}^-$ are the displacement, stress, temperature change, and heat flux column vectors at the top, $x_3 = d$, and the bottom, $x_3 = 0$, of the total array, respectively.

To establish the characteristic equation, relevant boundary conditions are introduced. A representative case is the medium composed of repeating a laminated unit cell. Hence, the periodicity condition can be introduced in the form of the following generalized Floquet condition:[30]

$$F^+ = F^- e^{i\xi d \cos\theta} \tag{31}$$

When Eq. (30) is combined with Eq. (31), the characteristic equation can be obtained as

$$\mathrm{Det}([A] - [I]e^{i\xi d \cos\theta}) = 0 \tag{32}$$

where $[I]$ is an 8×8 unit matrix.

V. Numerical Results

Numerical illustrations of the analytical characteristic equations are presented in the form of dispersion and attenuation or damping curves. These can be obtained in two ways. The first is to let the frequency be a real number with keeping the wavenumber ξ complex. In this case, the phase velocity of the waves c is defined as $c = \omega/Re(\xi)$, and the imaginary part of ξ is used as a measure for the attenuation of the waves in space. The second way is to keep ξ real, and let ω be complex. Then, the phase velocity is defined as $c = Re(\omega)/\xi$, and the imaginary part of ω is a measure for the damping of the waves with time (dissipation of energy). In order to find the solutions of a characteristic equation, Muller's method is used to solve it as an analytic complex function.

A representative orthotropic material with the following properties is used: $\rho = 2000$ kg/m^3, $C'_{11} = 128$ MPa, $C'_{12} = 7$ MPa, $C'_{13} = 6$ MPa, $C'_{22} = 72$ MPa, $C'_{23} = 5$ MPa, $C'_{33} = 32$ MPa, $C'_{44} = 18$ MPa, $C'_{55} = 12.25$ MPa, $C'_{66} = 8$ MPa, $k'_{11} = 100$ W/mK, $k'_{22}= 50$ W/mK, $k'_{33} = 25$ W/mK, $q'_{11} = 0.04$ MPa/K, $q'_{22} = 0.06$ MPa/K, $q'_{33} = 0.09$ Mpa/K, $T_0 = 300$ K, and $\tau_0 = 2 \times 10^{-7}$ sec, and $e_1 = 1 \times 10^{-2}$, where $e_1 = q_{11}T_0/\rho C_e C_{11}$. This material can be used to construct different layers, by rotating it by different angles in the $x_1{-}x_2$ plane. This choice of the material was stimulated by the way of manufacturing composite structures. Without any loss of generality, all of the layers, which constitute a unit cell, are assumed to have the same thickness in such a way that the total thickness of the unit cell is one.

The propagation direction is given by the representative angle $\theta = 45$ deg. The waves traveling in this direction are assumed to attack a trilaminated medium, with a combination of layers that makes angles of $-60, 0,$ and 60 deg with x_1.

Conventional dispersion curves are given in the form of variations of the phase velocity c with the wave number ξ. Next, the quasielastic modes are considered in detail by the presentation of the real part of ξ vs ω, and by monitoring the imaginary part of ξ with ω. The dispersive character of the mode can be seen in the first category, whereas the damping of the waves is shown in the other group. The effect of the inclusion of the thermal relaxation time in the formulation is also investigated. The results obtained when τ_0 is set equal to zero are depicted by dotted curves, and those which belong to the generalized formulation are drawn by dashed curves. The dispersion and attenuation of the quasithermal modes are given as functions of ω. For an assigned real value of ω, a complex wave number ξ is found and plotted vs ω. The wave-like behavior of the quasithermal modes is characterized.

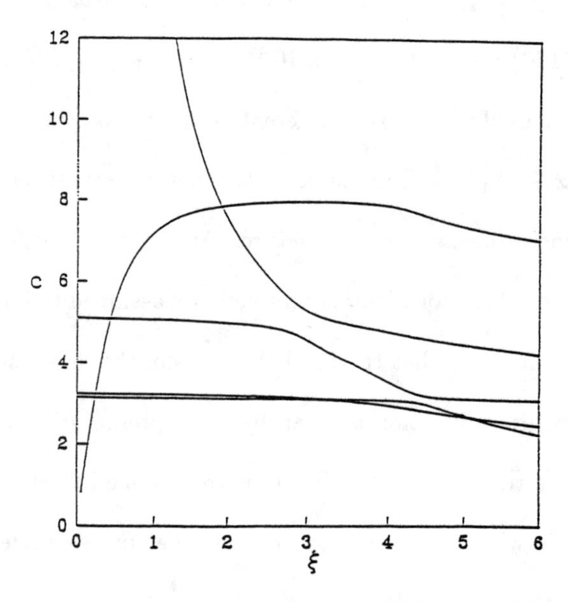

Fig. 3 Dispersion curves for the (-60, 0, 60) layup with $\theta = 45$.

Figure 3, shows the variation of the phase velocity c with the wave number ξ for the (-60/0/60) layup. In this figure, four fundamental modes are observed. Those are three quasielastic modes corresponding to the longitudinal, vertically polarized, and horizontally polarized waves, and a quasithermal generalized mode. The three quasielastic modes are far from each other, reflecting the anisotropic character of the medium. The quasithermal mode seems to show a rapid increase of the phase speed in the beginning, and then an asymptotic trend toward a finite value of the phase velocity.

The quasilongitudinal mode is considered in the $\xi - \omega$ plane in Fig. 4. The dissipation indicated by the imaginary part of ω is high for small wave numbers. The amount of dissipation dips down to local minima at certain values of the wave numbers. This behavior is similar to what was shown by Diamaruya and Naitoh[31] who studied coupled thermoelastic waves in infinite plates. The inclusion of the thermal relaxation time increases the amount of dissipation.

The quasishear mode includes the vertically polarized and the horizontally polarized modes. The vertically polarized mode is shown in Figs. 5. It shows less values of damping than the quasilongitudinal modes. The horizontally polarized mode is depicted in Fig. 6. The amount of damping is very small. The curves for these two cases show a maximum of damping corresponding to a steepest change in the dispersion curves. Thermal relaxation time increases the damping for both of the shear cases.

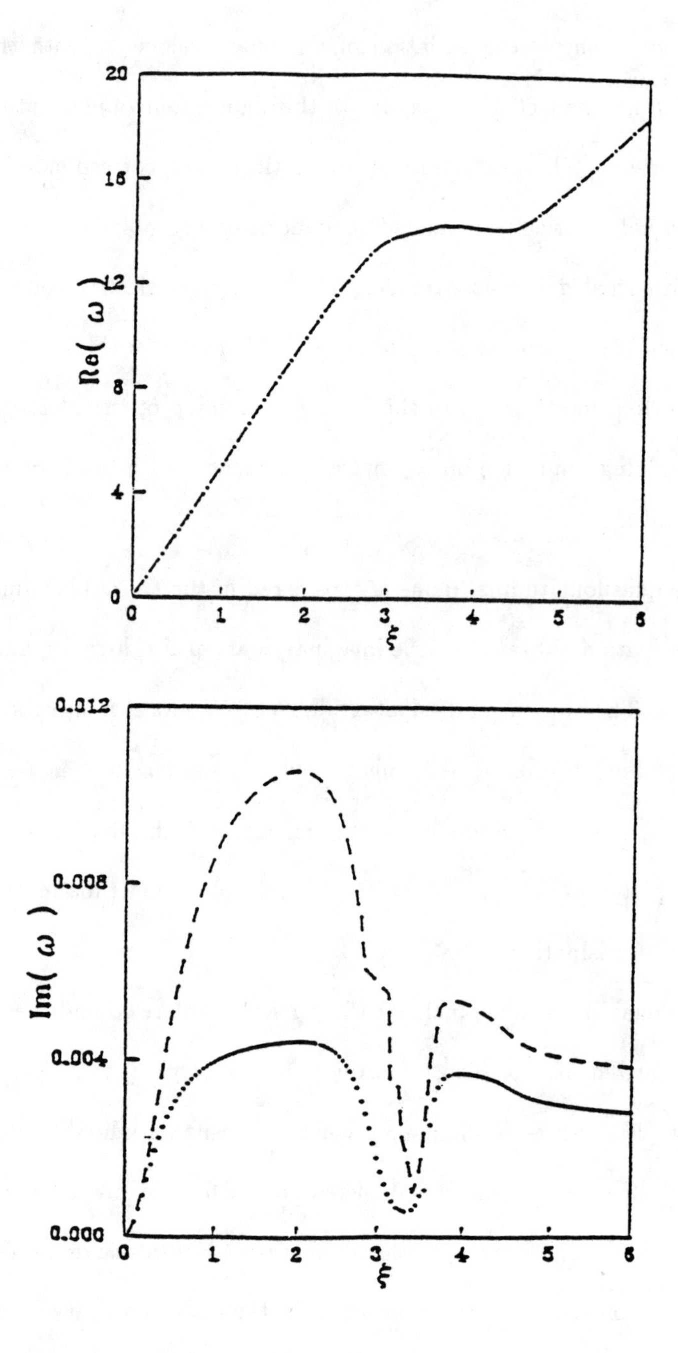

Fig. 4 Dispersion and damping of the quasilongitudinal mode; generalized theory (dashed), classical theory (dotted).

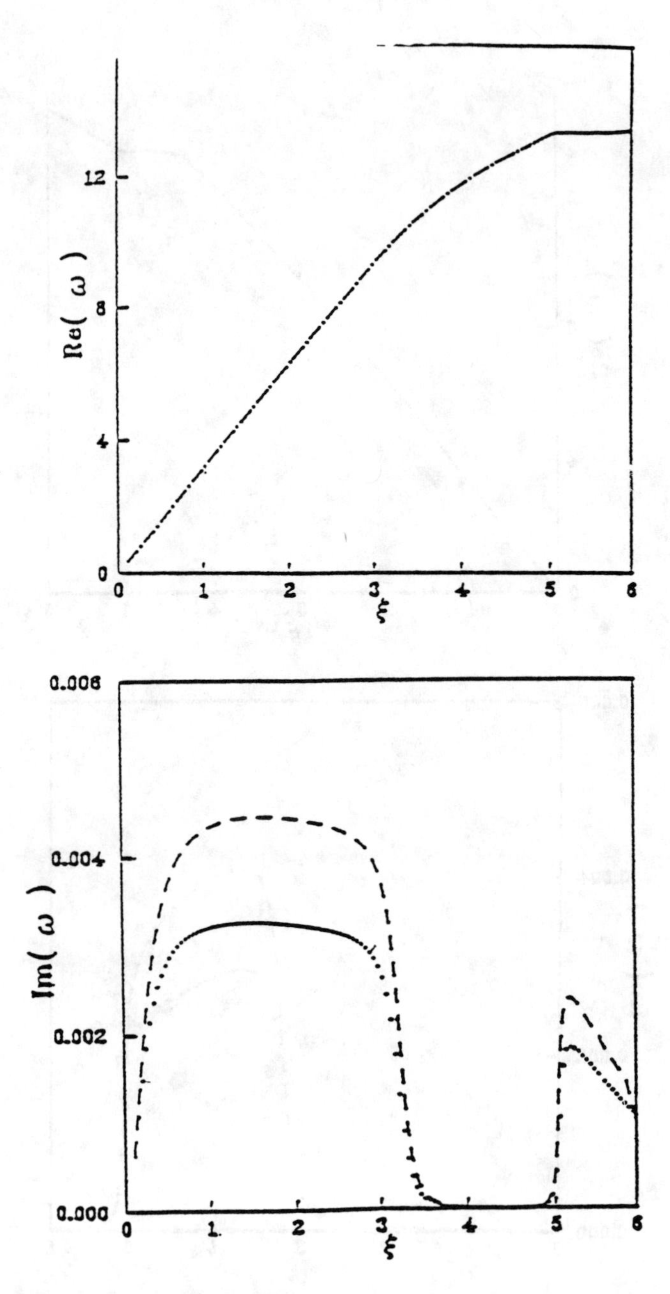

Fig. 5 Dispersion and damping of a quasishear mode; generalized theory (dashed), classical theory (dotted).

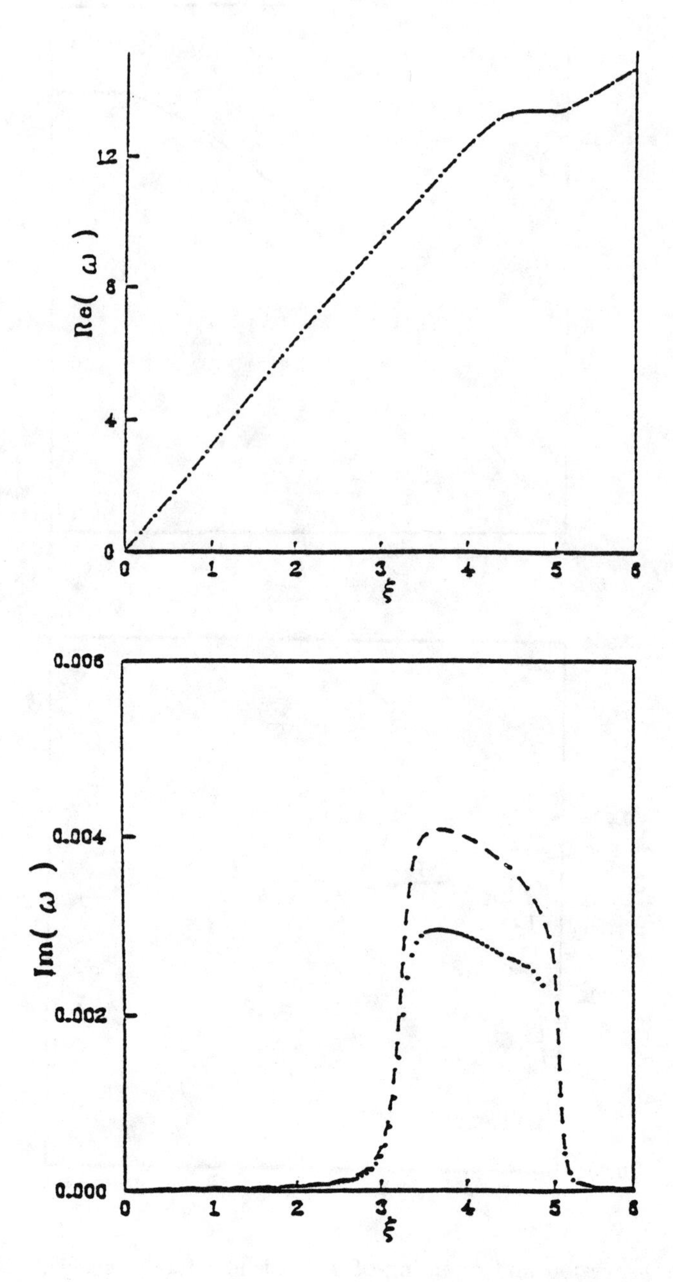

Fig. 6 Dispersion and damping of a quasishear mode; generalized theory (dashed), classical theory (dotted).

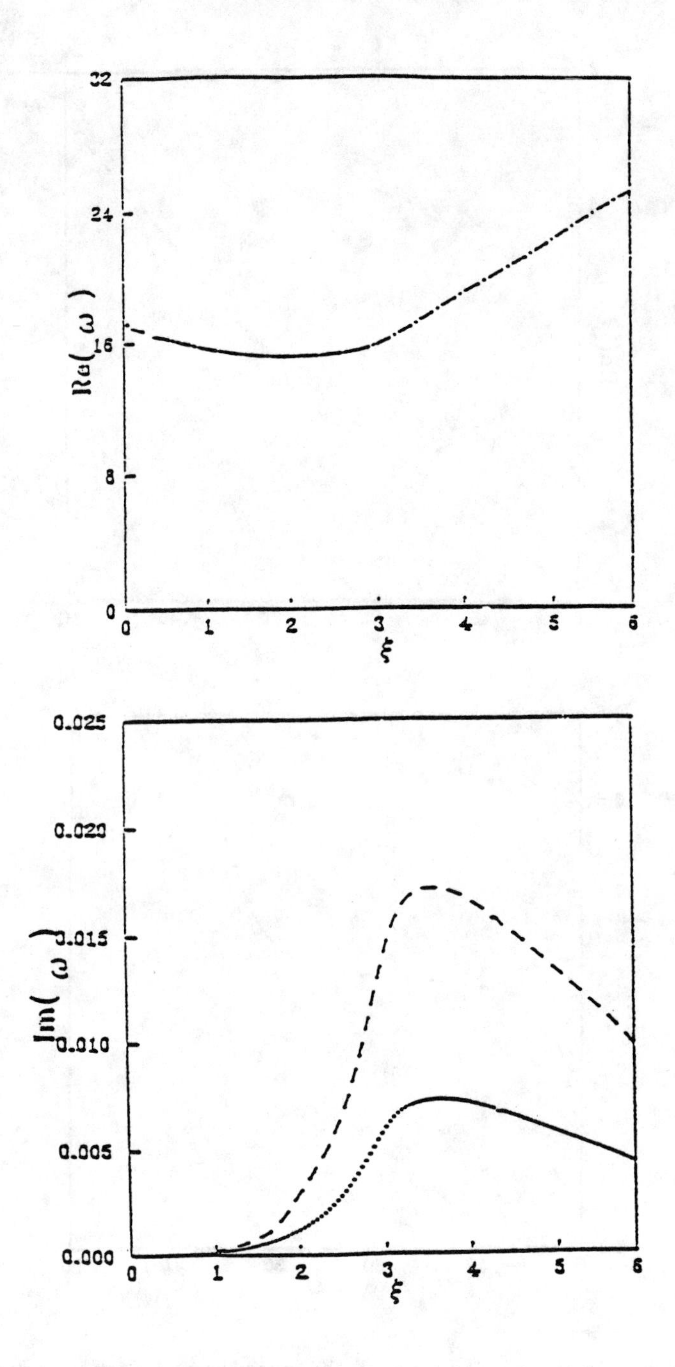

Fig. 7 Dispersion and damping of a higher-order quasielastic mode; generalized theory (dashed), classical theory (dotted).

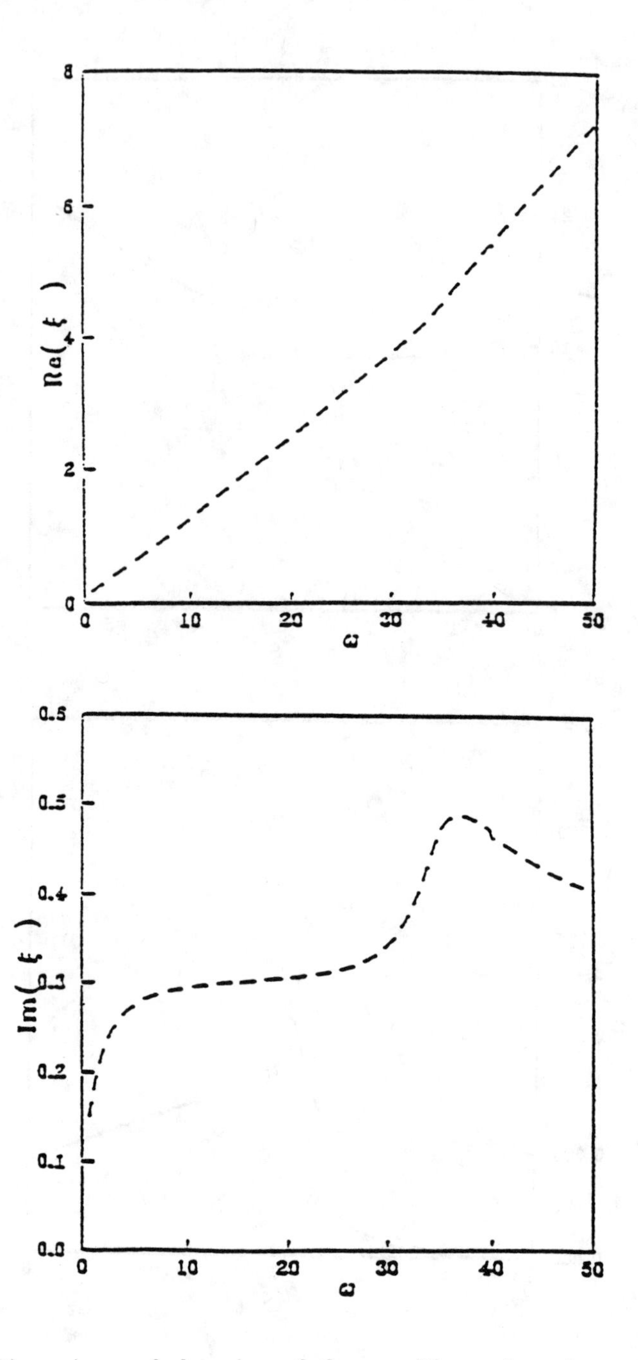

Fig. 8 Dispersion and damping of the quasithermal mode; generalized theory (dashed), classical theory (dotted).

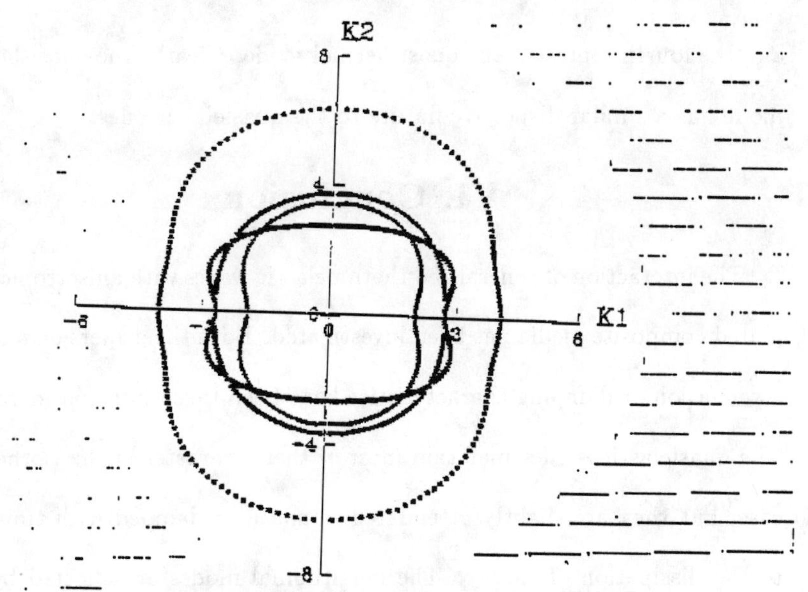

Fig. 9 Wave front for the (-60, 0, 60) layup with $\theta = 45$ and $\omega = 1$.

The higher order modes are represented by the curves in Fig. 7. In the plot, the amount of dissipation is high compared to those of the fundamental modes. They show maxima corresponding to the minima of the real part of ω. The thermal relaxation time also increases the dissipation.

The thermal mode is in Fig. 8. The dispersion curves are similar to any elastic dispersion one. The attenuation increases with ω, reaches a maximum corresponding to the steepest slope in the dispersion curve, and then starts to decrease.

Finally, the dispersive behavior of such media is investigated by plotting $(c \sin\theta)$ vs $(c \cos\theta)$ for a fixed value of the frequency ω. Figure 9 depicts the wave fronts for the (-60, 0, 60) layup at $\omega = 1$. Four wave fronts are shown, three of them correspond to the quasielastic fundamental modes, and the fourth represent the quasithermal mode. Clearly, the quasithermal mode has a similar dispersive nature to the quasielastic ones.

VI. Conclusion

The interaction of generalized thermoelastic waves with anisotropic laminated composite media has been investigated. Both the dispersion and the attenuation or damping characteristics have been taken into consideration. The quasielastic modes maintain most of their character in the isothermal case, but they are slightly attenuated in space or damped with time due to the dissipation of energy. The longitudinal modes are affected by the thermal coupling more than the shear modes. The amount of attenuation or damping is noticed to have large values when the slope of the dispersion curves is steep. Thermal relaxation time tends to increase the value of attenuation and damping of the different modes. Thermal mode has dispersion curves similar to those of the elastic waves. However, the values of attenuation for the thermal mode are of large order of magnitudes.

Appendix

Let $G = \cos\theta$ and $S = \sin\theta$.

The transformation relations of elastic constants are

$$C_{11} = C'_{11}G^4 + C'_{22}S^4 + 2(C'_{12} + 2C'_{66})S^2G^2$$

$$C_{12} = (C'_{11} + C'_{22} - 4C'_{66})S^2G^2 + C'_{12}(S^4 + G^4)$$

$$C_{13} = C'_{13}G^2 + C'_{23}S^2$$

$$C_{16} = (2C'_{66} + C'_{12} - C'_{11})SG^3 + (C'_{22} - C'_{12} - 2C'_{66})GS^3$$

$$C_{22} = C'_{11}S^4 + 2(C'_{12} + 2C'_{66})S^2G^2 + C'_{22}G^4$$

$$C_{23} = C'_{23}G^2 + C'_{13}S^2$$

$$C_{26} = (C'_{11} - C'_{12} - 2C'_{66})GS^3 + (C'_{12} - C'_{22} - 2C'_{66})SG^3$$

$$C_{33} = C'_{33}$$

$$C_{36} = (C'_{23} - C'_{13})SG$$

$$C_{44} = C'_{44}G^2 + C'_{55}S^2$$

$$C_{45} = (C'_{44} - C'_{55})SG$$

$$C_{55} = C'_{55}G^2 + C'_{44}S^2$$

$$C_{66} = (C'_{11} + C'_{22} - 2C'_{12} - 2C'_{66})S^2G^2 + C'_{66}(S^4 + G^4)$$

The transformation relations of thermal conductivities are

$$k_{11} = k'_{11}G^2 + k'_{22}S^2 - 2k'_{12}SG$$

$$k_{12} = (k'_{11} - k'_{22})SG + k'_{12}(G^2 - S^2)$$

$$k_{22} = k'_{11}S^2 + k'_{22}G^2 + 2k'_{12}SG$$

$$k_{33} = k'_{33}$$

The transformation relations for the γ's are obtained by replacing every k_{ij} in the above relations by γ_{ij}.

The coefficients of Eq. (22) are

$$A_1 = \{C_{55}C_{44}C_{33}b_{44} + C_{55}C_{44}k_{33}b_{44} + (C_{45}^2 - C_{55}C_{44})b_{34}b_{43} + C_{55}C_{33}k_{33}b_{22}$$

$$-C_{55}k_{33}b_{23}^2 + C_{44}C_{33}k_{33}b_{11} - C_{45}^2C_{33}b_{44} - C_{45}^2C_{33}b_{33} + 2C_{45}k_{33}b_{13}b_{23}$$

$$-2C_{45}C_{33}k_{33}b_{12} - C_{44}k_{33}b_{13}^2\}/\Delta$$

$$A_2 = \{(C_{55}C_{44}-C_{45}^2)b_{33}b_{44}+C_{55}C_{33}b_{44}b_{22}+C_{55}k_{33}b_{33}b_{22}-(C_{55}b_{22}+C_{44}b_{11})b_{34}b_{43}$$

$$-(C_{55}+b_{11})k_{33}b_{23}^2+C_{55}b_{42}b_{34}b_{23}+C_{55}b_{43}b_{24}b_{23}-C_{55}C_{33}b_{42}b_{24}+C_{44}C_{33}b_{44}b_{11}$$

$$+C_{44}k_{33}b_{33}b_{11}+C_{33}k_{33}b_{22}b_{11}-2C_{45}C_{33}b_{44}b_{12}-2C_{45}k_{33}b_{33}b_{12}-2C_{45}b_{44}b_{23}b_{13}$$

$$-C_{45}b_{41}b_{34}b_{23} - C_{45}b_{43}b_{31}b_{24} + C_{45}C_{33}b_{41}b_{24} + 2k_{33}b_{23}b_{13}b_{12} - C_{33}k_{33}b_{12}^2$$

$$-C_{45}b_{42}b_{34}b_{13} + C_{44}b_{13}^2b_{44} + C_{44}b_{41}b_{34}b_{13} - k_{33}b_{13}^2b_{22} - C_{45}b_{43}b_{23}b_{14}$$

$$+C_{45}C_{33}b_{42}b_{14} + C_{44}b_{43}b_{14}b_{13} - C_{44}C_{33}b_{41}b_{14}\}/\Delta$$

$$A_3 = \{C_{55}b_{44}b_{33}b_{22} - C_{55}b_{42}b_{33}b_{24} + C_{44}b_{44}b_{33}b_{11} + C_{33}b_{44}b_{22}b_{11}$$

$$+k_{33}b_{33}b_{22}b_{11} - (b_{22}b_{11} - b_{12}^2) - (b_{44}b_{11} - b_{41}b_{14})b_{23}^2 + b_{42}b_{34}b_{23}b_{11}$$

$$+b_{43}b_{24}b_{23}b_{11} - C_{33}b_{42}b_{24}b_{11} - C_{45}b_{44}b_{33}b_{12} + C_{45}b_{41}b_{33}b_{24}$$

$$-C_{45}b_{44}b_{33}b_{12} + (C_{33}b_{44} - k_{33}b_{33})b_{12}^2 + b_{44}b_{23}b_{13}b_{12} - b_{41}b_{34}b_{23}b_{12}$$

$$-b_{43}b_{24}b_{13}b_{12}+C_{33}b_{41}b_{24}b_{12}+b_{44}C_{23}b_{13}b_{12}-b_{42}b_{34}b_{13}b_{12}-(b_{44}b_{22}-b_{42}b_{24})b_{13}^2$$

$$+C_{45}b_{42}b_{34}b_{13} + b_{41}b_{34}b_{22}b_{13} - b_{41}b_{24}b_{23}b_{13} + C_{45}b_{42}b_{33}b_{41} - b_{43}b_{14}b_{23}b_{12}$$

$$+C_{33}b_{42}b_{14}b_{12}-C_{44}b_{41}b_{33}b_{14}+b_{43}b_{14}b_{13}b_{22}-C_{33}b_{41}b_{14}b_{22}-b_{42}b_{23}b_{14}b_{13}\}/\Delta$$

$$A_4 = \{b_{44}b_{33}b_{22}b_{11}-b_{42}b_{33}b_{24}b_{11}-b_{44}b_{33}b_{12}^2+b_{41}b_{33}b_{24}b_{12}+b_{42}b_{33}b_{14}b_{21}$$

$$-b_{41}b_{33}b_{14}b_{22}\}/\Delta$$

where

$$\Delta = C_{55}C_{44}C_{33}k_{33} - C_{45}^2C_{33}k_{33}$$

and
$$b_{11} = C_{11}\sin^2\theta - \rho c^2$$

$$b_{12} = C_{16}\sin^2\theta$$

$$b_{13} = (C_{13} + C_{55})\sin\theta$$

$$b_{14} = i\gamma_{11}\sin\theta c/\omega$$

$$b_{21} = b_{12}$$

$$b_{22} = C_{66}\sin^2\theta - \rho c^2$$

$$b_{23} = (C_{36} + C_{45})\sin\theta$$

$$b_{24} = i\gamma_{12}\sin\theta c/\omega$$

$$b_{31} = b_{13}$$

$$b_{32} = b_{23}$$

$$b_{33} = C_{55}\sin^2\theta - \rho c^2$$

$$b_{34} = i\gamma_{33}c/\omega$$

$$b_{41} = i\gamma_{11}T_0\sin\theta c(i + \tau_0\omega)$$

$$b_{42} = i\gamma_{12}T_0\sin\theta c(i + \tau_0\omega)$$

$$b_{43} = i\gamma_{33}T_0 c(i + \tau_0\omega)$$

$$b_{44} = \rho C_e c^2(i/\omega + \tau_0) - k_{11}\sin^2\theta$$

References

[1]Lord, H. W., and Shulman, Y., "A Generalized Dynamical Theory of Thermoelasticity," *Journal of Mechanics and Physics of Solids*, Vol. 15, 1967, p. 299.

[2]Achenbach, J. D., "The Influence of Heat Conduction on Propagating

Stress Jumps," *Journal of Mechanics and Physics of Solids*, Vol. 16, 1968, p. 273.

[3]Nayfeh, A., and Nemat-Nasser, S., "Thermoelastic Waves in Solids with Thermal Relaxations," *Acta Mechanica*, Vol. 12, 1971, p. 53.

[4]Muller, I., "The Coldness, a Universal Function in Thermoelastic Bodies," *Archives for Rational Mechanics and Analysis*, Vol. 41, 1971, p. 319.

[5]Green, A. E., and Lindsay, K. A., "Thermoelasticity," *Journal of Elasticity*, Vol. 2, 1972, p. 1.

[6]Suhubi, E. S., "Thermoelastic Solids," edited by A. C. Eringen, *Continuum Physics*, Vol. II, *Academic, New York, 1975*, Chap. 2.

[7]Puri, P., "Plane Waves in Generalized Thermoelasticity," *International Journal of Engineering Science*, Vol. 11, 1973, p. 735.

[8]Puri, P., "Plane Waves in Generalized Thermoelasticity-errata", *International Journal of Engineering Science*, Vol. 13, 1975, p. 339.

[9]Agarwal, V. K., "On Plane Waves in Generalized Thermoelasticity," *Acta Mechanica*, Vol. 31, 1979, p. 185.

[10]Tao, D., and Prevost, J. H., "Relaxation Effects on Generalized Thermoelastic Waves," *Journal of Thermal Stresses*, Vol. 7, 1984, p. 79.

[11]Banerjee, D. K., and Pao, Y. H., "Thermoelastic Waves in Anisotropic Solids," *Journal of the Acoustical Society of America*, Vol. 56, 1974, p. 1444.

[12]Dhaliwal, R. S., and Sherief, H. H., "Generalized Thermoelasticity for Anisotropic Media," *Quarterly of Applied Mechanics*, Vol. 38, No. 1, 1980, p.1.

[13]Singh, H., and Sharma, J. N., "Generalized Thermoelastic Waves in

Transversely Isotropic Media," *Journal of the Acoustical Society of America*, Vol. 77, 1985, p. 1046.

[14]Sharma, J. N., and Sidhu, R. S., "ON the Propagation of Plane Harmonic Waves in Anisotropic Generalized Thermoelasticity," *International Journal of Engineering Science*, Vol. 24, 1986, p. 1511.

[15]Sharma, J. N., and Singh, H., "Generalized Thermoelastic Waves in Anisotropic Media," *Journal of the Acoustical Society of America*, Vol. 85, 1989, p. 1407.

[16]Sinha, A. N., and Sinha, S. B., "Velocity of Rayleigh Waves with Thermal Relaxation in Time," *Acta Mechanica*, Vol. 23, 1975, p. 159.

[17]Agarwal, Y., "On Surface Waves in Generalized Thermoelasticity," *Journal of Elasticity*, Vol. 8, 1978, p. 171.

[18]Harinath, K. S., "On Certain Wave Propagations in Generalized Thermoelasticity," *Letters in Applied Engineering Science*, Vol. 4, 1976, p. 401.

[19]Harinath, K. S., and Muthuswamy, V. P., "A Note on Waves Generated at Liquid-Solid Interface-II (Viscous Effects)," *Defence Sciences Journal*, Vol. 28, 1978, p. 137.

[20]Mondal, S. G., "On the Propagation of a Thermoelastic Waves at a Thin Infinite Plate Immersed in an Infinite Liquid with Thermal Relaxation," *Indian Journal of Pure and Applied Mathematics*, Vol. 14, No. 2, 1973, p. 185.

[21]Mondal, S. G., and Jana, R. N., "On the Propagation of a Thermoelastic Plane Waves in a Thin Infinite Plate with Thermal Relaxation," *Acta Physica*, Vol. 33, 1983, p. 235.

[22]Massalas, C. V., "Themoelastic Waves in a Thin Plate," *Acta Me-*

chanica, Vol. 65, 1986, p. 51.

[23]Massalas, C. V., and Kalpakidis, V. K., "Thermoelastic Waves in a Thin Plate with Mixed Boundary Conditions and Thermal Relaxation," *Ingenieur Archiv*, Vol. 57, 1987, p. 401.

[24]Massalas, C. V., and Tsolakidis, G., "Propagation of Thermoelastic Waves in an Infinite Circular Cylinder with Thermal Relaxation," *Journal of Sound and Vibration*, Vol. 117, No. 3, 1987, p. 529.

[25]Erby, S., and Suhubi, E. S., "Longitudinal Wave Propagation in a Generalized Thermoelastic Cylinder," *Journal of Thermal Stresses*, Vol. 9, 1986, p. 279.

[26]Massalas, C. V., and Kalpakidis, V. K., "Thermoelastic Waves in a Waveguide," *International Journal of Engineering Science*, Vol. 25, 1987, p. 1207.

[27]Sve, C., "Thermoelastic Waves in Periodically Laminated Medium," *International Journal of Solids and Structures*, Vol. 7, 1971, p. 1363.

[28]Nayfeh, A. H., Taylor, T. W., and Chimenti, D. E., "Theoretical Wave Propagation in Multilayered Orthotropic Media," *American Society of Mechanical Engineers*, ADM-Vol. 90, 1988.

[29]Henneke, II, E. G., "Reflection-Refraction of a Stress Wave at a Plane Boundary Between Anisotropic Media," *Journal of the Acoustical Society of America*, Vol. 51, 1972, p. 210.

[30]Nayfeh, A. H., "The Propagation of Horizontally Polarized Shear Waves in Multilayered Anisotropic Media," *Journal of the Acoustical Society of America*, Vol. 86, 1989, p. 2007.

[31]Daimaruya, M., and Naitoh, M., "Thermoelastic Waves in a Plate," *Journal of Sound and Vibration*, Vol. 117, No. 3, 1987, p. 512.

Superplastic Deformation Behavior of Physical Vapor Deposited Ti-6Al-4V

J. WARREN † and H.N.G. WADLEY **

University of Virginia, Charlottesville, Virginia 22903

Abstract

The high temperature creep and microstructural evolution accompanying the deformation of an initially nanocrystalline, metastable, single phase, physical vapor deposited, Ti-6Al-4V alloy has been examined. At test temperatures between 760 and 900°C, the as-deposited structure transformed to a two phase, fine grain size ($\alpha + \beta$) microstructure and exhibited up to a 10 fold increase in the creep rate compared with conventional grain size Ti-6Al-4V. Models based on grain boundary sliding and dislocation creep were combined with both phase evolution and grain growth relationships to predict stress-strain rate responses at each test temperature. The analysis indicated that the experimentally observed creep rates can be adequately modeled by a diffusion accommodated grain boundary sliding mechanism. The constitutive model can be used to optimize the physical vapor deposition method of metal matrix composite manufacture.

I. Introduction

Physical vapor deposition (PVD) methods are currently being used to fabricate materials with novel microstructures and supersaturated and/or metastable phases.[1] The as-deposited grain size of these alloys, typically nanometer in scale, can significantly affect their mechanical properties and is presently an area of widespread investigation.[2] The enhanced creep rates observed in nanocrystalline alloys and the responsible mechanisms have not been widely studied primarily because of the difficulty in preparing samples with sufficient thickness to adequately conducted creep tests. Because of this, knowledge about the creep behavior of technologically important alloys (e.g.,

---

* Postdoctoral Research Associate, Department of Materials Science and Engineering.

† Professor, Department of Materials Science and Engineering.

alloys based on Al, Ti, Ni, and Fe) with nanometer (or submicron) scale microstructure is presently unavailable.[2] We are interested in the possibility of enhancing the superplastic behavior of the Ti-6Al-4V alloy which is used extensively for superplastic forming and, more recently, for coating ceramic fibers prior to their subsequent hot deformation processing to form metal matrix composite components (MMC's).[3]

Superplastic behavior is observed in conventionally fabricated, fine grain size Ti-6Al-4V between 750° and 950°C.[4-9] In this temperature range a grain boundary sliding mechanism (GBS) is thought responsible for the superplastic behavior.[10] The strain rate in this regime increases inversely with grain size and so maintaining a fine grain size during the forming process is essential for achieving large, uniform strains. The grain size limitation is in the 3-7 μm range of today's conventionally processed alloys.[11]

In the superplastic temperature range of Ti-6Al-4V, the two substitutional alloying elements (Al and V) partition between the hexagonal close packed (HCP) α and body centered cubic (BCC) β phases. This partitioning also retards grain growth kinetics and helps to extend the temperature for superplastic forming to 950°C.[9] Above 950°C the volume fraction of the α phase decreases allowing coarsening to occur in the β phase and the GBS mechanism to become less significant. At temperatures below 750°C superplasticity is again not observed in Ti-6Al-4V, but in this case it is because both the diffusivity and the volume fraction of the β phase are insufficient to accommodate the GBS mechanism responsible for superplastic behavior.[5,6] Consequently, superplasticity in Ti-6Al-4V occurs over a relatively limited range of temperatures for which the α and β phases are present in nearly equal volume fractions.

In this work we have attempted to systematically investigate the stress and temperature dependence of steady state creep rate for thick, initially nanocrystalline Ti-6Al-4V material produced by a high rate sputter deposition method. Concurrent grain growth and phase evolution accompany each creep test and this has been measured, fitted to a model and included in the analysis.

II. Experimental Procedures

A. Material Fabrication and Composition

The material used for this study was 0.4 mm thick, fully dense, argon plasma sputter deposited Ti-6Al-4V sheet (produced at the 3M Metal Matrix Composites Center, Mendota Heights, MN). A flat 303 series stainless steel plate (100 mm x 150 mm x 1mm) was used as a substrate for deposition of the sample material. The sputtering sources were conventionally processed Ti-6Al-4V alloy. Deposition was conducted in a high vacuum chamber capable of reaching a background pressure of a ~ 1 x 10^{-6} Torr using a 350 kW multicathode sputtering system. The deposition rate was ~ 0.8 μm/min. At the conclusion of deposition, the vacuum chamber was argon purged and allowed to cool. The thermal expansion mismatch between the deposited Ti-alloy coating and the stainless steel substrate conveniently caused the deposit to release. Chemical analyses of the source and sputter deposited materials were performed (Table 1) and indicated little influence of the vapor deposition process on the sheet composition.

B. Creep Testing Method

For tensile creep testing, strips 10 mm wide and 150 mm long were cut from the PVD sheet. The ends of the strips were sandwiched between 65 mm x 10 mm x 1.5 mm conventional Ti-6Al-4V tabs and spot-welded together. This resulted in a specimen gauge length of 20 mm. Specimens in the as-deposited condition were used to investigate the creep behavior.

Constant load, isothermal creep tests, were performed in a flowing argon atmosphere at three test temperatures of 760, 840, and 900°C. The load was applied directly to the bottom half of the specimen by a load train mechanism capable of applying small fixed loads (18-140 kN). Specimen temperature was monitored by two type K thermocouples attached to the upper and lower ends of the gauge section. A maximum temperature difference of 4°C was observed between the thermocouples at each of the three test temperatures. The specimen was heated to the desired test temperature at a rate of 5°C / min. After reaching the preprogrammed temperature the sample soaked for 3 min, the load applied, and the elongation measured using a molybdenum alloy extensometer outfitted with a linearly variable capacitive transducer (LVCT).

Table 1 Chemical composition of Ti-6Al-4V alloy, 1% = 10,000 ppm

Element	Source, % wt.	PVD Sheet, % wt.
Al	6.0	5.7
V	4.0	3.9
Ga[1]	0.20	0.19
Fe[2]	0.18	0.23
O_2[3]	0.176 / 0.178	0.188 / 0.189
N_2	0.0148 / 0.0163	0.0138 / 0.0139
Ti	balance	balance

(1) weak α-phase stabilizer, solid solution strengthener.
(2) β-phase stabilizer.
(3) Potent α-phase stabilizer, 0.2 maximum allowable concentration

Creep tests were conducted at progressively higher fixed loads to obtain isothermal strain rate data over a range of stresses. A new specimen was used for every test. The data were used to calculate the instantaneous true stress σ, true strain ε, and minimum true strain-rate $\dot{\varepsilon}_{min}$. Each isothermal creep test yielded a single data point on log σ vs log $\dot{\varepsilon}_{min}$ plot.

C. Microstructural Characterization

To characterize the microstructural evolution accompanying a creep test, samples from as-deposited sheet were annealed at each of the three creep test temperatures (760, 840, 900°C). The samples were first vacuum encapsulated in quartz ampoules (to avoid sample oxidation during heating), heated at 5° C/min to the desired annealing temperature, held for the required time, and quenched in water (at ambient temperature). The quenched specimens were then metallographically prepared and chemically etched to reveal microstructural features (with Kroll's reagent). Samples were then placed in the scanning electron microscope (SEM) and a series of micrographs taken for grain size measurement. The average grain size for each test temperature was determined using the Hilliard circle technique.[12] Specimens were also analyzed in the transmission electron microscope (TEM) to determine the arrangement of dislocations in the gauge sections and the effect of strain on the microstructure. X-ray diffraction patterns were also taken to determine the phase(s) present.

III. Physical Vapor Deposited Microstructures

A TEM micrograph of the PVD material, in the as-deposited condition, is shown in Fig. 1. As is evident in the figure, the dislocation density was very

Fig. 1 TEM micrograph of as-deposited Ti-6Al-4V material.

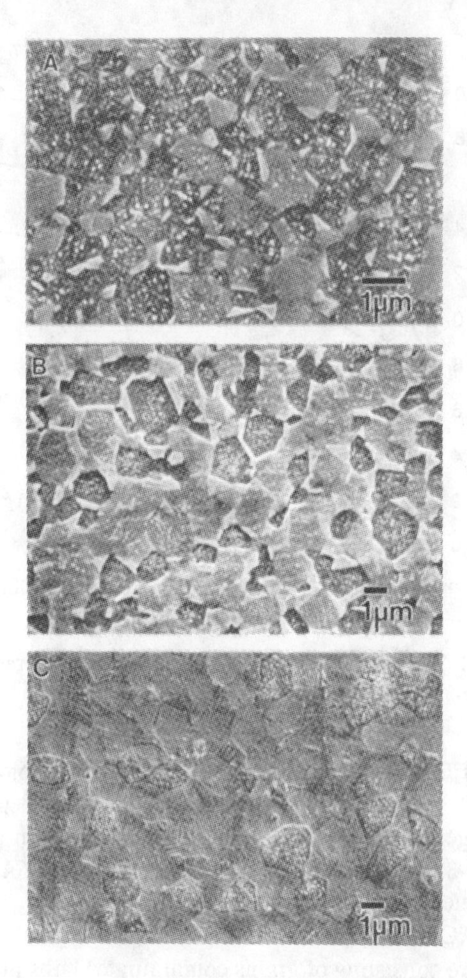

Fig. 2 Microstructures after a 60 s anneal (with subsequent water quench) from (a) 760, (b) 840, and (c) 900°C.

high and grain boundaries were not easily determined making an accurate grain size measurement difficult. We have estimated the grain size to be in the 30-100 nm range. X-ray diffraction of this sample indicated it had a HCP crystal structure. TEM analysis indicated the presence of two HCP phases; an α phase and a martensitic α' phase.

SEM micrographs of samples annealed at temperatures of 760, 840, and 900°C (for 1 min) followed by water quenching are shown in Figs. 2a-2c. A predominantly α phase microstructure with a small (0.09) volume fraction of polygonal β phase grains was obtained after a 760°C anneal. Large areas of the microstructure were comprised of α grains in nearest neighbor contact with α grains (α-α grain boundaries). The average grain size of this microstructure was 500 nm.

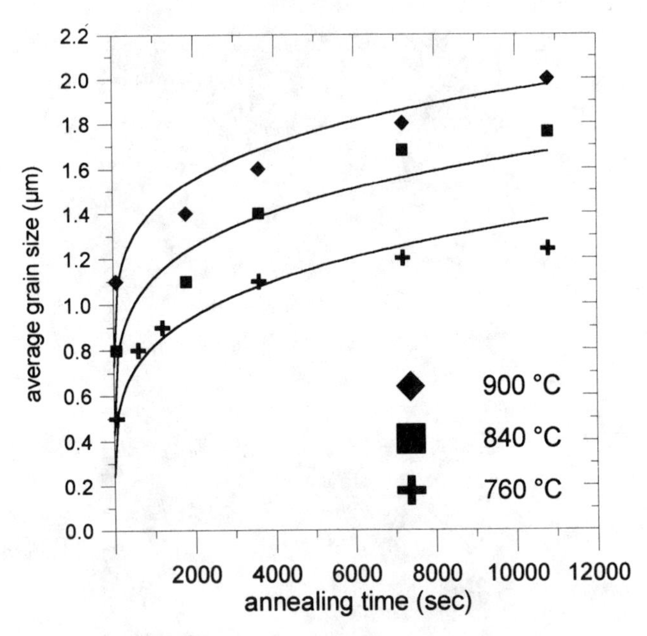

Fig. 3 The grain size vs time for various test temperatures.

A 1-min, 840°C, anneal (Fig. 2b) resulted in a morphology similar to that obtained at 760°C. However, the volume fraction of the β phase increased to 0.21. The average grain size was approximately 800 nm, still significantly smaller than the 3.0-μm grain size of conventional Ti-6Al-4V following a similar heat treatment.[4]

A 1-min 900°C anneal with a subsequent water quench, Fig. 2 c, resulted in a microstructure consisting of grains containing α' laths plus retained β and α grains. At 900°C the vanadium content in the β phase is insufficient to suppress the martensitic start temperature to below room temperature so prior β grains transform to a martensitic α' phase plus retained β microstructure during quenching.[13] The average grain size was 1.1 μm which is still significantly less than conventionally processed alloys (after a similar anneal at 900°C, have an average grain size of 4 μm[4]).

Figure 3 summarizes the measured grain sizes of samples annealed at the three test temperatures. The β phase volume fraction measured in the PVD alloy, shown in Fig. 4, is in reasonably agreement with the experimental data reported by Meier *et al.* for conventional alloys.[5]

IV. Creep Measurements

Figures 5a-5c show log true stress vs log minimum true strain rate behavior for the three test temperatures. The experimental data was fit to a

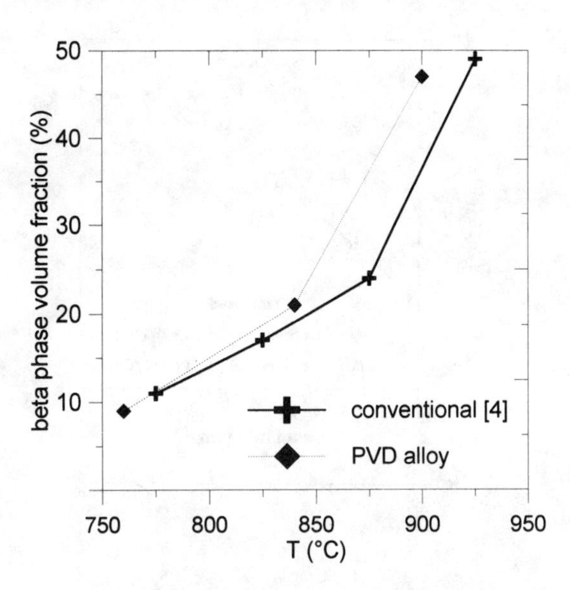

Fig. 4 The β-phase volume fraction of a conventionally processed Ti-6Al-4V alloy and the PVD alloy.

straight line using a least squares analysis. The slope of the line m defines the strain rate sensitivity

$$m = \frac{\partial \log \sigma}{\partial \log \dot{\varepsilon}_{min}}. \tag{1}$$

where $\dot{\varepsilon}_{min}$ is the minimum strain rate and σ is the applied true stress. The reciprocal of m is the creep stress exponent n in the Norton power law creep relationship

$$\dot{\varepsilon}_{min} = B\,\sigma^n \tag{2}$$

where B is a temperature dependent material parameter. The experimentally determined values of B, n, and m are given in Table 2.

Based on values of n greater than 0.5 and total engineering strains exceeding roughly two gauge lengths the PVD alloy deformed superplastically in the 760-900°C temperature range. For comparison, the superplastic

Table 2 Experimentally determined Norton power law creep parameters for the PVD alloy

Temperature, °C	Creep Stress Exponent, n (strain rate sensitivity, $m=1/n$)	B, 1/(s MPa^n)
760	1.70 (0.60)	1.6 10^{-6}
840	1.34 (0.75)	1.9 10^{-5}
900	1.70 (0.60)	1.2 10^{-5}

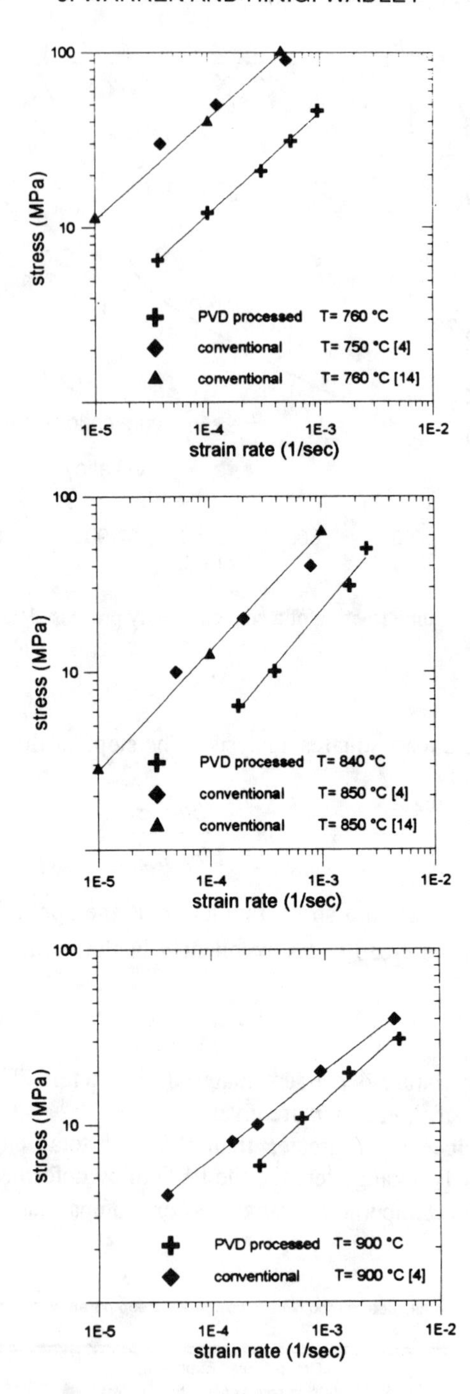

Fig. 5 Stress vs strain-rate relationships for the PVD sheet at various test temperatures, conventional alloy behavior is also shown.

behavior of conventionally processed Ti-6Al-4V is also shown.[4,14] For a given stress at 760°C, the PVD material exhibited up to a 10-fold increase in strain rate. A higher strain rate is also seen at the higher test temperatures but the difference becomes less. A list of stresses, associated minimum true strain rates, and grain sizes when the minimum creep rate was attained are given in Table 3. The duration of the transient creep stage (i.e., the time to reach the measured minimum strain rate) is also given in the table.

V. Deformation Microstructures

TEM analysis was performed on specimens prepared from strained gauge sections to obtain information into the possible mechanisms responsible for creep. Figure 6 is a TEM micrograph of a gauge section removed from a specimen tested at 840°C and a stress of 50 MPa. The total engineering strain to reach $\dot{\varepsilon}_{min}$ (after 2 min) was 0.45. The strained microstructure consisted of α-grains, of low dislocation density, and of smaller intergranular βgrains. Even though the strain was large in this particular sample, relatively few dislocations were present in the microstructure.

VI. Microstructural Evolution Relationships

A. Grain Growth

To extend conventional creep models to the phase distributions and submicron grain sizes observed in the PVD material the kinetics of grain

Table 3 Experimentally determined minimum strain rate values and measured times to reach the minimum strain rate for the test temperatures and stresses, estimated grain size at the time the minimum strain rate was measured is also shown.

T, °C	True stress, MPa	$\dot{\varepsilon}_{min}$, 1/s	Duration of transient creep stage, min(1)	Grain size at $\dot{\varepsilon}_{min}$, μm(2)
760	46.2	$9.5\ 10^{-4}$	0.5	0.6
760	31.2	$5.5\ 10^{-4}$	1.3	0.6
760	21.0	$3.0\ 10^{-4}$	1.7	0.7
760	12.0	$1.0\ 10^{-4}$	2.8	0.7
760	6.5	$3.7\ 10^{-5}$	9.6	0.8
840	50.2	$2.5\ 10^{-3}$	0.2	0.8
840	30.7	$1.8\ 10^{-3}$	0.3	0.8
840	10.1	$4.0\ 10^{-4}$	1.0	0.9
840	6.5	$1.2\ 10^{-4}$	1.3	0.9
900	30.7	$4.2\ 10^{-3}$	0.1	1.1
900	20.2	$1.6\ 10^{-3}$	0.3	1.1
900	10.9	$6.6\ 10^{-4}$	0.7	1.2
900	5.8	$2.6\ 10^{-4}$	1.3	1.2

(1) Each specimen held at test temperature for 3 minutes prior to start of test.
(2) Effect of strain enhanced grain growth is not included.

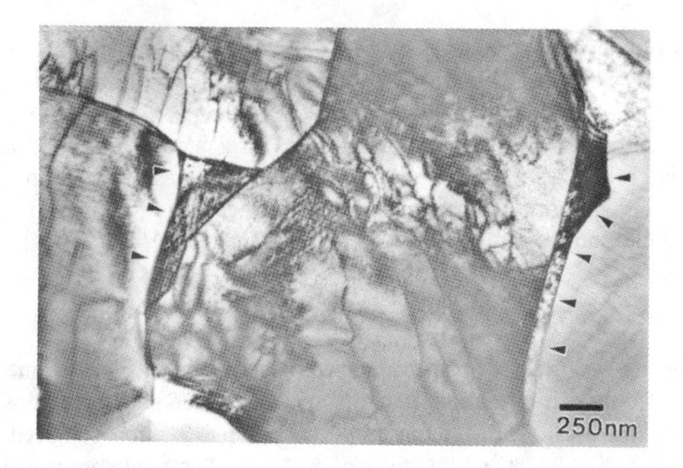

Fig. 6 Gauge section for a sample tested at 840°C and 50 MPa (TEM).

growth during the test must be established since the GBS mechanism exhibits an inverse dependence on grain size. When the applied stress was low, significant grain growth occurred during the relatively long primary creep stage. The measured strain rate for each creep test temperature is, therefore, not for a constant microstructural state but for one which changes with both with test conditions and test duration.

The solid lines that pass through the data sets in Fig. 3 are the best fits to an empirical grain growth relationship:

$$d = d_0 + kt^a \tag{3}$$

where d is the instantaneous grain size, d_o the initial grain size, t the time (s), a is the grain growth exponent and k is a constant. The best fit constants are given in Table 4. For the range of test temperatures studied here, the grain growth exponents were ~ 0.2. To determine the grain size at $\dot{\varepsilon}_{min}$ for each stress and strain rate (Table 3) we have assumed that the empirical relationship for grain growth given in Eq. 3 approximates the grain coarsening kinetics of the specimen during the transient creep stage of the test. We have neglected the effect of strain enhanced grain growth because of the small strains required to reach $\dot{\varepsilon}_{min}$ (~4%). Strain enhanced grain growth has been shown to increase the grain size by approximately 25% in gauge sections strained to 200%.[4]

Table 4. Experimentally determined empirical grain growth parameters for the PVD alloy.

T, °C	d_o, µm	a	k, µm / sec^a
900	0.50	0.20	0.23
840	0.20	0.20	0.23
760	0.11	0.24	0.14

Table 5 Material parameters used to model the creep behavior of the PVD Ti-6Al-4V alloy

Material parameter	α	β	Ref.
Burgers vector, *b*, m	$3\ 10^{-10}$	$3\ 10^{-10}$	23
room temperature shear modulus, μ_o, MPa *	$4.35\ 10^{4}$	$2.05\ 10^{4}$	23
atomic volume, Ω, m^3	$1.7\ 10^{-29}$	$1.7\ 10^{-29}$	23
grain boundary energy, Γ, MJ/m^2	$3.5\ 10^{-7}$	$3.5\ 10^{-7}$	23
pre-exponential lattice diffusion, D_{o_v}, m^2/s **	$6.6\ 10^{-9}$	$4.5\ 10^{-8}$	15
pre-exponential boundary diffusion, $D_{o_{gb}}$, m^2/s **	$1.3\ 10^{-8}$	$1.3\ 10^{-7}$	11
power law creep exponent, *n*	4.85 (1)	3.78 (2)	11
power law creep (Dorn) constant, A, (σ in MPa)	$3.6\ 10^{9}$ (1)	$1.2\ 10^{6}$ (2)	11
activation energy lattice diffusion, Q_v, kJ/mol	169	131	15
activation energy boundary diffusion, Q_{gb}, kJ/mol	101	77	11
grain boundary width, δ, m	$6\ 10^{-10}$ (3)	$6\ 10^{-10}$ (3)	11

* $\mu_\alpha (T) = \mu_o (1 - 1.2 ((T - 300) / 1933))$; $\mu_\beta (T) = \mu_o (1 - 0.5 ((T - 300) / 1933))$

** $D_v = D_{ov} exp(\frac{-Q_v}{kT})$; $D_{gb} = D_{o_{gb}} exp(\frac{-Q_b}{kT})$;

1. Values are for 100 % α-phase Ti-6Al.
2. Values are for 100 % β-phase Ti-6Al-4V (2300 ppm H_2 charged alloy).
3. Grain boundary width assumed to be equal to 2 *b*.

B. β-Phase Formation

The β-phase volume fraction has strong effects both on grain coarsening and the creep mechanisms in Ti-6Al-4V alloys. At 900°C, 47% of the alloy consisted of the β phase and decreased with decreasing test temperature. In addition to inhibiting grain growth, the BCC β phase has more slip systems and a diffusivity two orders of magnitude higher than the α phase (see Table 5). An empirical expression for the volume faction of β-phase present in the microstructure, as a function of test temperature, was determined from the experimental data and is equal to:

$$V_\beta [T(^\circ C)] = 11.870 - 3.088 \mathrm{x} 10^{-2} T + 2.024 \mathrm{x} 10^{-5} T^2 \quad (4)$$

where V_β is the β-phase fraction and *T* is the temperature in degrees centigrade.

VII. Creep Mechanisms

A. Activation Energy for Superplastic Flow

Figure 7 shows the relationship between $ln\ \dot{\varepsilon}_{min}$ and 1/T for the range of test temperatures and stresses studied. The average value of the activation energy was 150 kJ/mol, similar to the activation energy for volume self-diffusion in elemental β -Ti (131 kJ/mol). [15]

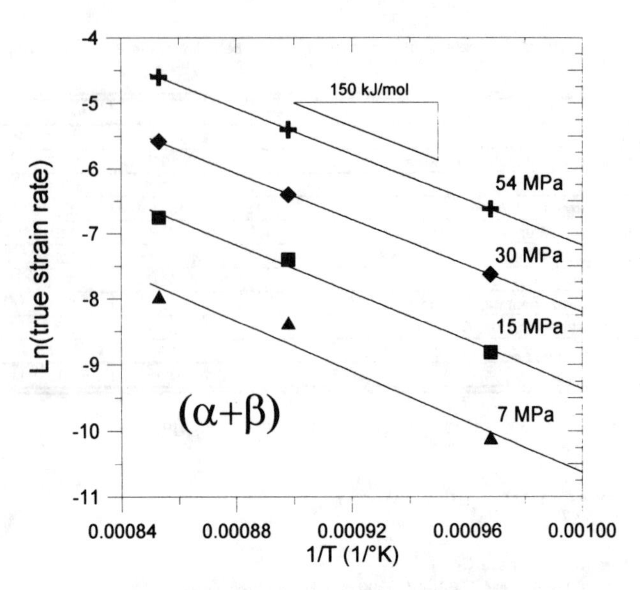

Fig. 7 A plot of strain-rate vs. $1/T$.

B. Creep Mechanisms

Table 2 indicates that the apparent n value was less than 2 for test temperatures ≥ 760°C. This, in addition to observed creep strains greater than about two gauge lengths, is normally indicative of superplastic behavior. It has been well established that GBS (accommodated by either dislocations or diffusion) is the primary mechanism responsible for superplastic deformation of Ti-6Al-4V in the 3-7 μm grain size range.[10, 14, 16] Various workers have reported that the optimum superplastic forming temperatures (where the maximum value of m is measured) are between 870 and 940°C with the variations resulting from alloy compositional differences.[4-6, 17, and 18]

The rate equation for a single-phase, superplastically deforming material is a *modified* Dorn relationship:

$$\dot{\varepsilon} = A\frac{\mu b}{kT}D\left(\frac{\boldsymbol{b}}{d}\right)^{p}\left(\frac{\sigma}{\mu}\right)^{n} \tag{5}$$

where A is the power law creep constant (Dorn constant), μ the temperature compensated shear modulus, $\boldsymbol{b}$ the Burger's vector, p the grain size exponent (experimentally in the 1.5 to 3.0 range) and D a diffusion coefficient (equal to D_v for lattice diffusion accommodated GBS or D_{gb} for grain boundary diffusion accommodated GBS) and d is the grain size and σ the applied stress and k and T are Boltzman's constant and the temperature, respectively.[19]

For a single-phase alloy, Eq. (5) shows that reducing the grain size *d* increases the attainable strain rate during superplastic forming. However, Equation (5) does not account for the complexities of the two phase alloy encountered in Ti-6Al-4V above 760°C. Experimentally, it has been found that to achieve superplasticity a fine grain size and a substantial volume fraction of the significantly less creep resistant β phase are necessary in the Ti-6Al-4V alloy system. In both conventional and PVD Ti-6Al-4V alloys tested at 760°C the β phase is located at grain boundary triple point junctions and does not form a fully interconnected structure. Deformation at 760°C must, therefore, involve either GBS along α-α grain boundaries or creep (dislocation and/or diffusion accommodated) of the α grains.[6] Hence, the α phase (at low β phase fractions) is expected to dominate the deformation behavior.

At 760°C the GBS mechanism could be accommodated by either dislocation motion or grain boundary diffusion.[20] Ignoring any microstructural evolution effects, it has been found that in Eq. 5, $n = 2$ ($m = 0.5$) for GBS accommodated solely by dislocation motion and $n = 1$ ($m = 1$) for sliding accommodated solely by grain boundary diffusion. The experimentally measured creep stress exponent was determined to be $n = 1.7$ at 760°C, suggesting a dislocation accommodated mechanism. However this analysis fails to take account of the effects of grain coarsening on the creep stress exponent. At 840°C, we note that the apparent *n* value of the experiments was 1.3, a value more indicative of diffusion accommodated GBS. Further support for this can be found in the relatively low density of dislocations adjacent to the α grain boundary in the specimen tested at 840°C, Fig. 6.

Hamilton *et al.*[11] successfully used the Ashby-Verrall [21] (A-V) mechanism for diffusion accommodated GBS to model the superplastic behavior of a conventional Ti-6Al-4V alloy (at a test temperature of 870°C). The predicted strain rate of the specimen was assumed to be modeled by the summation of the diffusional GBS mechanism $\dot{\varepsilon}_{gbs}$ and the dislocation creep mechanism $\dot{\varepsilon}_{d}$

$$\dot{\varepsilon} = \dot{\varepsilon}_{gbs} + \dot{\varepsilon}_{d} \tag{6}$$

with $\dot{\varepsilon}_{gbs}$ in Eq. (6) represented by the A-V relationship

$$\dot{\varepsilon}_{gbs} = \frac{100\Omega D_v}{kTd^2}\left(\sigma - \frac{0.72\Gamma}{d}\right)\left(1 + \frac{3.3\delta D_{gb}}{dD_v}\right)$$

where δ is the grain boundary width, Γ the grain boundary energy and Ω the atomic volume of the diffusing species (the remaining parameters were defined earlier). The dislocation creep relationship in Eq. (6) was assumed to be independent of the grain size and was represented by Eq. (5) (with the grain size exponent p equivalent to 0). The material parameters used in the analysis are given in Table 5.

To model the deformation behavior of a conventionally processed, two-phase Ti-6Al-4V alloy, Hamilton *et al.*[11] proposed and experimentally verified that each phase in the alloy experienced the same strain rate (an iso-strain rate assumption). Thus

$$\dot{\varepsilon} = \dot{\varepsilon}_{\alpha} = \dot{\varepsilon}_{\beta} \tag{7}$$

The applied stress, σ, was also assumed to partition according to a rule of mixtures

$$\sigma = V_{\alpha}\sigma_{\alpha} + V_{\beta}\sigma_{\beta} \tag{8}$$

where V_{α} and V_{β} are the volume fractions of the α and β phases (obtained from Eq. 4) at the creep test temperature, and σ_{α},and σ_{β} are the stresses in each phase.

A predicted strain rate was calculated, by solving Eqs. (6-8) simultaneously, for each experimental creep stress test, temperature, β-phase fraction, Eq. (4) and grain size, Eq. (3). The predicted $\dot{\varepsilon}$ points were then plotted on a log-log plot, fitted with a straight line (using a least squares analysis) and compared to experiment.

Figures 8a - 8c compare the experimental results and the model predictions for the test temperatures of 760, 840, and 900°C, respectively. At 760°C, Fig. 8a, the dominant mechanism was the A-V GBS mechanism whose contribution was nearly three orders of magnitude greater than the dislocation creep contribution even at the highest test stress of 46.2 MPa. Grain growth had the largest effect on the creep test strain rates conducted at 760°C because a relatively large time difference was required to reach $\dot{\varepsilon}_{min}$ (~ 10 min) for the highest and lowest test stresses (Table 3). Equation (3) predicts that this results in a 140% increase in grain size. The effect of concurrent grain growth is to increase the predicted creep exponent to n = 1.5, Fig. 8a, which is in good agreement with the experimental value of n = 1.7. Thus, when concurrent grain growth is included in the analysis the predicted creep stress exponent is shifted towards a value of 2; a value more usually associated with a dislocation accommodated GBS mechanism which again emphasizes the need for careful incorporation of coarsening when seeking to ascribe a mechanistic significance to creep data.

The model predictions for a test temperature of 840°C are also in good agreement with experimental data, Fig. 8b. At 840°C the model predicted a creep exponent of n = 1.1, which was quite close to the experimentally determined value of n = 1.3. The A-V strain rate was the dominant deformation mechanism at all stresses: it was nearly 80 times greater than the dislocation creep strain rate contribution at the highest test stress (50.2 MPa). The β-phase volume fraction at 840°C increased from 0.09 at 760°C to 0.21

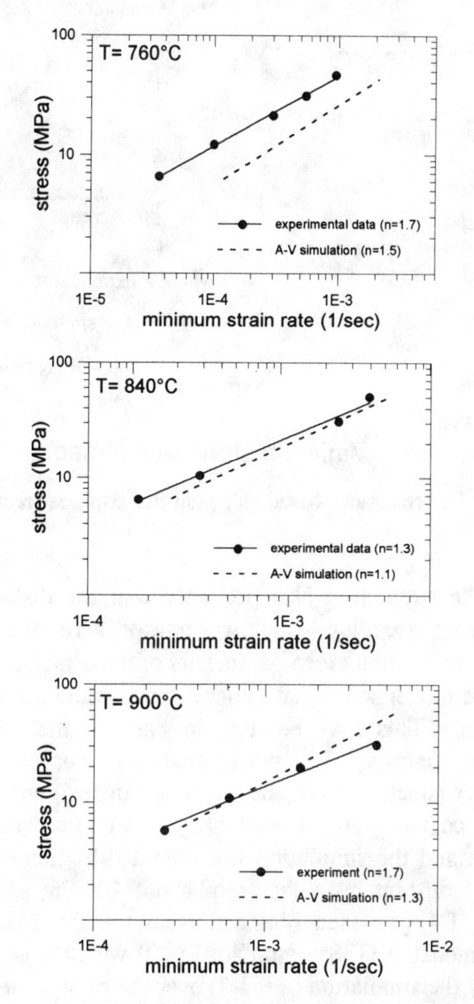

Fig. 8 Comparison of the Ashby-Verral model predictions and experimental results.

and this may have been responsible for the decrease in the value of n for either or both of two reasons: 1) The minimum β-phase volume fraction required to "break up" the skeletal network of α grains and achieve fully accommodated superplastic flow is approximately 0.26[22], 2) The β-phase volume fraction more effectively inhibited concurrent grain growth during the creep tests.

At 900°C, Fig. 8c, the experimentally determined value of n had increased to 1.7. The model simulation, which incorporated the effect of grain growth, also predicted an increase, although somewhat lower in value (n = 1.3). Based on the material properties given in Table 5 the modeling results predicted that the A-V mechanism was again the dominant creep mechanism at all test stresses (the strain rate contribution due to the A-V mechanism at the highest stress of 30.7 MPa was about 18 times greater than the dislocation creep

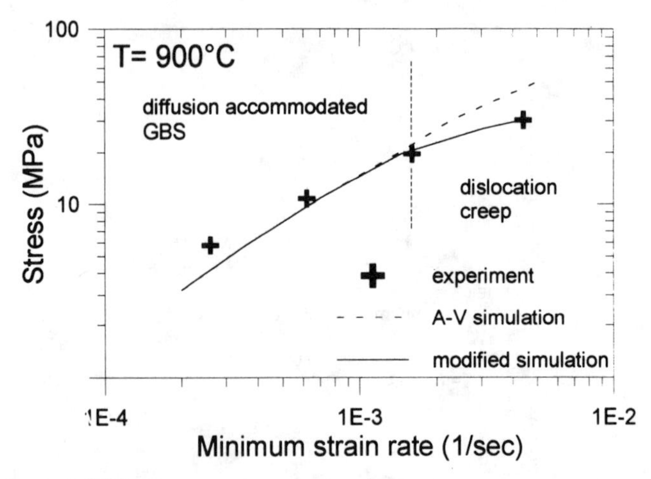

Fig. 9 Model predictions based on a modified α-phase creep constant.

contribution). The simulation also predicted that the dislocation creep rate behavior of the two-phase alloy, which was now of increasing significance, was dominated by the dislocation creep parameters of the α phase.

At 900°C, the model somewhat underestimates the measured strain rate at the highest stress. This may be due, in part, to the choice of material parameters in the analysis. The power law creep constant A is a material sensitive parameter (stacking fault energy, subgrain size) and the value used in Table 5 is for a coarse grain, Ti-6Al alloy. An improved fit between the experimental data and the simulation was obtained by increasing the α phase power law creep constant by a factor of about 10, Fig. 9. A creep stress exponent of $n = 1.4$ was then obtained from the three data points in the "diffusion accommodated GBS region" of Fig. 9 which was quite close to the value predicted by the simulation ($n = 1.3$) over the same region.

The enhanced superplastic strain rate of PVD Ti-6Al-4V has significant practical importance. During superplastic forming, shorter, and therefore more economical forming cycles can be used to shape and diffusion bond components. During the manufacture of composites it will facilitate much lower and/or high strain rate consolidation (e.g. roll bonding) and promise to improve both the quality and affordability of this promising new materials class.

VIII. Conclusion

Enhanced superplastic behavior was observed when the PVD alloy was creep tested between 760° and 900°C. The dominant creep mechanism was shown to be diffusion accommodated GBS. Grain growth during testing has resulted in subtle, but important, changes in the creep stress exponent n. The

Ashby-Verrall superplastic flow model, solved by simultaneously considering the flow stress-strain rate behavior of each phase, the phase fractions present at test temperature, and the effect of concurrent grain growth has accurately predicted the superplastic behavior of the two-phase, PVD alloy. The enhanced superplasticity of PVD Ti-6Al-4V will significantly improve the processibility and afordability of titanium matrix composites.

Acknowledgments

This work has been supported by the Advanced Research Projects Agency (W. Barker, Program Manager) and NASA (D. Brewer and R. Hayduk, Program Monitor) through grant NAGW-1692 and through the ARPA/ONR funded University Research Initiatives (URI) program at the University of California, Santa Barbara. The authors are grateful to Dr. J. Storer (3M) for supplying the material used in this study and Dr. L.M. Hsuing (UVa) for supplying the TEM micrographs. We also thank Prof. J. A. Wert (UVa) for many helpful discussions about superplasticity in the Ti-6Al-4V system.

References

1. Ward-Close, C.M and Partridge, P.G., *Journal of Materials Science*, No. 25, 1990, pp. 4315-4323.
2. Shull, R.D., *Nanostructured Materials*, Vol. 2, No. 3, 1993, pp.213-216.
3. P.G. Partridge and Ward-Close, C.M., *International Materials Reviews*, Vol. 38, 1993, pp. 1-23.
4. Arieli, A. and Rosen, A., *Metallurgical Transactions A*, Vol. 8, 1977, pp. 1591-1596.
5. Meier, M.L. , Lesuer, D.R., and Mukherjee, A.K., *Material Science and Engineering*, Vol. 154 A, 1992, pp. 165-173.
6. Meier, M.L. , Lesuer, D.R., and Mukherjee, A.K., *Material Science and Engineering*, Vol. 136 A, 1991, pp. 71-78.
7. Lee, D. and Backofen, W.A., TMS-AIME, Vol. 239, 1967, pp. 1034-1040.
8. Wert, J.A., *Superplastic Forming of Structural Alloys*, N.E. Paton and C.H. Hamilton,eds., TMS-AIME, Warrendale, PA., 1982, pp. 169-189.
9. Wert, J.A. and Paton, N.E., *Metallurgical Transactions A*, Vol. 14, 1983, pp. 2535-2545.
10. Sherby, O.D. and Wadsworth, J., *Progress in Material Science*, Vol. 33, 1989, pp. 169-221.
11. Hamilton, C.H., Gosh, A.K., and Mahoney, M.W., *Advanced Processing Methods for Titanium Alloys*, D.F. Hasson and C.H. Hamilton, eds., TMS-AIME, Warrendale, PA., 1982, pp. 129-144.
12. *Metals Handbook Eighth Ed.*, Vol. 8, ASM, Metals Park, OH, pp. 42-43.
13. Brooks, C.R., *Heat Treatment, Structure and Properties of Nonferrous Alloys*, ASM, Metals Park, OH, 1982, p. 366.

14. Pilling, J., Livesey, D.W., Hawkyard, J.B., and Ridley, N., *Metal Science*, Vol. 18, 1984, pp. 117-122.
15. Dyment, F., *Proceedings of the Fourth International Conference on Titanium*, H. Kimuraand, O. Izumi, eds., TMS-AIME, Warrendale, PA., 1980, pp. 519-528.
16. Takeuchi, S. and Argon, A.S., *Journal of Material Science*, Vol. 11, 1976, pp. 1542-1566.
17. Boyer, R.R., and Magnuson, J.E., *Metallurgical Transactions A*, Vol. 10, 1979, pp. 1191-1193.
18. Paton, N.E. and Hamilton, C.H., *Metallurgical Transactions A*, Vol. 10, 1979, pp. 241-250.
19. Mukherjee, A.K., Bird, J.E., and Dorn, J.E., TMS-AIME, Vol. 62, 1969, pp. 155-179.
20. Kashyap, B.P., and Mukherjee, A.K., *International Conference on Superplasticity*, B.Baudelet and M. Suery, eds., Editions du C.N.R.S., Paris, France,
21. Ashby, M.F. and Verrall, R.A., *Acta Metallurgica*, Vol. 21, 1973, pp. 149-163.
22. Gifkins, R.C., *Metallurgical Transactions A*, Vol. 7, 1976, pp. 1225-1232.
23. Ashby, M.F. and Frost, H.J., *Deformation-Mechanism Maps*, Pergamon Press, 1982, pp. 44-45.

Chapter 4. Performance of Aircraft Materials

Aluminum Alloys for Subsonic Aircraft

J. T. Staley*

Aluminum Company of America
100 Technical Dr.
Alcoa Center, PA 15069

Abstract

The current use of aluminum alloy products in subsonic airframe structure is reviewed and future trends are predicted. The paper describes the performance requirements for different components including skins, frames, and stringers for fuselages plus covers, spars, ribs, stringers, and leading edges for upper and lower wings and empennages. The relationship between these performance requirements and the properties and characteristics needed in aluminum alloy products for these components is discussed. Particular emphasis is placed on the recent shift from performance-driven material development to emphasis on reducing the cost of ownership. To meet this need, much of the structure now made by riveting together formed sheets could be replaced by a simpler structure which consolidates the functions of individual pieces. The tasks of the aluminum producer include the following: 1) improve the properties of thick plate so that performance of structure machined from the plate matches that of the current built-up structure, 2) increase the forming rate and mechanical properties of superplastic sheet to make this method of producing structure both cost effective and structurally efficient, and 3) develop alloys and processes to make the use of castings for primary structure technically and economically feasible.

I. Introduction

Aluminum alloy products have been the material of choice for airframes for about 70 years because of their high specific strength, fabricability, durability, and damage tolerance. They have been replaced by polymer matrix composites for many components of high-performance military aircraft although they continue to dominate in military transports and jetliners. Although this paper

---

* Chief Scientist, Aerospace and Commercial Products, Alcoa Technical Center.

describes the performance requirements for components in subsonic aircraft, components in supersonic aircraft must meet the same performance requirements in addition to those imposed by aerodynamic heating. In the following sections, the performance criteria for different portions of the airframe are reviewed, materials which meet these criteria and are currently being employed are noted, and new materials in the development stage are discussed.

II. Fuselage

A. Performance Criteria and Property Requirements

The fuselage is a semimonocoque structure made up of skin to carry cabin pressure and shear loads, stringers or longerons (longitudinal) to carry the longitudinal tension and compression loads, frames (circumferential) to keep the fuselage in shape and redistribute loads into the skin, and bulkheads to carry concentrated loads. Hoop loads are always in tension, and predominant axial loads during flight are tension in the crown, compression in the bottom, and shear in the sides. Specific strengths, modulus, fatigue crack initiation, fatigue crack growth, fracture toughness, and corrosion resistance must all be considered in design of a fuselage. Depending on the component and location in the fuselage, some of these characteristics are more important than others.

1. Materials for Skins

Fracture toughness and resistance to the growth of fatigue cracks at high levels of stress intensity are probably the most important criteria used in evaluating a material for this application because of their effect on damage tolerance.[1] The most recent fuselage skin material, Alcoa C-188, is a proprietary alclad Al-Cu-Mg alloy product which has been specified for the Boeing 777 jetliner.[2] It replaces alclad 2024-T3, the standard of the industry since the DC-3. This material was developed to meet the fracture toughness targets established by Boeing for an improved fuselage skin sheet.[3] Moreover, fatigue cracks grow at a lower rate in C-188 than they do in an alclad 2024-T3 sheet, and the advantage increases with increasing level of peak stress intensity factor, Fig. 1. The higher toughness and greater resistance to the growth of fatigue cracks save weight and decrease manufacturing cost through simplification of structural detail. Stringent controls on both chemical composition and processing are required to meet the rigorous toughness and fatigue requirements without sacrificing strength.[4] To achieve the fatigue and fracture characteristics, the microstructure has a low volume fraction of insoluble constituent particles and almost no sparsely soluble constituent particles. The K_c of C-188 measured using 16-in. wide panels is about 22–23% higher than that of 2024-T3, Table 1. Because K_c values depend on geometry as well as on material characteristics, however, toughness measurements using this size panel give an underestimate of the advantage of C-188 when used in the fuselage. For example, Boeing demonstrated a 40%

advantage in K_C over 2024-T3 using 60-in. wide panels, Fig. 2, and this translates to a 24% increase in residual strength in a hypothetical two-bay crack in a fuselage, Fig. 3. The toughness advantage of C-188 also translates into an advantage in small aircraft structure, Fig. 4. Because of this, Canadair has chosen C-188 for the fuselage skin of its new Global Express long-range executive jet.

A candidate for the next generation jetliners, GLARE®, a family of glass fiber-aluminum alloy laminates, offers weight savings competitive with Gr/Ep nonetheless retaining the advantages of aluminum at a lower total cost than Gr/Ep. Compared to 2024-T3, GLARE® offers lower density, higher strength and damage tolerance, and much greater resistance to the growth of fatigue cracks. Performance on the barrel test, which measures fatigue resistance of a section of a fuselage under simulated service, is far superior to that of any aluminum alloy product. The nonmetallic (prepreg) layers in GLARE® are about 0.1 mm thick and consist of S2 glass fibers embedded in an epoxy resin. The aluminum sheet in current versions is either 2024-T3 or 7475-T76 and is about 0.3 mm thick. The high strength of GLARE® laminates is attributed to the strength of the glass fibers whereas the high damage tolerance is attributed to the aluminum alloy sheet. The resistance to the initiation of fatigue cracks is control-

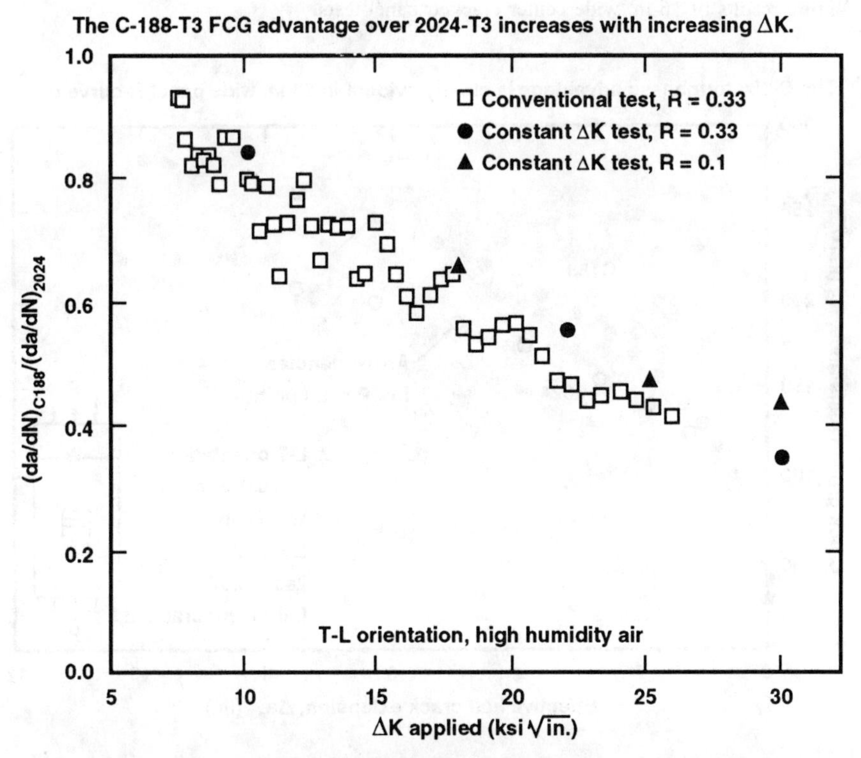

Fig. 1 Ratio of C-188 to 2024 fatigue crack growth rates vs ΔK.

Table 1 Typical strength-toughness properties of Alcoa C-188 and 2024 skin sheet (source: Alcoa Aerospace Technical Fact Sheet)

Property	Test orientation	C-188-T3 alclad, 0.060–0.249 in. thick	2024-T3 alclad,[a] 0.060–0.249 in. thick
FTU, ksi	L	67	67
	LT	66	66
FTY, ksi	L	50	50
	LT	45	45
El, %	LT	18	18
K_C ksi $\sqrt{\text{in.}}$ [b]	L-T	148	126
	T-L	158	128

[a]Typical of current production.

[b]From results of 16-in. wide center cracked panel tests.

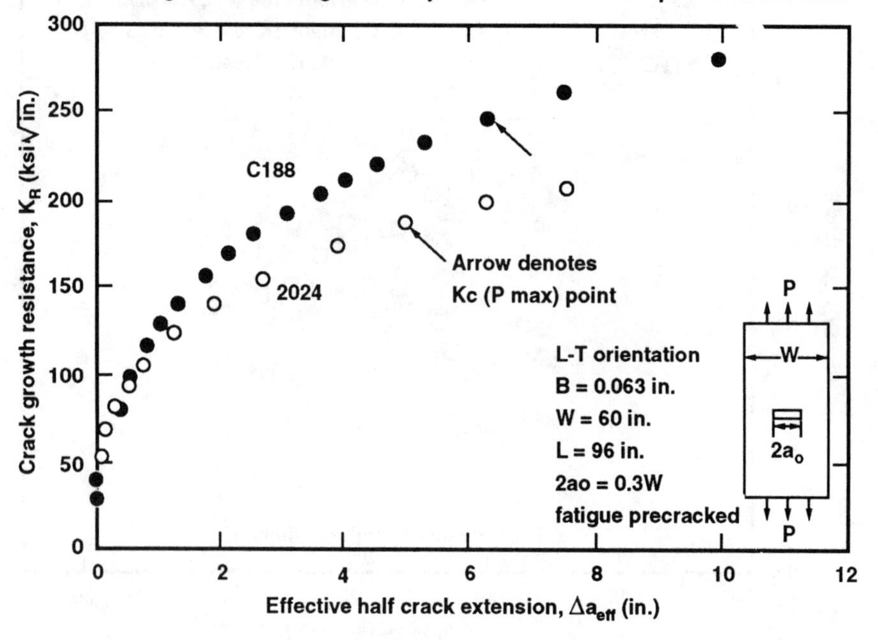

Fig. 2 K_R curves for 0.063-in. 2024-T3 and C-188 T3 clad sheet (courtesy of Boeing).

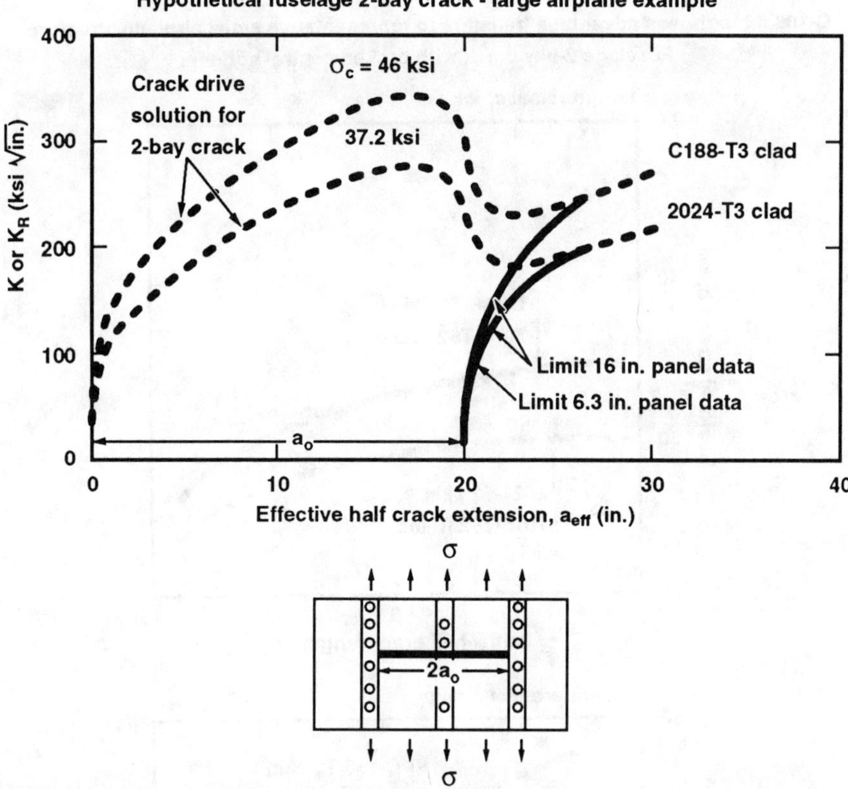

Fig. 3 Residual strength prediction for 0.063-in. 2024 and C188 cracked stiffened panels.

led by the resistance of the aluminum sheet whereas the resistance to the growth of the fatigue cracks is attributed to the bridging effect of the fibers and to some controlled debonding at the epoxy-aluminum sheet interface. Although GLARE® is more expensive than aluminum alloy sheet, economies of construction promise a cost effective structure. Based on expressed interest by manufacturers of airframes, the next generation jetliners may well have at least a portion of the fuselage skin fabricated from this hybrid material.

2. Materials Under Deve' ɔment for Fuselage Skin

To address concerns with the corrosion resistance of existing aluminum alloy products for this application and to provide additional weight savings to lower fuel costs, Al-Mg-Sc alloy sheet is being developed. Laboratory tests of specially processed sheet indicate that this material is capable of developing

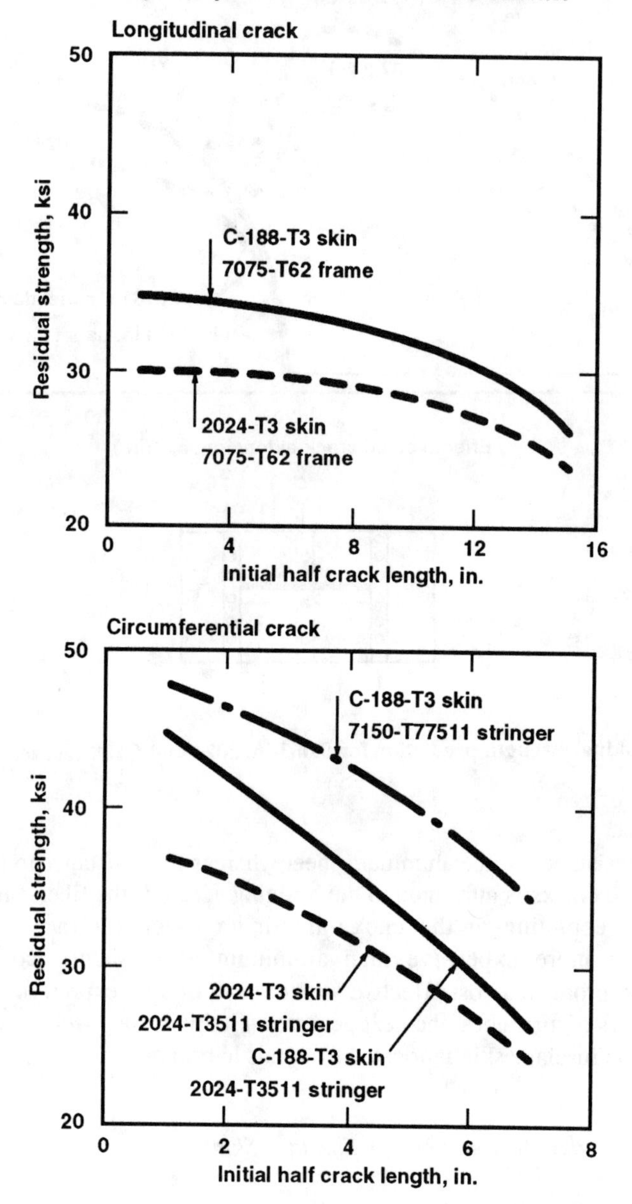

Fig. 4 Panel residual strength vs initial damage size.

strength and fracture toughness similar to that of C-188 with the corrosion resistance of Al-Mg alloys used for marine applications. Weight savings are possible because of the lower density. Al-Mg alloys have not been considered heretofore for aircraft applications because strain hardening to increase strength is accompanied by unacceptable ductility and toughness, and thermal treatments to recover ductility result in unacceptably low strength. The addition of scandium helps in two ways. First, scandium increases strength by both solid solution and precipitation strengthening of Al_3Sc particles. Second, these dispersoids pin the dislocation cell structure which forms during thermal treatments of strain-hardened material. This pinning mechanism minimizes the loss in strength which usually accompanies the recovery of ductility and toughness.

3. Materials for Fuselage Frames and Stringers

Alloy 7075-T62 sheet is usually used for frames, whereas alloy 2024-T3 and alloy 7075-T6 in sheet or extrusions are used for stringers, depending on strength requirements and on manufacturing considerations. Some 7075 stringers are provided as O temper sheet which is taper rolled along the length to various thicknesses at the manufacturers. This taper-rolled sheet is then roll formed transversely before heat treatment to the T62 temper. This material has two potential liabilities. First, the stringent taper rolling and forming requirements can lead to low recovery due to cracking during fabrication. Second, the grain shape after this operation is more equiaxial than that in 7075-T6 sheet supplied by the aluminum producer. Consequently, stringers fabricated in this manner may be susceptible to stress-corrosion cracking. Alloy 7055-O sheet which is under development solves both of these liabilities. Experimental lots have exhibited higher formability than 7075-O and developed higher strength, fracture toughness, and resistance to stress-corrosion cracking than 7075-T6 when heat treated to a T7 type temper.

Alloy 7150-T77 extrusions offer higher strength than 7075-T6 with durability and damage tolerance characteristics matching or exceeding those of 7050-T76. These extrusions are being used by Boeing as fuselage stringers for the upper and lower lobes of the new 777 jetliner because of the superior combination of strength, corrosion and stress corrosion cracking characteristics, and fracture toughness.

Recent analysis indicates that the new, higher strength 7XXX alloy products for frames and stringers may be required to take full advantage of the higher fracture toughness of the new fuselage skin Alcoa C-188. Increasing the strength of the frames and the stringers significantly increases the residual strength of stiffened panels containing a crack. For example, the residual strength of a damaged structure of C-188 skin and 7150-T77 stringers is significantly higher than the residual strength of a similar structure composed of C-188 skin and the 2024-T3511 stringer which is usually used with 2024-T3 skin, Fig. 5. Alloy 7150-T6 extrusions would supply the same weight savings as 7150-T77, but the

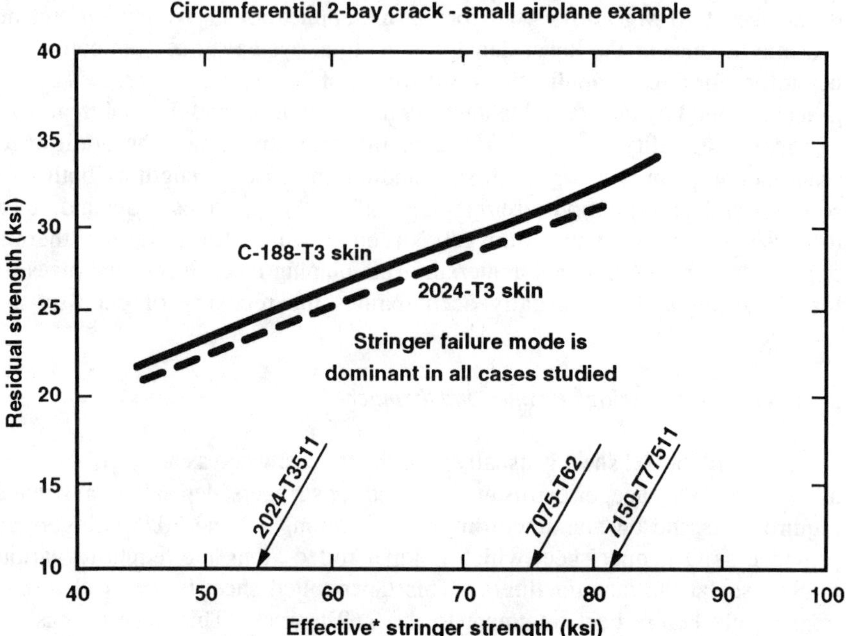

Fig. 5 Effect of stringer failure strength on panel residual strength.

higher corrosion resistance offered by 7150-T77 provides greater durability. The effects of different combinations of frame and stringer materials on the residual strength and resultant opportunity to save weight are presented in Table 2. These results demonstrate that a structure consisting of C-188 skin and 7150-T77 frames and stringers could save over 17% weight compared to the incumbent design. These weight savings compare favorably with the 25% weight savings claimed by graphite-epoxy at significantly lower cost and risk. This analysis is significant because it demonstrates that simpler analyses such as that made by Ekvall *et al.*[5] are inadequate to predict the weight savings possible by increasing fracture toughness. That analysis would predict a weight savings of less than 4% based on the tests of 60 in.-wide panels. The foregoing illustrates that close ties between the developer of aluminum alloy products and the structure designer can lead to more effective structures by concurrently designing the material and the structure.

4. Bulkheads of High-Performance Aircraft

These structures are fabricated from plate and forgings because stringent requirements for structural strength mandate a thick section. High levels of

Table 2 C-188/7150 weight saving opportunity 2-bay crack scenario, small airplane example

			Residual Strength, ksi		Weight saving opportunity,[a] %			
		Extruded			Uncoupled		Coupled	
Skin sheet	Frame	Stringer	Hoop	Axial	Hoop	Axial		Remarks
2024-T3	7075-T62	2024-T3511	24.3	24.0	0	0	0	incumbent
		7150-T77511	24.3	31.4	0	20.7	0	
2024-T3	7150-T77511	2024-T3511	24.7	24.0	1.3	0	0	
		7150-T77511	24.7	31.4	1.3	20.7	1.3	
C-188-T3	7075-T62	2024-T3511	28.5	24.5	13.2	1.4	1.4	
		7150-T77511	28.5	32.6	13.2	24.1	13.2	
C-188-T3	7150-T77511	2024-T3511	29.5	24.5	16.4	1.4	1.4	
		7150-T77511	29.5	32.6	16.4	24.1	16.4	
C-188-T3	Upgauged[b] 7150-T77511	7150-T77511	33.0	32.6	17.5	24.1	17.5	
C-188-T3	Upgauged[b] 7150-T77511	Upgauged[c] 2024-T3511	33.0	32.6	17.7	7.7	7.7	

a Over the incumbent by skin downgauging.
b Upgauged by 42% (to optimum frame strength).
c Upgauged to meet 7150-T77511 strength.

cyclic load also impose requirements for durability and damage tolerance. The F14 and F15 use 2124-T8, whereas F16 and F18 use either the highest toughness candidate, 7475-T73, or the highest strength one, 7050-T74, respectively. New alloys containing about 1.5% Li are being developed for evaluation as thick structure for high-performance military aircraft. The goals are: 1) strength and toughness at least equivalent to 2124-T8, 2) no loss in toughness with elevated temperature exposure, 3) resistant to exfoliation corrosion and to SCC in the short-transverse direction at 172 MPa, 4) minimal anisotropy, 5) 5% lower density and 7% higher modulus than 2124, and 6) significantly higher resistance to the growth of fatigue cracks compared to 2124-T8.[6] The new materials would save weight because of their lower density and are anticipated to provide greater durability because of higher resistance to the growth of fatigue cracks.

III. Wing

A. Performance Criteria and Property Requirements

A wing is essentially a beam that transmits all of the applied air load to the central attachment to the fuselage. The wing box, the part of the wing that carries the loads, consists of top and bottom covers (each cover is composed of a skin and stringers that stiffen the skin and run lengthwise), spars that create the sides of the wing box and run lengthwise, and ribs. The covers take the bending and the torsional loads. Specific compressive yield strength and modulus of elasticity in compression are the static material properties that influence the design of the top skin-stringer which is primarily in compression. Specific tensile strength, tensile yield strength, and tensile modulus are the static material properties that influence the design of the lower skin-stringer which is primarily in tension. The front and rear spars, as well as the covers, take the torsion loads as the wing tries to twist during flight. Ideally, the top of the spars should have the same material properties as that of the upper wing skin-stringer, whereas the bottom should have the same material properties as that of the lower wing skin-stringer. The ribs keep the covers in shape and keep buckling of the covers and stringers to a minimum. Static strength properties are important for ribs (tensile strength, compressive yield, modulus). Fatigue resistance and fracture toughness are most important on the lower wing structure, but must be considered in the whole wing structure. Commercial jetliners have successfully used 7XXX-T6X aluminum alloy products for wing structure despite their low resistances to exfoliation corrosion resistance and stress-corrosion cracking in the short-transverse direction. Potential exfoliation problems are managed by the use of corrosion protection systems. Stresses are relatively low in the short-transverse direction of wing skins because the starting plate stock is less than 2 in. thick, so stress in the short-transverse direction is minimized. Military aircraft, however, must use 7XXX material in corrosion resistant T7 tempers.

Leading and trailing edges are used to increase the aerodynamics of the airplane and reduce the take-off and landing speed. These elements of the wing are not primary load carrying parts. Fatigue, bending and torsional stiffness, and resistance to buffeting and hail damage are some of the items considered in their design.

1. Materials for Upper Wing Skin and Stringers

In the late 1980s, builders of military aircraft desired a higher strength material than that provided by 7050-T76 to save weight while maintaining corrosion characteristics. In addition, builders of civil aircraft desired a material with the strength of 7150-T6 along with the corrosion characteristics of 7050-T76 to minimize costs of coatings and special procedures. In response to these needs, Alcoa developed a proprietary temper for 7150. Alloy 7150-T77 plate and extrusions develop the same mechanical properties as 7150-T6 with significantly improved resistances to both exfoliation corrosion and SCC. The first application of 7150-T77 for wing structure was by McDonnell-Douglas on the C17. This saved a considerable amount of weight because corrosion performance of 7150-T6 was deemed to be inadequate by the Air Force for this application, and strength of 7050-T76 is considerably lower.

The implementation of the T77 temper for 7150 was followed by development of 7055-T77 plate and extrusions for compressively loaded structure. Alloy 7055-T77 plate offers a strength increase of about 10% relative to that of 7150-T6 (almost 30% higher than that of 7075-T76). It also provides a high resistance to exfoliation corrosion similar to that of 7075-T76 with fracture toughness and resistance to the growth of fatigue cracks similar to that of 7150-T6.[7] In contrast to the usual loss in toughness of 7XXX products at cryogenic temperatures, fracture toughness of 7055-T77 at -65°F (220 K) is similar to that at room temperature. Resistance to SCC is intermediate to that of 7075-T6 and 7150-T77. The attractive combination of properties of 7055-T77 is attributed to its high ratios of Zn/Mg and Cu/Mg. When aged by the proprietary T77 process this composition provides a microstructure at and near grain boundaries that is resistant to intergranular fracture and to intergranular corrosion. The matrix microstructure resists strain localization while maintaining a high resistance to the passage of dislocations.

2. Materials Under Development for Upper Wing Skin

To compete with Gr/Ep for applications such as a high aspect ratio wing which requires high specific modulus, aluminum alloy products must be reinforced. Several ways to develop the target specific modulus are possible, but all have problems of either a technical or an economic nature which must be addressed.

Aluminum products which match or even exceed the specific modulus of Gr/Ep are available today. These products are reinforced with discontinuous particles or whiskers (DRA) of silicon carbide and manufactured either by powder metallurgy or ingot metallurgy techniques. Toughness levels of current

DRA products are below target values, however, and size limitations and high costs are additional deterrents to their use.

Special DRA products have been produced in the laboratory which show promise of meeting the specific modulus and fracture toughness goals.[8] Plate products composed of alternate layers which are reinforcement free and reinforcement rich as well as extruded products composed of rod-shaped reinforcement rich areas embedded in reinforcement free areas are both promising. They provide high toughness by extrinsic toughening mechanisms which cause crack deflection and bifurcation. These products are not cost effective, however, because of the processing costs. Moreover, scale up to the size needed for wing skins of jetliners presents problems. Products produced using spray metallurgy techniques, however, have the potential of solving technical, scale up, and economic problems. For wing skins, alternate reinforcement-free and reinforcement-rich layers will be sprayed to form a rectangular billet. This billet will be hot rolled to gauge, then heat treated. The alloy content and the percentage of reinforcement particles will be determined by the strength and modulus targets. Because the combination of properties of this new material will be so unlike those of current aluminum alloy products, the structure designer must work closely with the material developer so that the property set and the structure design are developed concurrently.

3. Lower Wing Materials

Lower wing covers for jetliners and transports are usually made from plate and extrusions in a skin-stringer construction. They have also been designed successfully as integrally stiffened panels using either plate or extruded panels. High-performance aircraft use unstiffened lower wing covers machined from plate.

4. Materials for Lower Wing Skins/Stringers

Alloy 2024-T3 plate and extrusions are the standard for McDonnell-Douglas jetliners and the C17 and for the jetliners built by Airbus Industries jetliners. Boeing, however, has been using 2324-T39 plate and 2224-T3 extrusions since the 757 and 767 because these higher strength variants of 2024 save weight.

Alloy 7475-T73 has found use in the wing structure of high-performance military aircraft because of its superior combination of strength and fracture toughness, but it has not been used for this application in commercial aircraft. This material is being considered for the lower wing skin of jetliners because of potential cost savings if the wing is age formed.[9] Questions regarding fracture toughness at the subzero temperatures encountered at high altitudes are being resolved.

Al-Li alloy 8090-T86 plate and extrusions are also being considered as materials for the lower wing skin of jet liners. They promise weight savings because of the lower density of 8090 and the products are age formable.

Concerns regarding deviation of fatigue cracks from a path 90 deg to the applied load, thermal stability, and fracture toughness of the plate product are being addressed to determine their relevance and impact. All technical concerns regarding the use of 8090-T8 extrusions have been resolved, and application depends on resolution of the technical concerns with the plate. Aluminum producers are working on next generation Al-Li alloys, but sufficient information is not available to allow a valid assessment of these developmental materials.

The high durability and damage tolerance of fiber-metal laminates make them strong candidates for the lower wing. Fokker has been performing "in-service" evaluations of portions of a lower wing on one of their commercial aircraft. The ARALL® versions (aramid fibers) offer lower density than GLARE® (glass fibers), but some manufacturers have expressed concern over possible problems with degradation of the aramid fibers even though they are only exposed at the edges of the laminate.

5. Materials for Spars and Ribs

The standard alloys 7075-T6 and 2024-T3 continue to be specified in most cases, but Boeing selected 7150-T77 extrusions for the spar chords of the 777 jetliner. The Airbus models use a spar integrally machined from thick 7050-T76 and 7010-T76 plate.

Ribs are usually constructed using stiffened sheet of 7075-T6. The newer materials do not appear to be cost effective.

6. Fixed Leading Edge

Fixed leading edges are usually fabricated from 2014-T6 sheet in Europe and 7075-T76 sheet in the USA. Alloy 2090-T84 sheet has been specified on several new commercial aircraft. It saves weight because of its low density combined with ultimate longitudinal tensile strength equaling that of 7075-T76, superior resistance to the growth of fatigue cracks, and acceptable fracture toughness.

IV. Empennage

A. Performance Criteria and Property Requirements

The empennage is the tail of the airplane and consists of a horizontal stabilizer, a vertical stabilizer, elevators, and rudders. The horizontal stabilizer is like an upside down wing whose span is roughly 50% that of the wing. Structural design of both the horizontal and vertical stabilizers is essentially the same as for the wing. Both upper and lower surfaces of the horizontal stabilizer are often critical in compression due to up and down bending. Consequently, the modulus of elasticity in compression is the most important property.

1. Materials for the Empennage

The vertical stabilizers of the newer jetliners are made from composites, and the newest Airbus and Boeing jetliners also use composite horizontal stabilizers. These components in older aircraft are fabricated from sheet, plate, and extrusions of 7075-T6 and 2024-T3. Aluminum alloys developing higher strength and fracture toughness generally offer less advantage in the empennage than in wings or fuselage because much of the structure is sized by elastic modulus, and requirements for damage tolerance are not as critical. McDonnell-Douglas, however, found 7150-T77 to be cost effective in the horizontal stabilizer of their new MD-11.

2. New Materials for the Empennage

A recent study evaluating potential weight savings with a next generation Al-Li alloy and ARALL® aramid fiber-aluminum alloy composites indicates that they offer considerable weight savings relative to horizontal stabilizer structure fabricated from 2024-T3 and 7075-T6 at an affordable cost premium.[10] The properties of the developmental Al-Li alloy products were the most cost effective, so they are strong candidates for horizontal stabilizers of future jetliner and transport aircraft.

V. Future Trends

The U.S. airlines lost more money in the last few years than they made in their entire history. Consequently, they have little money to spend to replace their aging fleets. As a result, airframe builders have shifted from a technology driven environment to a market driven one. Boeing has reported that they are focusing on helping the airlines save costs by improving maintainability/reliability and reducing airline acquisition and operational costs.[11] They point to 7055-T77 for the upper wing covers and 7150-T77 for upper wing spar chords, body stringers, and seat tracks as providing reduced operational cost (fuel and maintenance) for the new 777 aircraft because of their combinations of high strength and corrosion resistance. They also reduced the use of forgings significantly in the 777 by substituting stress-relieved plate to save machining costs. Airbus Industries is sending a similar message. Deutsche Aerospace Airbus reports that improved economy in connection with high reliability is the overriding requirement for the development of aircraft.[12] The structure should have a long life, low weight, and require minimum maintenance costs. These requirements strongly indicate that the next generation jetliners will contain less built-up structure and considerably more 1) structure machined from thick plate, 2) structure fabricated by superplastic forming, and 3) large area components made as thin walled castings.

These needs can be met by: 1) thick plate which will develop the fatigue and fracture characteristics equivalent to that of thin plate, 2) sheet which is superplastically formable at rates which are orders of magnitude higher than that of 7475 with strength and durability equaling or bettering that of 7475-T6 and 2024-T3, respectively, and 3) casting alloys and processes which will provide properties similar to those of today's wrought materials in section sizes desired by customers.

VI. Summary And Conclusions

The requirements of the airframe industry for structural materials have changed over the years, and the aluminum producers have developed new products to meet these needs. The requirement for materials with improved resistance to exfoliation corrosion with no loss in strength led to 7150-T77 plate and extrusions, whereas the requirement for higher strength in compressively loaded structure with good resistance to corrosion provided the incentive to develop 7055-T77 plate and extrusions. The need for a fuselage skin material with improved fracture toughness and fatigue resistance while maintaining strength led to the development of Alcoa C-188, a new 2XXX alclad sheet and plate product for fuselage skins. Al-Li alloys 2090, 8090, and 2091 were developed to slow the advance of polymer matrix composites in aircraft structure because of their low density, but technical problems and costs have, thus far, relegated them to niche applications. Alloy 2090 looks like it will become the standard material for leading edges, however, and alloy 8090 may yet become a material for the lower wing covers of jetliners because of its low density and ability to be age-formed. GLARE®, a structural laminate of glass fibers and aluminum alloy sheet, is being evaluated for fuselage skin, and ARALL®, a laminate of aramid fibers and aluminum alloy sheet materials, is being evaluated for lower wing skins. The high specific modulus and low density of the next generation Al-Li alloy products and ARALL® make them strong candidates to regain the empennage which has been lost to graphite-epoxy composites for some jetliner applications. Next generation Al-Li alloy plate is also being developed for bulkheads of high-performance aircraft to replace 2124-T8. Other materials under development include DRA composites which develop acceptable toughness by extrinsic toughening mechanisms and corrosion resistant Al-Mg-Sc sheet. The challenges posed by the need for lower acquisition and maintenance costs and for higher reliability will drive future developments. Performance targets for a thick plate which is fatigue resistant, high-performance superplastic forming process (SPF) sheet, and large-area high-quality castings are being established.

References

[1] Staley, J. T., and Rolf, R. L., "Trends in Alloys for Aircraft," *Proceedings of International Symposium on Light Metals*, The Metallurgical Society of the Canadian Institute of Mining, Metallurgical, and Petroleum Engineers, Montreal, 1993, pp. 629-642.

[2] Bray, G. H., Bucci, R. J., Kulak, M., Rolf, R. L., Witters, J. J., and Yeh, J. R., "Damage Tolerance Attributes of New Fuselage Skin Alloy C-188," AeroMat '94, Anaheim, CA, June 1994.

[3] Hyatt, M. V., and Axter, S. E., "Aluminum Alloy Development for Subsonic and Supersonic Aircraft," *Science and Engineering of Light Metals, RASELM 91*, Japan Inst. Light Metals, Tokyo, Oct. 1991, pp. 273-280.

[4] Colvin, E. L., Petit, J. I., Westerlund, R. W., Magnusen, P. E., U.S. Patent 5, 213, 639, May 25, 1993.

[5] Ekvall, J. C., Rhodes, J. E., and Wald, G. G., "Methodology for Evaluating Weight Savings from Basic Material Properties," *Design of Fatigue and Fracture Resistant Structures*, edited by P. R. Abelkis and C. M. Hudson, American Society for Testing and Materials, Philadelphia, PA, 1982, pp. 328–341, (ASTM STP 761).

[6] Balmuth, E., "Aluminum-Li Alloy for Thick Section Applications," The Minerals, Metals, and Materials Society of the American Society of Mining, Metallurgical, and Petroleum Engineers Meeting, Denver, CO, Feb. 1993.

[7] Lukasak, D. A., and Hart, R. M. "Aluminum Alloy Development Efforts for Compression Dominated Structure of Aircraft," *Society of Allied Weight Engineers Paper 1985*, S.A.W.E., Inc., Chula Vista, CA, 1991.

[8] Hunt, W. Jr., Osman, T. M., and Lewandowski, J. J., "Micro- and Macrostructural Factors in DRA Fracture Resistance," *JOM*, Vol. 45, No. 1, Jan., 1993, pp. 30–35.

[9] Meyer, E. S., and Brooks, C. L., "New Aircraft Lower Wing Skin Trade Study," 1992 USAF Structural Integrity Program Conf., San Antonio, TX, Dec. 1-3, 1992.

[10] Warren, C. J., "Advanced Metallic Horizontal Stabilizer Structure," AeroMat '94, Anaheim, CA, June, 1994.

[11] Lovell, D. T., and Miller, A., "Engineering Needs in Materials and Processes for Commercial Airplanes," AeroMat, Anaheim, CA, June, 1994.

[12] Rendigs, K., "Metallic Structures Used in Aerospace During 25 Years and Prospects," *50 Years of Advanced Materials or Back to the Future, Proceedings of the 15th International European Chapter Conference of the Society for the Advancement of Material and Process Engineering*, Toulouse, France, edited by J. Hognat, R. Pinzelli, and E. Gillard, 1994, pp. 49-65.

Materials Requirements for Aircraft Engines

James C. Williams*

General Electric Aircraft Engines, Cincinnati, Ohio 45215

Abstract

Each successive generation of aircraft gas turbine engines have had improved performance. These improvements are the result of component improvements, better materials, and improved design methods. Specific metrics that track these performance improvements are described and data are presented to demonstrate the performance improvements. It is suggested that the rate of improvement is slowing due to increasing product maturity and that future improvements will be more difficult to achieve. In part, this is because there are fewer new metal-based materials to support future product improvements. Some of the current business realities in the aircraft engine industry are discussed because these affect the investment level available for materials development. The new reality is product improvements driven by cost rather than performance. The implications of this on future product and materials requirements is outlined. In particular, the requirements for future subsonic and supersonic propulsion systems will be

---

* General Manager, Engineering Materials Technology Laboratories.

described and contrasted. Some suggestions for meeting these requirements also will be offered.

Introduction

There are number of quantitative parameters that can be used as measures of performance of gas turbine engines. Perhaps the most important of these are thrust normalized by weight (thrust to weight or *T/W*) of aircraft engines and thrust normalized fuel consumption [specific fuel consumption (SFC)]. During the past 25 years or so, each new generation of engine has exhibited dramatic improvement in these characteristics. One result of these combined improvements in efficiency and performance is the capability for modern commercial aircraft to fly $\geq$ 7000 miles nonstop at subsonic cruising speeds. Another is the availability of commercial supersonic flight made possible by the introduction of the Concorde. The improvements in *T/W* and SFC have been accomplished through increased engine operating temperatures that are the result of increased compression ratios in the engine cycle, as measured by overall pressure ratio (OPR). The increases in OPR result in improved thermodynamic efficiency. Improved designs that utilize proportionately more of the energy in the hot gas stream and the use of stronger and/or lighter materials for increased structural efficiency have permitted steady increases in OPR. The best indicators of engine operating temperatures are the compressor exit temperature T_3 and the turbine inlet temperature T_{41}. In commercial and military transport engines, additional gains in *T/W* and SFC have been achieved through the use of front fans of ever increasing diameter relative to that of the core engine. The best measure of fan size is bypass ratio (BPR), which is the ratio of the air flow through the core to that moved by the fan.

At the same time that *T/W* and SFC have improved, the durability, as judged by mean time between overhauls (MTBO), and reliability, as measured by the frequency of unscheduled engine removals (UER) and in-flight shutdowns (IFSD), also have improved. Today, the excellent reliability of modern turbofan engines permit twin engine commercial aircraft to operate routinely on long, overwater routes. Further, for commercial engines, the environment-related operating characteristics also have improved, particularly with regard to reduced emissions (both hydrocarbons and oxides of nitrogen) and noise.

As already mentioned, improved materials have enabled increases in T_3 and T_{41}, but this is only part of the basis for improved engine performance. Commensurate improvements both in component performance and in design and analysis methods also have made major contributions to the overall gains realized in engine performance. In combination, these three elements account for much of the performance improvements and the durability increases. There is a strong degree of interdependence between these elements and, because of this, the sum of the gains exceeds the individual contributions.

The object of this paper is to outline some of the major materials improvements and future materials requirements, but the foregoing should make it clear that there has been a highly productive synergy between materials, component performance, and design and analysis. Nevertheless, this paper will address materials improvements because this area is a focus of the volume and comprises the expertise of the author. The focus of the discussion will be commercial aircraft propulsion for two reasons; first, because this sector is the major factor for the future of the engine business and, second, because this sector more closely approximates a free market economy. As the aircraft engine industry matures and becomes largely a commercial business, the precepts of a free market economy will begin to dominate the business decisions that are made. The paper also describes some of the future product requirements and how these shape the opportunities and challenges that lie ahead for design and materials engineers in the aircraft engine industry.

The paper will conclude with a short discussion of the business environment in the aircraft engine industry. This discussion is important because it directly affects both the availability of resources for futher product improvements and the willingness of the materials suppliers to invest in capital equipment needed to produce new classes of materials needed to support future generations of engines. A more detailed account of this subject has recently been published elsewhere,[1] but this paper summarizes and updates this earlier paper.

Improvements in Materials for Gas Turbine Engines

Historically, materials have been a major enabling technology for gas turbine engine performance improvements but, until recently, the emphasis has been placed on incremental improvements in metals, particularly Ni and Ti base alloys. Further, the introduction of new or greatly improved materials processing methods in many cases have allowed component operating stresses to be increased without any attendant material modifications that increase the material strength. These processing methods range from improved melting processes that yield cleaner metals to near net shape manufacturing methods such as precision casting and forging. Some of these new or improved processes have allowed the production of cleaner, more sophisticated materials that can be used at a higher fraction of the ultimate strength with an equal or higher degree of confidence. Clean metals made with these processing methods usually are higher cost but still affordable, in a cost-benefit sense. That is, the overall payoff from these materials in terms of durability, reliability, and performance is substantial over the life of an engine, even though the initial cost is higher.

Table 1 shows the time-based progression of the key temperature parameters along with OPR and BPR for typical aircraft gas turbines placed into service at the approximate time indicated in the table. This table shows that the temperatures, pressure ratios, and bypass ratios have increased with each successive generation of machine.

The engine temperature parameters T_3 and T_{41} are significant because these set materials temperature capability requirements even though, strictly speaking, they are gas temperatures and not materials temperatures. In the compressor, the gas temperatures are roughly equal to the maximum metal temperature. In the turbine section, the gas temperatures are considerably higher than can be accommodated by the metals used for the turbine airfoils and disks. The difference between gas temperatures and metal temperatures is sustained through the use of cooling technology. Cooling is achieved by bleeding cooler air from the compressor and directing it through a series of passages to the turbine section. This air is used to cool the turbine rotor and the airfoils. Large temperature differences can be achieved with this technique, but the cooling air

Table 1 Representative gas turbine characteristics as a function of time

Service date	T_3, °C (°F)	T_{41}, °C (°F)	OPR	BPR
1955	379 (715)	871 (1600)	10	<2
1965	427 (800)	938 (1720)	12-13	2-3
1975	593 (1100)	1343 (2450)	14-16	5-6
1995	693 (1280)	1427 (2600)	35-40	8-9
2015 (est.)	816 (1500)	1760 (3200)	65-75	12-15

that is used reduces the overall fuel efficiency because it does not participate in the propulsive process. Cooling air can be thought of as a form of overhead, the amount of air used is proportional to the reduction in efficiency. Thus, increases in fuel efficiency of the engine can be achieved through more efficient use of cooling air and/or through the use of higher temperature capability turbine materials that require less cooling.

The introduction of investment cast turbine airfoils created new materials temperature capability horizons because alloys could now be selected on the basis of creep strength rather than manufacturability. In the cast airfoils, the cooling air passages are created during casting and this permits more intricate designs than could be produced when the passages were put in wrought blades by drilling, which is a line-of-sight process. Later, controlled solidification processing led to the introduction of directionally solidified castings followed by monocrystal castings. These processing improvements permitted further tailoring of alloys to increase the maximum temperature capability and eliminated creep strength reductions due to the presence of grain boundaries. Current generation turbine airfoil alloys are capable of operating at temperatures $\geq$ 150° C higher than the early wrought alloys. In summary, the steady increases in T_{41} have been possible because of the introduction and subsequent improvements in cooling technology and because of advances in materials temperature capability and materials processing technology. In combination, these improvements have allowed a nearly twofold increase in T_{41} during the past 40 years as seen in Table 1.

Although there is still room for continued improvement in temperature capability of turbine alloys, the point of diminishing returns is beginning to be called into question. This is because some characteristics of the materials, such as melting temperature and density, cannot be circumvented and these ultimately limit the improvement opportunities for metals. At some point in the not too distant future, achievement of further materials temperature capability will require the introduction of new materials concepts or classes such as composites and intermetallic compounds.

There is another method of inceasing the gas-metal temperature difference in cooled airfoils. This is the use of thick (75-150 μm) ceramic coatings known as thermal barrier coatings (TBCs). The TBC is effective in reducing metal operating temperature because the ceramic layer has much lower thermal conductivity than the metallic airfoil substrate. The TBC thus reduces the heat flux across the airfoil wall under a constant temperature gradient. The use of TBCs, thus, either reduces the cooling air flow required to maintain a constant metal temperature or enables an increase in the temperature difference between the hot gas and the metal at a constant cooling air flow. In either event, the use of TBCs increases turbine efficiency in the same way that better turbine airfoil alloys do. The practical challenge for using TBCs is ensuring the long-term integrity of the ceramic layer. If the TBC spalls locally during service, the underlying metal temperature rises locally. Some spalling is inevitable, thus, it is important to calculate the extent of the temperature rise as a function of spalled region size. These calculations have now been made, and the results show that local spalling is not a barrier problem for TBC use. Successful applications of TBCs are now common. As a result, the use of TBCs is more extensive with each new generation of engine. TBCs are relatively expensive to apply, but the durability and SFC benefits justify the initial expense from a life cycle cost standpoint.

In addition to the progress made over the past 10-15 years in improving the turbine section, major progress also has been made in the fan and compressor sections. This progress is measured by the temperatures at the back of the compressor (higher T_3), by weight reductions, and by more efficient aerodynamic designs that improve performance while reducing the tendency for compressor stalls to occur. In the fan, the steady increases in BPR, shown in Table 1, are attributable to improved capability of the low- pressure turbine to drive a large fan and the ability to increase the fan size without unacceptable increases in weight. The steady increase in fan size (increasing BPR) has been possible because of the availability of high-quality rotor grade Ti alloys for the fan disk and fan blades and the early (lower temperature) stages of the compressor. Lightweight static structures such as the fan frame and outlet guide vanes also have been major contributing factors.

An example of the improved melting practice mentioned earlier started in the late 1960s with the use of triple melting of Ti alloys to further refine the material. By the early to mid 1970s triple-melt Ti was used throughout the engine industry for rotors. Presently, hearth melting of rotor grade Ti alloys is being introduced to further reduce the frequency of melt related inclusions. Although development of these improved melting technologies is too complicated to relate in detail here, it is important to recognize that high quality Ti alloys are primarily responsible for the present capability to design and manufacture large rotor structures, such as fan disks, with confidence. The continuous improvement of Ti alloys including the manufacturing practices for Ti has occurred over a period of about 30 years. In addition to improved melting practices, processing to allow higher sensitivity ultrasonic inspection has been developed for increased confidence in the reliability of the material. Also, higher strength alloys have been developed and introduced into service. Further, hot die isothermal forging has been reduced to practice as a means of reproducibly achieving the desired microstructural characteristics that provide the proper balance of properties in large forgings such as fan disks.

In all high BPR turbofan engines, the weight of the fan section and its frontal area become practical limitations for fan sizes. Lightweight materials and lightweight construction methods will be essential in moving the limiting fan size to much larger diameters. As the BPR of newer engines increases, the fan blades have become larger, both in length and width (chordwise). The increases in width are intended to improve the fan efficiency and robustness for events such as bird strikes. These larger blades would be unacceptably heavy if they were made of solid, forged Ti like the smaller, narrower blades used on most current generation large engines today. Two new types of lower weight fan blades have emerged. The first is a hollow Ti blade presently made by diffusion bonding (DB) or superplastic forming and diffusion bonding (SPFDB). The second is a carbon fiber reinforced polymer matrix composite blade. Both of these new blade types are more expensive to produce, but they are necessary to meet the acceptable weight limits and durability requirements imposed by newer, large, high BPR engines. There is no doubt that both types of blades will work, and the ultimate discriminator between the two technologies will be the cost of the articles. As discussed earlier, the fan rotor weight in higher BPR engines is intrinsically important, but the blade weight (mass) also affects the weight of the fan disk and the static fan structures. i.e., the frame, case, and containment. Thus, the overall weight benefit of lightweight fan blades greatly exceeds that of the blades themselves.

Ultimately, T_3 limits due to materials capability will pace further increases in OPR. This is because, unlike the turbine section, there is no provision in any current production engines to cool the compressor rotor. The higher OPR machines currently under consideration serve to underscore this limitation. That is, machines with OPR values ≥ 50 will have corresponding T_3 levels of ≥ 1375°F (746°C). Today there are no proven, production ready materials for compressor rotors that will operate at such high temperatures for the lifetimes expected from a commercial transport gas turbine. Machines that operate with T_3 values above ≈ 1375-1400°F will require substantially different materials of construction and different methods of construction, e.g., the airfoil attachments. Use of cooling technology to ameliorate the effects of higher T_{41} values for very high OPR machines also becomes more challenging. This is because the air from the high-pressure compressor stages is now hotter and loses its cooling effectiveness. Yet, extracting air from earlier stages is not feasible because there is insufficient pressure at these points to effectively pump the air through the cooling circuits without incurring reverse flow because the external pressure in the turbine section is higher than that of the cooling air. In light of these limitations, the BPR, OPR, T_3, and T_{41} levels shown in Table 1 as representative of 1990 appear to be approximate barrier values for engines with metallic rotors. That is, an OPR of ≤ 40 is set by T_3 and T_{41} values that are near the current limits for rotor materials available today.

There also has been substantial progress in combustor technology, some of which impacts SFC and some of which allows reduced emissions. Until now the combustor has not been limited by special materials requirements. Thus, improved materials developed for other engine components have been successfully adopted for the combustor as the needs dictate. The recent introduction of dual annular combustor designs has allowed the CO and NO_x emissions to be reduced dramatically. Two factors are essential to control emissions from an aircraft engine: minimizing the residence time of the burning fuel in the hot zone of the combustor and controlling the fuel-air ratio in this zone. The former is achieved by combustor design. The second is more difficult to control in today's combustor because of the use of a metallic combustor liner. Acceptable liner temperatures are maintained by impinging cooling air on both the inner and outer liner surfaces. Reduction of the amount of cooling air to more precisely control the fuel air ratio and minimize the emissions would cause the currently used metallic combustor liner to overheat. Whereas the progress in lower emissions made through design is impressive and meets all standards under discussion today, further progress could be made if higher temperature liner materials were available. The prospects for this will be discussed later in this paper.

Future Development Needs and Opportunities

Continued product improvements in future generation gas turbines depend on the availability of improved materials and on innovative component design concepts. These concepts must allow increased component efficiency but also must constrain the additional demands placed on these materials to the achievable levels of strength, temperature, and durability. The previous virtual quantum leaps in performance of each successive generation of engines have been achieved through large increases in rotor operating temperatures and stresses. Similar increases seem less likely in future generation engines because it appears there are few or no remaining major untapped opportunities for large improvements in turbine rotor materials performance. Certainly, the advances realized from the introduction of monocrystal turbine airfoils and powder metallurgy superalloy disks do not appear repeatable.

It, thus, appears that a greater portion of the efficiency improvements in future generation high bypass ratio engines will come from the use of larger fans and from weight reduction of the fan, compressor, and low pressure turbine (LPT) sections of the engine. Lightweight and/or high strength materials such as composites and intermetallic compounds will play important roles in realizing these weight reductions. Lightweight metallic structures such as thin walled castings and hollow, fabricated structures also will play a role in weight reductions, provided they can be produced at an affordable cost.

There is an additional, nontechnical factor that will affect the pace and nature of new technology developments. This is the shift in emphasis for future products away from the traditional single focus on technology driven performance improvements to much greater consideration of reduced cost. The reasons for this are the maturation of aircraft engines, the overcapacity of the engine industry, the depressed business climate, and the intense competition in the airline industry which are forcing operators to emphasize reductions in operating costs. Achieving these cost reductions appears at least as probable through reduced ownership cost and improved reliability and durability than through better fuel efficiency. Recent studies show that, during periods of steady or declining oil prices, fuel costs are not as important as ownership costs. Thus, the benefit of increased utilization factors and lower maintenance costs is becoming the central concern for the air lines. As a result, the improvements in fuel efficiency will only be welcomed by the operators if these do not come at higher initial equipment cost or at the expense of reduced operability and durability.

This change in focus will slow the rate of introduction of new materials, but still there are many opportunities for introduction of incrementally improved materials (evolutionary materials) in selected components. There also are areas where new component efficiencies only can be attained through the use of new, specialty materials that have large improvements in selected properties, such as stiffness (these often are called revolutionary materials). The needs and opportunities for introduction of both incrementally improved materials and revolutionary materials will be described in this section.

Rotor Materials

Perhaps the most pressing materials need for future generation, high OPR engines is the higher temperature disk material described earlier in connection with current limitations on T_3 . There are two principal material characteristics needed to meet this requirement. First, the creep strength of the material at the higher operating temperatures must be adequate to prevent rotor growth due to creep during service. Second, the rate of fatigue crack growth must be low enough to give acceptable service life, even when the crack growth rates are measured in tests where the peak load is held for an arbitrary period [hold time crack growth or (HTCG)]. All Ni-base alloys exhibit acceleration of crack growth rate under hold time conditions. The variation in HTCG rate between alloys, although significant, is smaller than can be observed in a single alloy if the grain size is varied. Fortunately, coarser grain sizes reduce the HTCG rates and also improve the creep strength. This leads to the conclusion that coarser grain sizes are beneficial for higher T_3 applications. Although this is encouraging, it must be recalled that coarser grained materials also exhibit shorter low cycle fatigue (LCF) lives and lower yield stress values. These trades in property balance must be accommodated in component design. This can be done by designing to lower maximum stresses to accommodate the lower yield strength and reduced LCF life characteristics of coarse grained materials. Such designs realize the benefits from improved HTCG and creep behavior without unacceptable reductions in LCF life. It is not clear how much improvement in HTCG behavior is possible beyond that achievable with coarse grained materials. The near-term solution will be to use coarse grained materials and to keep the stress amplitudes low in the highest temperature regions of the disk, since HTCG rate depends on both temperature and stress amplitude.

Current estimates of the temperature limitations for disk alloys are based on limited data, but these suggest that any of the uniform coarser grain size materials examined so far have a maximum service temperature of about 1350°F (732°C). It appears that a bi-metallic or dual alloy disk (DAD) will be necessary to achieve the 1375-1400°F capability required by engines with OPR ≥ 50. The benefit of DAD technology is the ability to independently alter and, to a greater extent, control specific properties in those regions of a disk that are limited by different requirements. For example, the creep strength of the high-temperature regions of the rim of the disk can be increased independently from the bore and web. Concurrently, the LCF life and strength of the highly stressed but lower temperature regions of the web and bore can be improved relative to the coarse grained rim. This is very advantageous since these properties limit the life and performance of these respective regions of a disk. As an important step in reducing the DAD concept to practice, it has been demonstrated that a high strength, LCF resistant fine grained alloy can be forge bonded at the disk midradius to a lower yield strength but highly creep resistant coarser grained (ASTM 2 or coarser) alloy. Enough work has been done on the bonding process to create confidence that a high-integrity joint can be created reproducibly. Thus, the technological feasibility of DAD has been established. The open issue with respect to implementation of DAD technology is cost. The new cost conscious environment emerging in the aircraft engine industry will require large life cycle benefits to be demonstrated before a disk of this type, with higher first cost will be considered for introduction into a production application.

Airfoil Materials

The benefits of turbine airfoil alloys with higher temperature capability has been described earlier. Each of the last several generations of Ni-base alloys for airfoils have produced a 50-75°F (30-40°C) capability increase. Each generation also contains increasing amounts of high density refractory elements such as Ta and Re. These elements improve the creep strength of the alloy because they have lower diffusivity, but airfoils made from these alloys are heavier, which increases the disk stresses at the airfoil attachment points. It is essential to balance the increase in temperature capability against increases in density because, at some point, the increased stresses offset the benefit of any further increases in creep strength. However, within these density constraints, it appears that at least one more generation of improved creep strength turbine blade alloys can be identified and developed. At present, the question of alloy stability appears to be of greater concern for the next generation alloys. Alloy stability means that the constitution of the alloy must remain essentially constant over the life of the airfoil. Previously, unsuccessful alloys have exhibited the tendency to form brittle, solute rich constituents known as topologically close-packed (TCP) phases. Alloys that form TCP phases are known to have time-dependent reductions in creep behavior, manifested as increased creep rate and reduced rupture ductility.

A possible alternate material for high-pressure turbine airfoils is NiAl. This intermetallic compound is attractive because it has a high melting point, has about three times the thermal conductivity of superalloys and is only two-thirds as dense. The limitations of NiAl are modest high-temperature strength and very low-room-temperature ductility. There has been a large investment made in developing alloys based on NiAl for turbine blades and vanes. If NiAl can be made to work, analysis shows the weight reduction percentage of the rotor exceeds the density difference. This is because lighter static structures (frames, bearing supports, etc.) also are possible when the rotor weight is reduced. The biggest unanswered questions regarding the viability of NiAl are its toughness and ductility. Already, ductility-related difficulties in manufacturing experimental engine test hardware have been encountered. On the other hand, blade attachments have been successfully re-designed to accommodate the lower ductility. Worker training may resolve these ductility-related manufacturing concerns. However, there are real questions regarding thermal shock resistance. Component tests designed to examine this issue have yielded encouraging results. As a minimum, new designs that minimize thermal stresses will be required. The next step is a combustor rig test. If the rig test results are acceptable, an engine test will be conducted in 1995. The effort to reduce NiAl to practice is an expensive, very high risk activity, but it also has a very high payoff if it works. This risk-reward relationship typifies the stakes involved in reducing revolutionary materials to practice for turbine engines.

There also are other material possibilities for improved LPT airfoils. Improvement in the LPT performance is more likely to be derived from lower density materials for weight reductions than from increased temperature capability. Alloys from the family of intermetallic compounds based on γ titanium aluminide (TiAl) have < 0.5 the density of Ni-base superalloys and have the potential to replace them with substantial weight saving in the LPT. Data shows that these alloys have adequate strength and creep resistance to directly replace conventionally cast equiaxed Ni-base alloys in LPT blades. An additional advantage of γ TiAl is that the elastic modulus has a weaker temperature dependence than for superalloys. In the longer LPT airfoils, this makes the stresses that are generated by the natural harmonic modes of the rotor easier to analyze and manage across the entire flight envelope. The principal limitation of γ Ti aluminides is the low ductility at temperatures ≤ 450-500°C. As with NiAl, it is hard to specify a minimum required ductility.

Setting ductility requirements is troublesome because it cannot be done analytically with adequate accuracy. This is because the ductility limitations are localized at attachments, etc. Therefore, an empirical approach, drawing on past experience with metals is the historical norm. Examination of this approach shows prior experience is relevant in deciding what ductility levels have been adequate, but it provides much less useful guidance about how little ductility may be acceptable. An alternate empirical approach is to make test hardware, perform component tests and, if these results are acceptable, run the components in a factory engine test. Here, it is prudent to use a low risk introduction strategy. That is, components should be selected that have modest consequence of failure. For example, last stage LPT blades are relatively low risk because there is no rotating machinery behind them and, should a blade failure occur, the pieces can benignly exit the exhaust without inflicting harm on the rest of the engine. Recently, CF6 factory engine tests have been completed using cast last stage LPT airfoils about 15 in. (38 cm) in length, made from the alloy Ti-48Al-2Nb-2Cr (atomic %). These tests already have successfully run ≥ 1500 cycles. This is an excellent result which represents a significant demonstration of the capability of this material. There is more to do before γ alloys are introduced into production engines, but such an introduction appears to be technically feasible. The major barrier may be cost. These alloys are intrinsically less expensive than superalloys, e.g., they contain no Co and Ni, the prices of which are climbing. The cost issues seem to center on lack of production experience in casting them, which translates into concerns about yields. If a decision is made to introduce this material, there is sufficient capability in the casting industry that negotiation of a mutually acceptable supply agreement should be possible.

At the higher T_3 values characteristic of higher OPR machines, the fine grained wrought compressor airfoils currently used will have inadequate creep strength. The castable alloys used for turbine airfoils have more than enough temperature capability for this application. However, the cross-sectional shape of compressor airfoils is much thinner and, therefore, different from turbine airfoils. In particular, the casting of airfoils with thin leading and trailing edges is at the limit of current casting technology. Moreover, compressor airfoils have been typically limited by high cycle fatigue (HCF) strength. Conventionally cast superalloys have coarser equiaxed grains than equivalent wrought material; this lowers the yield stress and the HCF run out stress. Cast directionally solidified or monocrystal alloys have still lower yield strengths but considerably improved creep strength. The tradeoff here is between creep strength improvements and HCF reductions due to lower yield strengths. There are a number of questions to be answered before it will be known if cast superalloy compressor blades represent a suitable means of accommodating the higher temperatures required for future generation higher OPR engines. However, recent results are promising.

Two classes of materials that appear to be alternatives to cast superalloys for higher T_3 requirements are the intermetallic compounds of Ti (Ti_3Al or α_2) and Ni (Ni_3Al or γ'). The investment in developing the Ti_3Al- and Ni_3Al-based alloys was initiated before the new market realities were clear. The uncertain future of these materials serves to illustrate the implications of these realities for a new strategy and a set of guidelines for materials development and introduction. The current thinking at GE is to strongly favor product pull and move away from technology push to guide investment in future materials and process development. New projects that culminate in more costly technology when it is implemented will be carefully analyzed in terms of life cycle cost benefit and competitive advantage before any investment is made. As our product matures, this disciplined approach is necessary to meet the needs of our business with the resources available.

Static Structures

There are three major opportunities to improve the efficiency of static structures by combining component design with the availability of new materials. As the sheer size of next generation engines increases, substantial weight savings can be realized from static structures having increased structural efficiency.

The first of these opportunities is the increased use of polymer composites (PMCs) in the fan case, fan containment structure, and core cowl of the engine. These applications would result in significant weight savings, but at a cost increase using current manufacturing methods. There currently are several programs under way to improve the efficiency of manufacturing methods for polymer composites. There also are new materials under development that will be easier to fabricate and some that are less toxic for the factory workers and are easier to dispose of from an environmental standpoint. Alleviating these environmental, health, and safety issues also will reduce the cost of using PMCs.

The second opportunity is the use of titanium matrix composites (TMCs) for weight reduction of stiffness limited static structures. Examples include fan frames, casings, and actuator links. These are attractive because the risk is low and the benefit is large. There has been a large investment in TMCs over the past 8-10 years. As a result, the technical feasibility of TMCs has been proven in terms of their properties, but the cost of these materials is still considerably too high. The properties of current generation TMCs are shown in Table 2, from which it can be seen that these materials have about twice the structural capability of Ti alloys but have the same density. The reproducibility of these properties also has been demonstrated, thus technical feasibility seems to be at hand. Indeed, TMCs are an interesting class of material that is limited in its potential only by cost. Based on detailed analysis, it now is believed that the intrinsic cost of this material can be reduced through process improvement

Table 2 One example of reproducible TMC[a] properties

Property	Value
Ultimate Tensile Strength, ksi (MPa)	276 (1.9×10^3)
Young's Modulus, msi (MPa)	32.8 (226×10^3)
Fracture strain, %	0.95
Density (equivalent to Ti alloys), lbs/in.^{3} (gms/cm^3)	0.16 (4.43)

[a]System: Ti 6Al 2Sn 4Zr 2Mo+Si; about 150 μm SiC fibers

and the emergence of sufficient demand to permit economies of scale to be realized. A large cooperative effort that includes both material users and producers has been initiated to address these issues. This effort will focus both on intrinsic cost reduction and increasing demand through commonalty of materials requirements and product forms.

As is the case with most revolutionary materials, cost is the only real, remaining barrier to introduction of this material for low risk applications. In the case of the fan frame, analysis shows the use of selective reinforcement by TMCs can achieve the desired stiffening result with a modest quantity of the higher cost TMC material. Innovative designs such as this that combine creative use of materials and novel manufacturing concepts materials will be required to expand the introduction of revolutionary materials in a cost effective manner. If the TMC cost reduction program succeeds, it is likely that these materials will be introduced in production systems before the end of the century. If not, TMCs likely will be remembered as a technological success that did not reach commercialization because of cost.

The third example of static structures that can be improved by introduction of revolutionary materials is the liner for very low-emissions combustors. The need for higher temperature liner materials was described previously. Calculations show that liners which operate without cooling air on the inner surface will experience temperatures in excess of 2600°F (1425°C). This exceeds the temperature capability of any metals that could be realistically used. Thus, it will be necessary to make these new liners from ceramic matrix composites (CMCs). These temperatures are higher than any of the currently available CMC systems are capable of withstanding for the long times expected of a component such as a combustor liner. A very large materials development program is underway to provide new materials for this purpose.

Even in a summary, it must be made clear that the requirements for a successful CMC combustor liner material include the ability to maintain fiber dominated behavior for at least 10,000 h, much of which is at an operating temperature of ≈1425°C. Fiber dominated behavior is the characteristic of CMCs that toughens them and differentiates their fracture behavior from that of monolithic ceramics such as hot pressed Si_3N_4. The challenge is to develop a material that has a fiber-matrix interface that provides toughening behavior. To retain the required toughness during service, the behavior of this interface must remain largely unchanged during service exposure for the times and temperatures mentioned earlier. To maintain liner wall temperatures of ≈1425°C, the material also must have thermal conductivity about equivalent to SiC or Si_3N_4 because the cooling air is only applied to the outside of the liner wall. Analysis shows that the lower conductivity values characteristic of other ceramics, e.g. oxides, will cause the wall temperatures to rise by another ≈175-250° C, reaching 1600°C or higher. There are no CMC systems today that can withstand such high temperatures and meet the long time toughness requirement. No matter which system is chosen, the requirements are very challenging. Added to the basic materials behavior challenge is the requirement to manufacture combustor liners from the CMC material.

All in all, the ability to further reduce emissions through the use of new combustors that do not use cooling for the inner liner wall represent, perhaps, the greatest challenge of any material or component development effort described in this paper. The program underway is examining the most promising systems and

Benefits Estimation of New Engine Technology

Edward R. Generazio[1] and Christos C. Chamis[2]
NASA Lewis Research Center
Cleveland, Ohio 44135

Abstract

A technology benefit estimator system has been developed to provide a formal method to assess advanced technologies and quantify the benefit contributions for prioritization. The technology benefit estimator may be used to provide guidelines to identify and prioritize high-payoff research areas, help manage research and limited resources, show the link between advanced concepts and the bottom line, i.e., accrued benefit and value, and to credibly communicate the benefits of research. An open-ended, modular approach is used to allow for modification and addition of both key and advanced technology modules. The technology benefit estimator has a hierarchical framework that yields varying levels of benefit estimation accuracy that are dependent on the degree of input detail available. This hierarchical feature permits rapid estimation of technology benefits even when the technology is at the conceptual stage. The technology benefit estimator's software framework, status, novice-to-expert operation, interface architecture, analysis module addition, and key analysis modules are discussed. Representative examples of the technology benefit estimator's analyses are shown.

---

1 Head of the Non-destructive Evaluations Science Branch.

2 Senior Aerospace Scientist, Structures Division.

developing shape making manufacturing processes. This is a high risk activity. Even if technical feasibility is demonstrated, cost of these combustor liners will initially be very high, and achieving adequate cost reductions also will be a challenge. Even so, the payoff from success is enormous and may even control the future of high OPR aircraft engines as emission requirements in the 21st century become more stringent.

Business Environment

After several decades of continuous improvement, as outlined, the jet engine is approaching a degree of maturity that has and will continue to affect the nature, pace, and extent of future developments. The current business environment is shaped by a depressed state of the airline industry, the rapid contraction of military spending throughout the world, and a protracted period of flat fuel prices. These factors only serve to accelerate recognition of the new business realities for an industry that previously has been driven almost exclusively by technological advancements. The new realities are cost, not performance centered, and this relatively abrupt change tends to redefine the term: affordable performance improvements. Going forward, the engine manufacturers will find it more difficult to effectively differentiate their products in the market place on the basis of technology unless they also are either the low cost engine provider or at cost parity with their competitors.

The commercial aircraft and engine industry deals with a very expensive product in terms of development cost and has a long product cycle. One of the current objectives of all manufacturing companies is to reduce by as much as 50% the cycle time required to develop a new or derivative product and introduce it into service. For aircraft engine companies, it also is important to shorten their product development cycle in absolute terms but also in relative terms. It is important for the engine development cycle to be shorter than that for an airplane. Achieving this will permit the engine development to focus on a more sharply defined set of requirements. This is economically advantageous because it minimizes the amount of redesign due to changing requirements of the airplane. To achieve this objective an integrated product development (IPD) approach is required. The IPD approach requires that all technology that is to be included in the new product must be ready at the initiation of the product development effort. This requires a greater degree of planning than historically has been the case. Previously, the longer development times allowed much of the required technology to be developed as the need arose. This will no longer be the case. It, thus, is essential to focus on the correct set of products and to invest in the relevant technologies that can provide benefit to these products. This investment must be started before the need is well defined in actual product requirement terms. Technology development costs can only be controlled by selection of the correct projects. This requires a level of discipline that often has not been apparent. In commercial products all development costs must be recovered in the selling price. This suggests that greater selectivity in technology investment will be required in the future.

The author believes that cost will be the principal discriminator in the market place and this represents a paradigm shift for both the engine and aircraft industries. This shift requires a culture change in the engine companies for many of the reasons outlined earlier. In the future, the most successful companies will be those that make the change most rapidly and most completely. Those that do not respond to this new reality will lose control of their destiny and, perhaps, disappear.

Summary

This paper has described the advances in aircraft engine development since their inception. The advances described are those related to either component efficiency, design methods, or materials capability. The progress made is measured in terms of several fundamental turbofan engine parameters, such as compressor exit temperature, turbine inlet temperature, overall pressure ratio, and bypass ratio. The article focuses on commercial instead of military engines because this is where the business is concentrated now and in the future. The requirements for future products and the prospects for achieving these are discussed using past progress as a reference point. It is suggested that new measures of business performance will be needed to assess the viability of engine companies and their products in the future. The essence of these measurements is product cost and product development cycle time, which is related to product development cost. The requirement to meet both relative and absolute cost targets is the basis for ongoing viability of aircraft engine companies. This must be balanced with retention of a product line that has competitive performance and meets the operators' expectations. Thus, the future challenges for the industry will be different than those of the past but, possibly, even more daunting.

References

[1]Williams, J. C., "The Development of Advanced Gas Turbine Engines: the Technical and Economic Environment," *Materials for Advanced Power Engineering 1994,* Kluwer Academic, Norwell MA, 1994, pp. 1831-1846.

Introduction

Over the past several years the costs and risks of introducing dramatically new technologies into aerospace propulsion systems have been perceived to be very high, whereas other benefits, e.g., noise emissions, weight, and reliability are generally unknown. This perception and insufficient knowledge of benefits of new technologies is an effective barrier for introducing new technologies into propulsion systems. A good example of perceived high-risk technology is composite materials. Here the manufacturing costs, durability, and material properties are often unknown until detailed analyses are performed. That is, an engineer cannot simply turn to a handbook of standards on composites to obtain the required material data for designing a composite component. The engineer needs to perform a detailed multidisciplinary analysis, including effects of fiber orientation and external loads, etc., on the engineered component. Also, since the durability is not represented in a handbook, a component life analysis needs to be done in conjunction with the multidisciplinary analysis. Here the internal makeup of the component, the component geometry, and the component's life can greatly affect the life cycle costs of the component. It is costs, such as these, that need to be quantitatively determined before new technologies are introduced into aeropropulsion systems. The over arching goal of this work is to credibly communicate the benefits of research in new technologies. A formal method is needed to assess technologies and quantify and prioritize benefit contributions.

Such a method can be used for providing guidelines to identify and prioritize high-payoff research areas, to help manage research and limited resources for greatest impact, and to show the link between advanced concepts and the bottom line, i.e., accrued benefit and value. Credible determination of benefits can only be obtained from an optimization, with respect to industry specific objective functions, of multidisciplinary analyses that includes analyses such as mission, engine cycle, weight, life, emissions, noise, manufacturing, and cost. The following work describes the technology benefit estimator (T/BEST) analysis simulation system that is being developed for credibly communicating the benefits of introducing new technologies into aerospace propulsion systems. Typical benefits needed to make investment decisions are range, speed, thrust, capacity, city pairs reached, component life, noise, emissions, specific fuel consumption, component and engine weights, precertification test, engine cost, direct operating cost, life cycle cost, manufacturing cost, development cost, risk, development time, and return on investment.

Approach

A hierarchical, modular, open-ended approach has been used to interface a wide variety of disciplines: mission analysis, thermodynamic engine cycle analysis, engine sizing or flowpath analysis, weights, analysis, structural analysis, cost analysis, and noise and emissions analyses. The integration and control of these software analyses is the T/BEST system. The hierarchical arrangement permits credible estimation of benefits even when there is a limited amount of data. Here a package of statistical correlations is relied upon for projecting both the interpolated and extrapolated data that is needed for completing a benefits analysis. Increasing hierarchical detail yields increased accuracy of the benefits (Fig.1) determined. The hierarchical details span the range from speed, capacity, and range of the first estimation level to the constituent components of the subcomponent of the twelfth estimation level (Fig. 2). The open-ended feature allows for system growth, whereas the modular aspect permits easy addition, updating, and replacement of analysis modules (Fig. 3). The T/BEST system is designed to be operated by beginners, intermediate, and experts users. The beginner has control of basic parameters, e.g., mission, range, component materials, and airfoil geometry selection from the

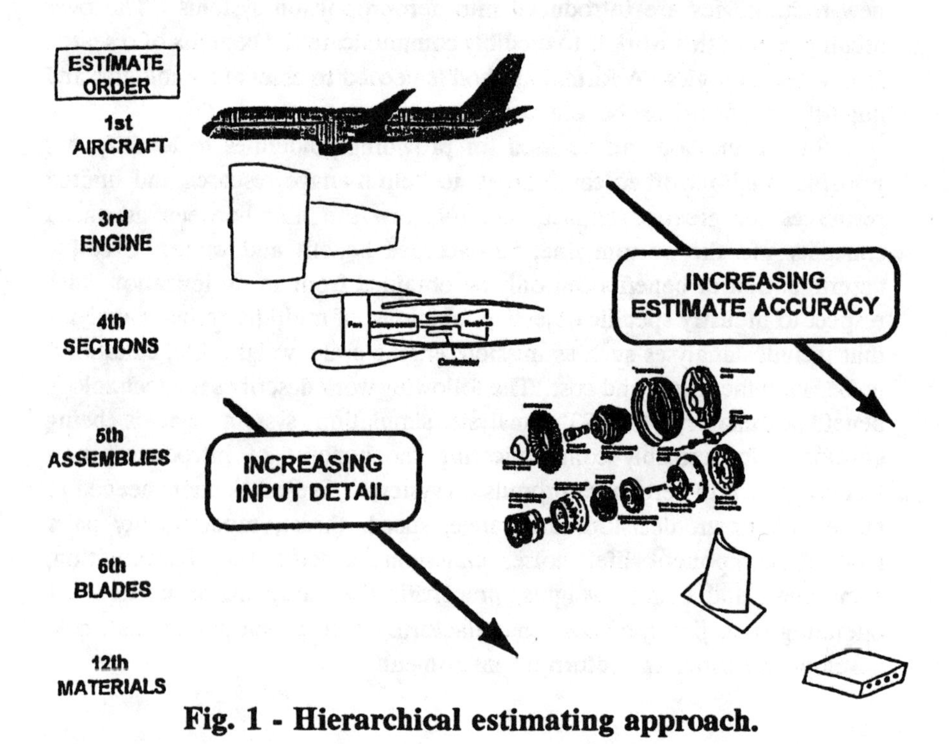

Fig. 1 - Hierarchical estimating approach.

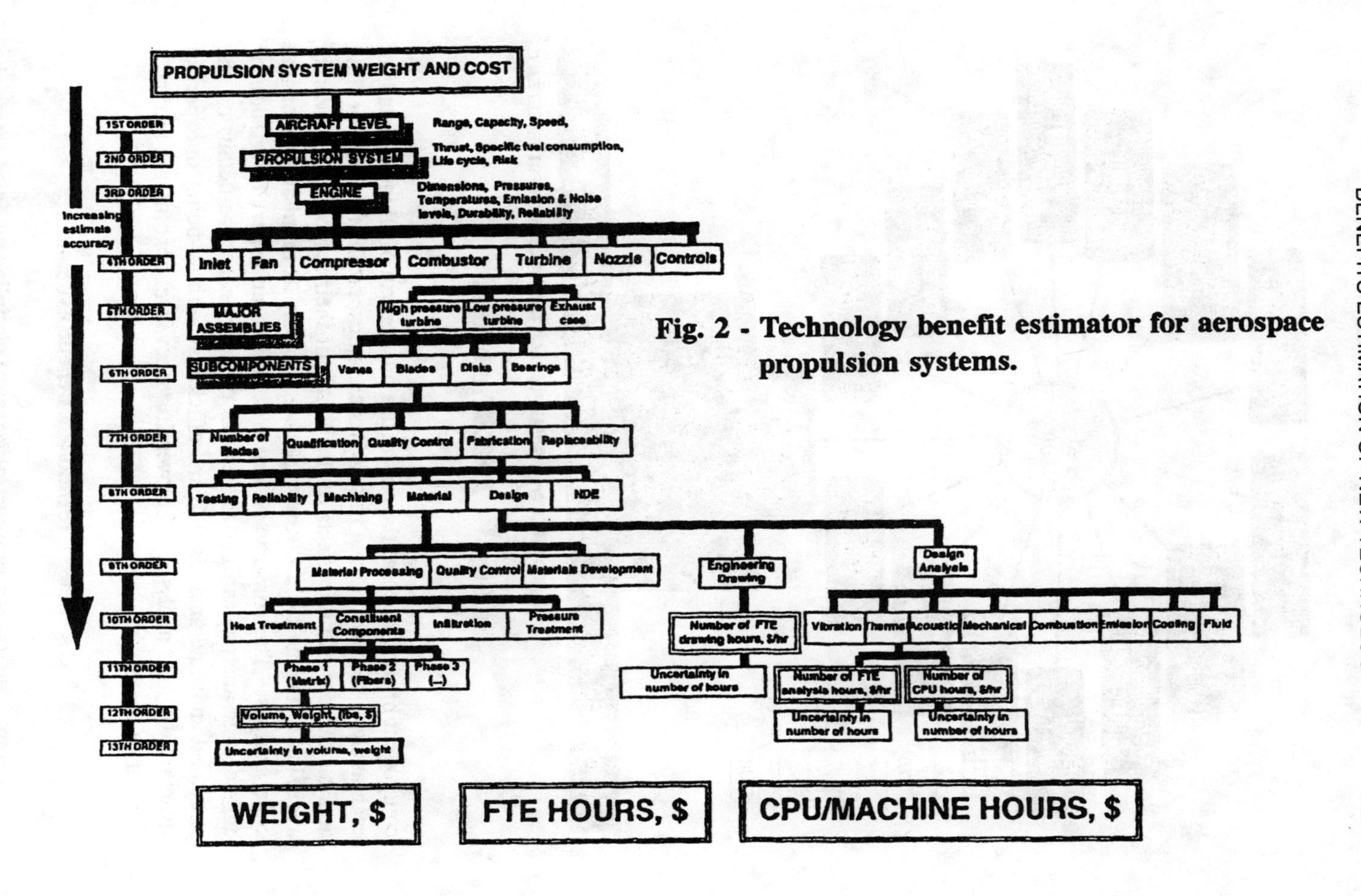

Fig. 2 - Technology benefit estimator for aerospace propulsion systems.

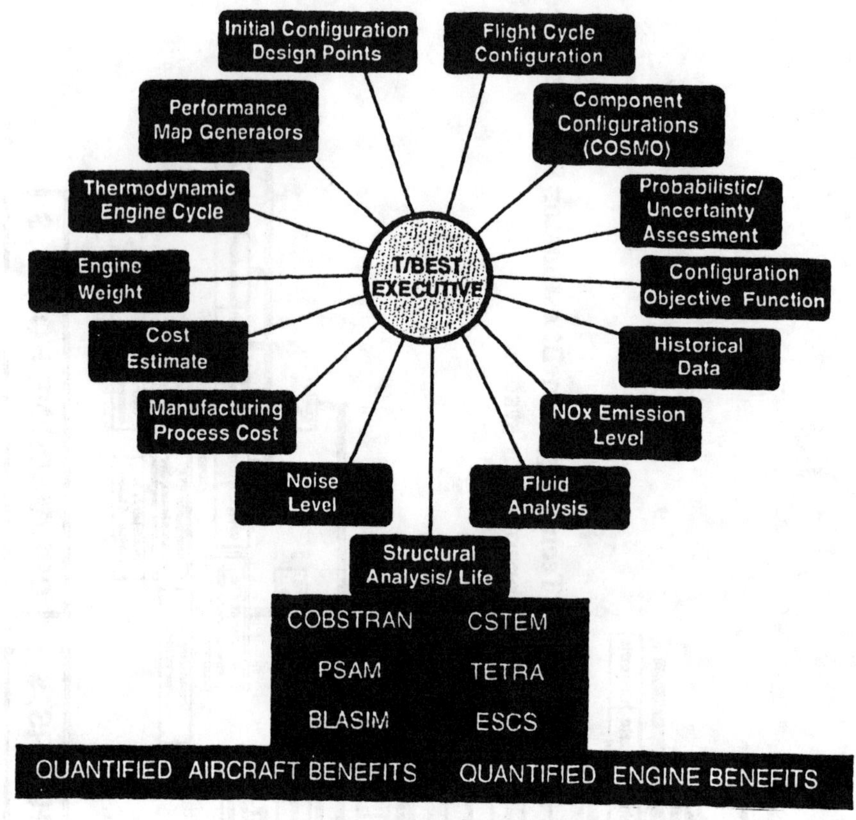

Fig. 3 - T/BEST modules.

component library, whereas the expert has full control of all input details of each analysis module. Using this approach, an expert in structural or fluid analysis can observe the effects of the details of their respective analysis on a mission or engine cycle, etc. without being an expert in mission or engine cycle analysis. The converse is also true, where an expert in mission or engine cycle analysis can see the effects of their respective details on the structural and fluid analysis (life and efficiency) without being an expert in structural or fluid analysis. The overall system level benefits are available to all user levels.

Overview of Technology Benefit Estimator

Figure 3 indicates a representative set of discipline modules within the T/BEST system. The T/BEST executive, a UNIX shell, controls the

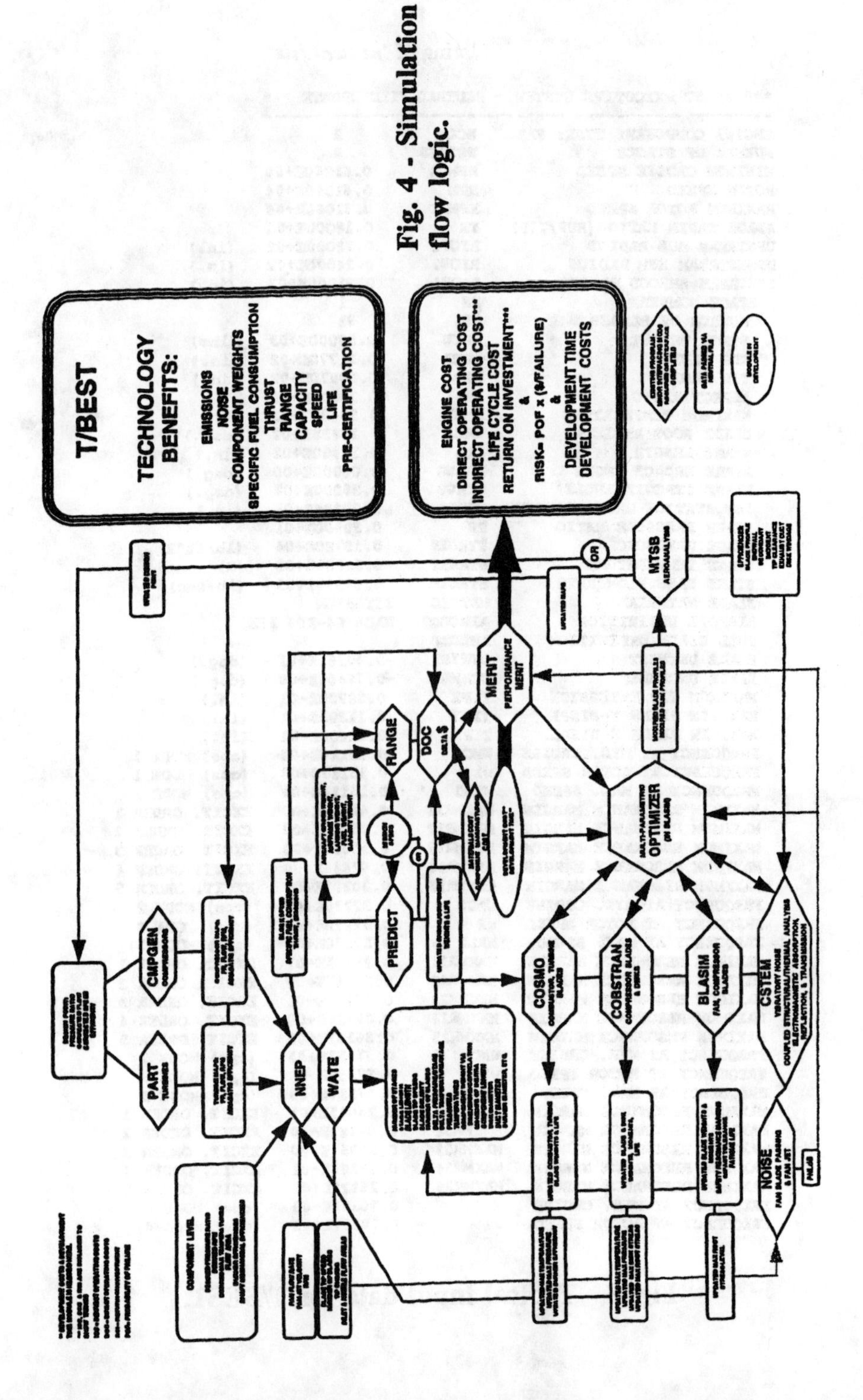

Fig. 4 - Simulation flow logic.

Listing of "*neutral.file*"

```
*** TBEST EXECUTIVE SYSTEM - NEUTRAL FILE UPDATE ***
-----------------------------------------------------------
ENGINE COMPONENT TYPE: FAN     NCC          2
NUMBER OF STAGES               NSTAGE       2
MINIMUM CRUISE SPEED           RPMCR        0.61060E+04
ROTOR SPEED                    RPM          0.61060E+04
MAXIMUM ROTOR SPEED            RPMAX        0.61060E+04
BLADE TAPER RATIO (HUB/TIP)    TR           0.18000E+01
UPSTREAM HUB RADIUS            RIUP1        0.12000E+02   (in.)
DOWNSTREAM HUB RADIUS          RIDW1        0.24000E+02   (in.)
UPSTREAM SHROUD RADIUS         ROUP1        0.31000E+02   (in.)
  STAGE NUMBER                 NS           1
   NUMBER OF BLADES            NB           33
   STAGE WEIGHT                WSTW         0.53000E+03   (lbs)
   HUB RADIUS                  RHBA         0.11770E+02   (in.)
   TIP RADIUS                  RTBA         0.30970E+02   (in.)
   ASPECT RATIO                AR           0.30000E+01
   MAXIMUM TEMPERATURE         TMAX         0.85400E+03   (R)
   BLADE ROOT ANGLE            THER         0.18435E+02   (deg.)
   STAGE LENGTH                STL          0.13900E+02   (in.)
   BLADE BROACH ANGLE          BRANG        0.00000E+00   (deg.)
   BLADE STAGGER ANGLE         STAGG        0.35000E+02   (deg.)
   1ST STATION CHORD LENGTH    CHORD(1)     0.82286E+01   (in.)
   STAGE PRESSURE RATIO        PR           0.20900E+01
   STAGE PRESSURE              STAGEP       0.19720E+04   (lb/ft^2)
   STAGE TEMPERATURE           STAGET       0.51900E+03   (R)
   STAGE MASS FLOW RATE        STAGEF       0.69390E+03   (lb/sec)
   BLADE MATERIAL              MATSLC       TITANIUM
   AIRFOIL DEFINITION          AIRCODE      NACA 64-206 FAN
   FULL BLADE DEFINITION       ABLDEF
   BLADE UNTWIST               UTWIST      -0.40261E+01   (deg.)
   BLADE UNCAMBER              UCAMB       -0.76460E+00   (deg.)
   MAXIMUM TIP EXTENSION       TIPX         0.32700E-01   (in.)
   MAX. IN PLANE Y-DISPL.      TIPY         0.11305E+01   (in.)
   MAX. IN PLANE Z-DISPL.      TIPZ         0.31400E-01   (in.)
   FREQUENCY AT MIN. CRUISE    WMC1         0.15116E+03   (cps) MODE 1
   FREQUENCY AT ROTOR SPEED    w1           0.15117E+03   (cps) MODE 1
   FREQUENCY AT MAX. SPEED     WRL1         0.15116E+03   (cps) MODE 1
   MAXIMUM RESONANCE MARGIN    MAXMR11      0.48536E+00   EXCIT. ORDER 1
   MAXIMUM RESONANCE MARGIN    MAXMR12      0.25732E+00   EXCIT. ORDER 2
   MAXIMUM RESONANCE MARGIN    MAXMR13      0.50488E+00   EXCIT. ORDER 3
   MAXIMUM RESONANCE MARGIN    MAXMR14      0.62866E+00   EXCIT. ORDER 4
   MAXIMUM RESONANCE MARGIN    MAXMR15      0.70293E+00   EXCIT. ORDER 5
   FREQUENCY AT MIN. CRUISE    WMC2         0.32270E+03   (cps) MODE 2
   FREQUENCY AT ROTOR SPEED    w2           0.32270E+03   (cps) MODE 2
   FREQUENCY AT MAX. SPEED     WRL2         0.32270E+03   (cps) MODE 2
   MAXIMUM RESONANCE MARGIN    MAXMR21      0.21709E+01   EXCIT. ORDER 1
   MAXIMUM RESONANCE MARGIN    MAXMR22      0.58547E+00   EXCIT. ORDER 2
   MAXIMUM RESONANCE MARGIN    MAXMR23      0.56982E-01   EXCIT. ORDER 3
   MAXIMUM RESONANCE MARGIN    MAXMR24      0.20726E+00   EXCIT. ORDER 4
   MAXIMUM RESONANCE MARGIN    MAXMR25      0.36581E+00   EXCIT. ORDER 5
   FREQUENCY AT MIN. CRUISE    WMC3         0.37537E+03   (cps) MODE 3
   FREQUENCY AT ROTOR SPEED    w3           0.37538E+03   (cps) MODE 3
   FREQUENCY AT MAX. SPEED     WRL3         0.37537E+03   (cps) MODE 3
   MAXIMUM RESONANCE MARGIN    MAXMR31      0.26885E+01   EXCIT. ORDER 1
   MAXIMUM RESONANCE MARGIN    MAXMR32      0.84427E+00   EXCIT. ORDER 2
   MAXIMUM RESONANCE MARGIN    MAXMR33      0.22951E+00   EXCIT. ORDER 3
   MAXIMUM RESONANCE MARGIN    MAXMR34      0.77866E-01   EXCIT. ORDER 4
   MAXIMUM RESONANCE MARGIN    MAXMR35      0.26229E+00   EXCIT. ORDER 5
   FREQUENCY AT MIN. CRUISE    WMC4         0.70356E+03   (cps) MODE 4
   FREQUENCY AT ROTOR SPEED    w4           0.70357E+03   (cps) MODE 4
```

Fig. 5 - Typical input data for T/BEST.

$$\text{MERIT}_i = \sum_j A_{ij} * (X_{ij}/N_{ij})$$

A = PROPRIETARY COEFFICIENTS

N = NORMALIZATION FACTORS

X = DELTA T/BEST OUTPUTS (e.g., Δ WEIGHT, Δ COSTS, ETC.)

i = PERFORMANCE MERIT RELATIONSHIP NUMBER

j = NUMBER OF PARAMETERS IN RELATIONSHIP

$$\text{TOTAL MERIT} = \sum_i \text{MERIT}_i$$

Fig. 6 - Merit function for evaluating benefits.

OBJ = 0.43 * (% DELTA TSFC) + 0.40 * (DELTA WEIGHT/1000) + 0.21 * (DELTA COST/100,000) + 0.32 * (DELTA MC/10)

where DELTA TSFC = 6 * (1 - η) for rotor 6

= 60 * (1 - η) for fan

η - Efficiency

DELTA WEIGHT = No. of Blades * Blade Weight

DELTA COST = Material Cost

DELTA MC = Maintenance and Material Cost

Fig. 7 - Numerical values for specific merit function.

operation of the entire system. All critical data, e.g., specific fuel consumption, number of stages, subcomponent weight, stress levels at critical areas of the components, etc., is passed through a neutral file so that other modules can easily assess and share this data. Copious data, such as airfoil profile geometries, nodal pressure, and temperature loads are passed via pointers that point to the appropriate files. The flow of data is T/BEST is graphically shown in Fig. 4 and a section of the neutral file is shown in Fig. 5. The figure of merit or objective function varies (Fig. 6) with the problem being addressed, however, a typical objective function is shown in Fig. 7. Here the changes in specific fuel consumption, weight, material, and

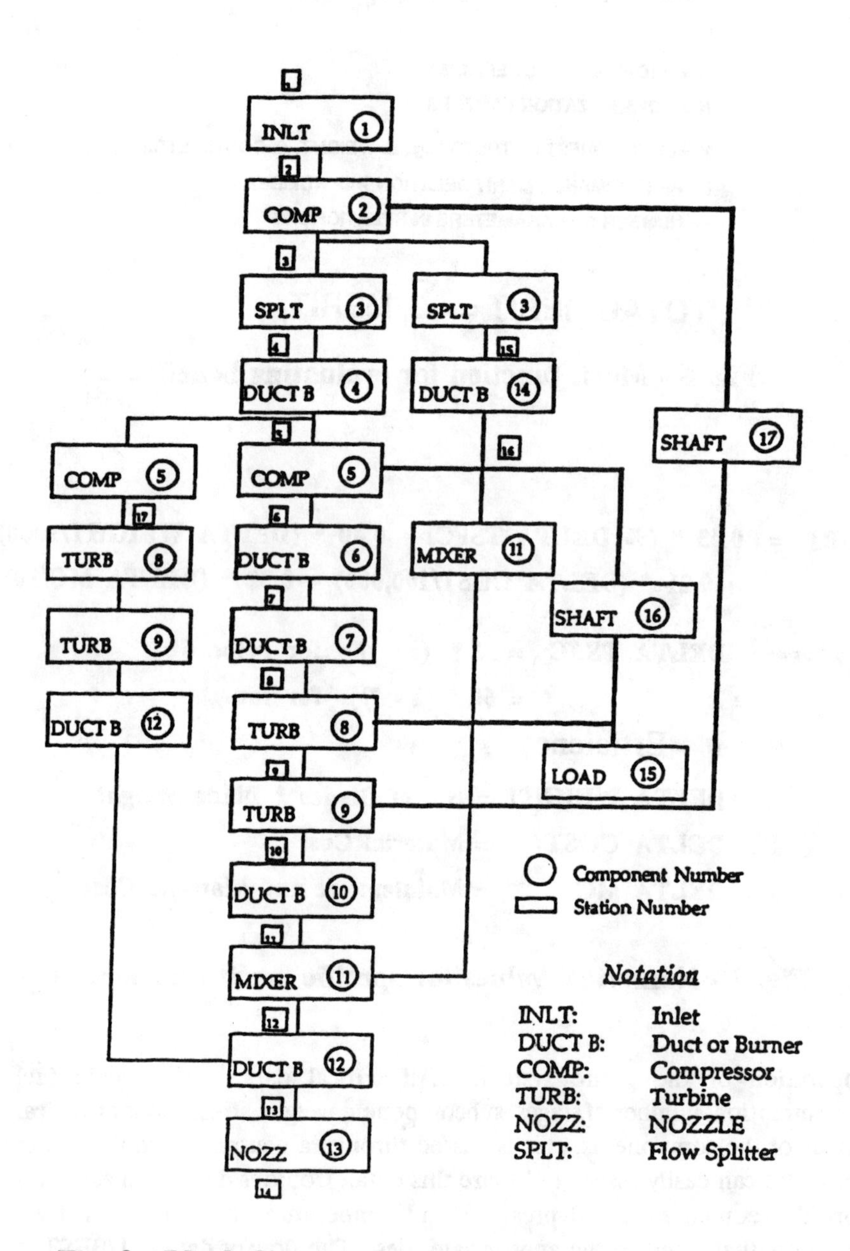

Fig. 8 - Block diagram for supersonic research engine.

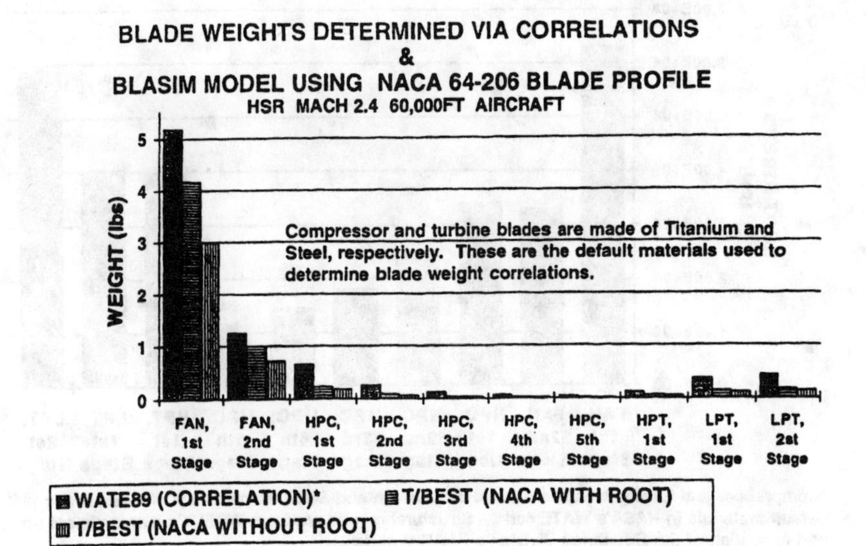

* The blade weights determined by the configuration evaluation weight code are based on correlations developed from historical data.

Fig. 9 - Blade weights comparison

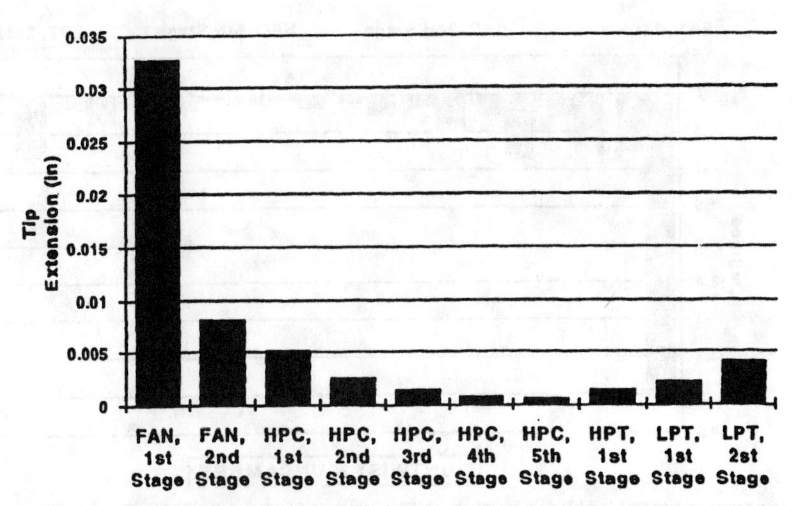

*Compressor and turbine blades are made of Titanium and Steel, respectively. These are the default materials in NASA's WATE code. Structural results are from T/BEST's automatic set up and execution of the Structures Division's BLASIM code.

Fig. 10 - Blade structural response - tip extension.

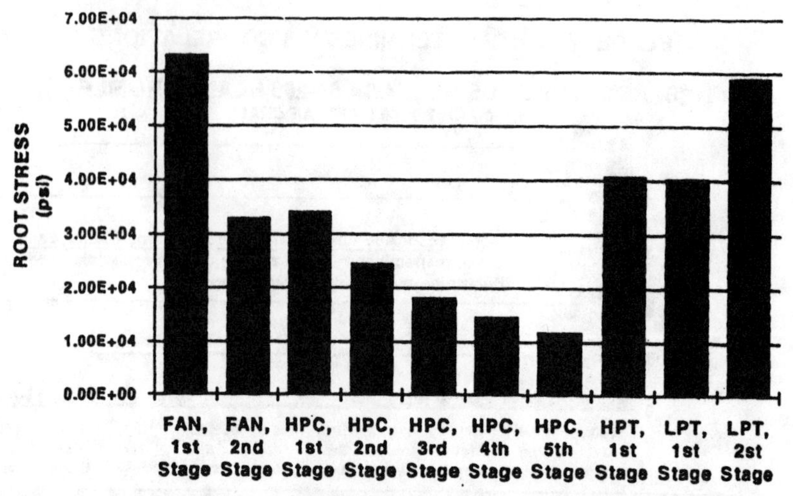

*Compressor and turbine blades are made of Titanium and Steel, respectively. These are the default materials in NASA's WATE code. Structural results are from T/BEST's automatic set up and execution of the Structures Division's BLASIM code.

Fig. 11 - Blade structural response - root stress level.

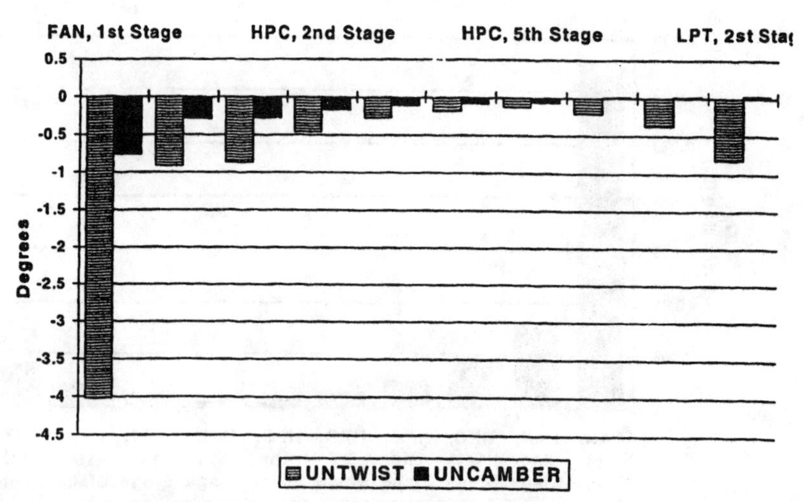

*Compressor and turbine blades are made of Titanium and Steel, respectively. These are the default materials in NASA's WATE code. Structural results are from T/BEST's automatic set up and execution of the Structures Division's BLASIM code.

Fig. 12 - Blade structural response - untwist and uncamber.

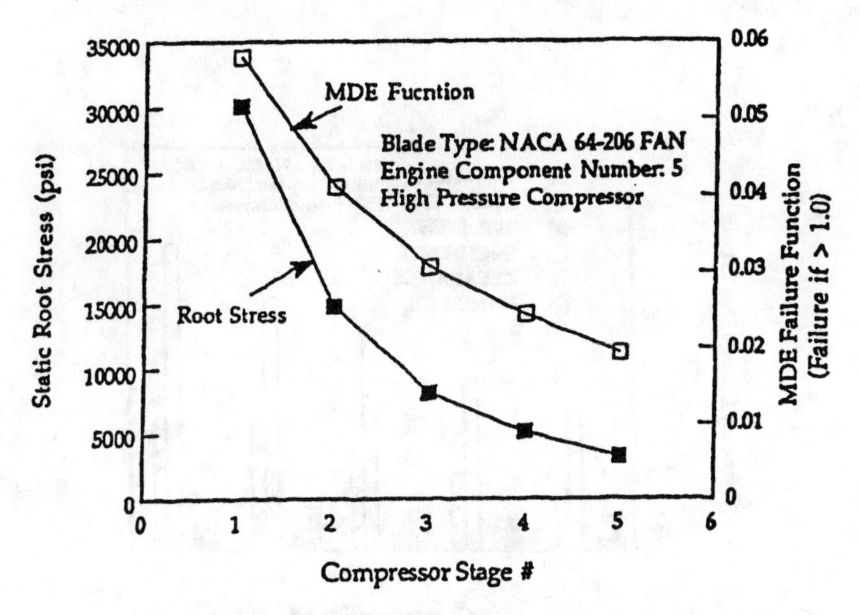

Fig. 13 - Blade structural response - static root stress level and MDE failure function.

maintenance costs are combined with coefficients and normalization factors are usually proprietary.. The security of the T/BEST system is assured, in that the entire system is transportable to a workstation. Since the analysis codes are all in portable Fortran and UNIX shell, the transportability feature has come easily. It is pointed out here that a software module developed using other languages, for example, object-oriented languages, can also be easily attached to T/BEST. After a complete simulation cycle or run, a wide range of data is available for display.

Example

Figure 8 shows an initial engine configuration for a supersonic aircraft engine. During a simulation T/BEST performs the thermodynamic engine cycle analysis, flight cycle optimization, engine sizing or flowpath determination, generates components and applies thermomechanical loads, structural analysis, weights analysis, fluid analysis, noise analysis, emission analysis, mean time between repair analysis, and direct operating cost analysis. Some typical outputs from T/BEST is shown in Figures 9 -- 14. The blade's weights, tip extensions, static blade root stress levels, uncamber and untwist, failure function, and rotor efficiencies are available, stage-by-stage throughout the engine. Figures 9 -- 14 show these results for the high-pressure compressor. The flight cycle mission is shown in Fig. 15, and the

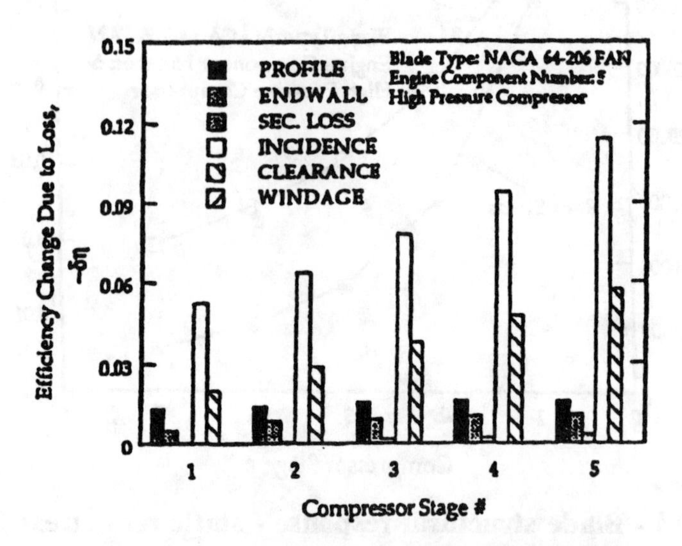

(a) Compressor Efficiency Changes Due to Loss

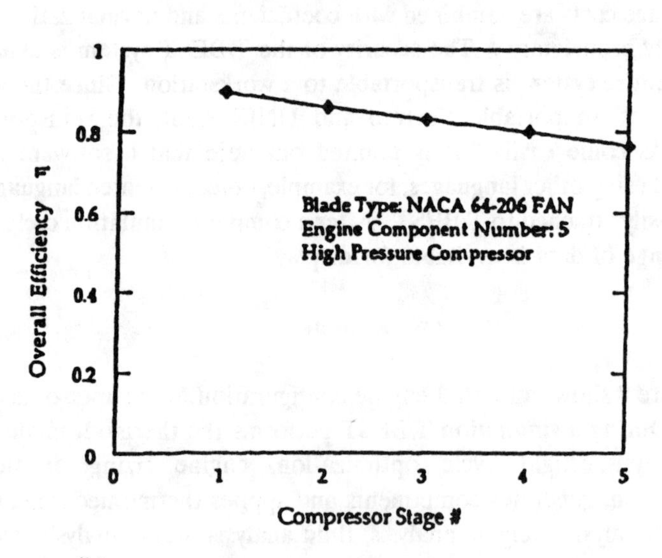

(b) High Pressure Compressor Stages Overall Efficiency

Fig. 14 - Compressor efficiencies.

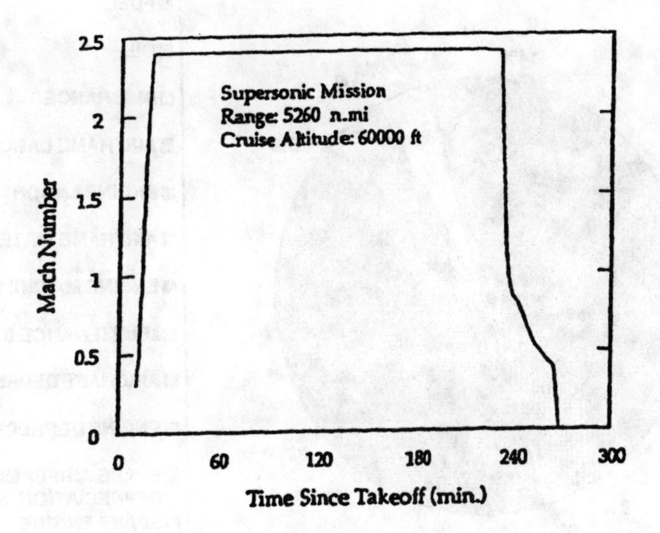

(a) Mission Performance: Mach Number at Climb, Cruise, and Descent.

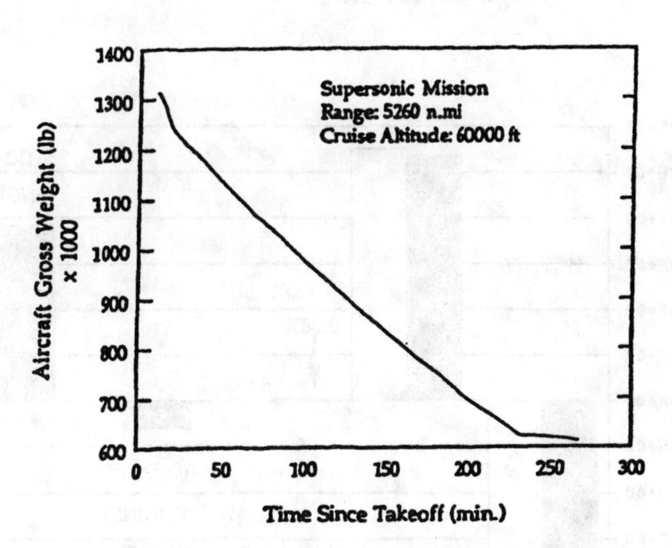

(b) Mission Performance: Aircraft Gross Weight at Climb, Cruise, and Descent.

Fig. 15 - Mission performance.

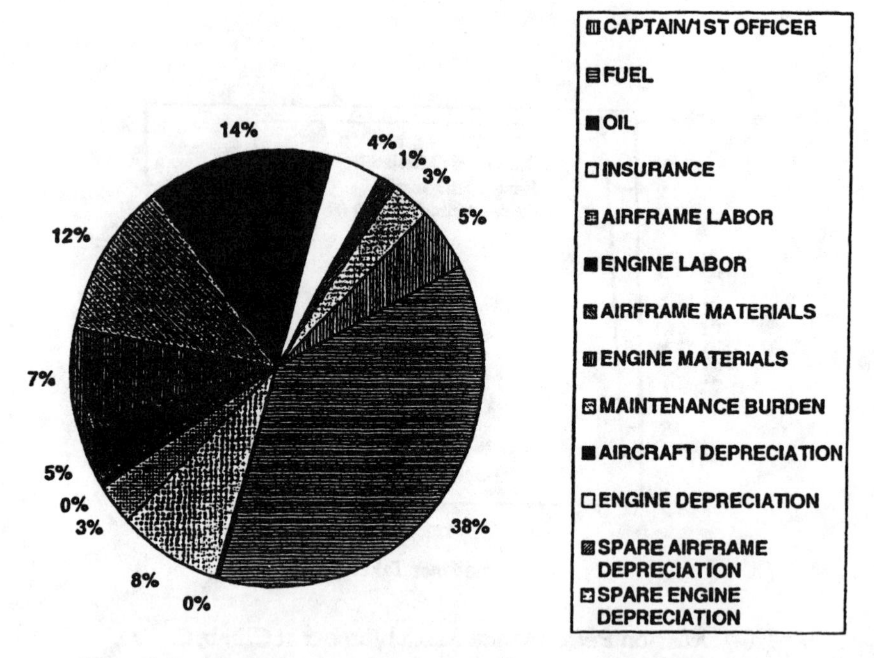

Fig. 16 - Turbine engine aircraft cost (%) mile and design 60,000 ft.

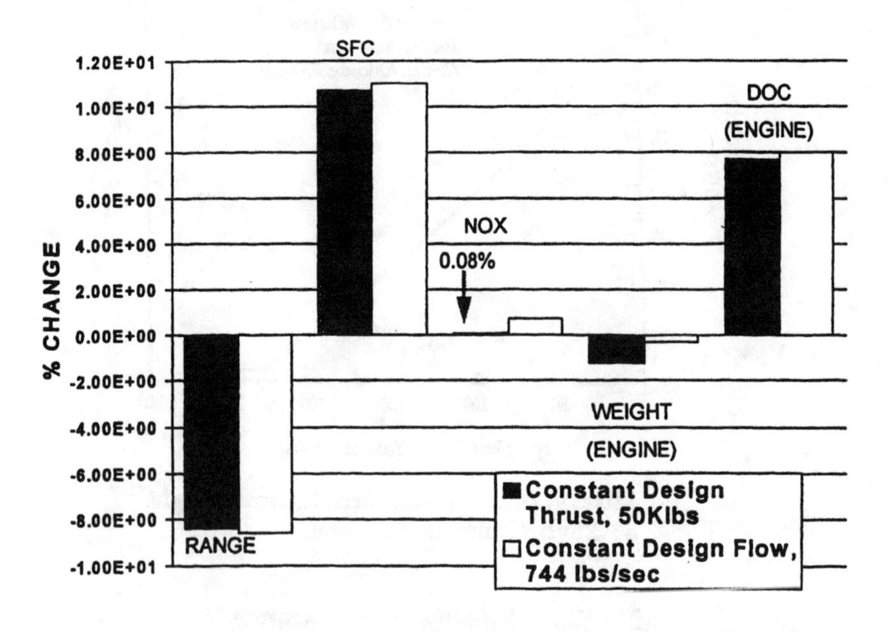

Fig. 17 - Combustor efficiency is 10% less than 0.99.

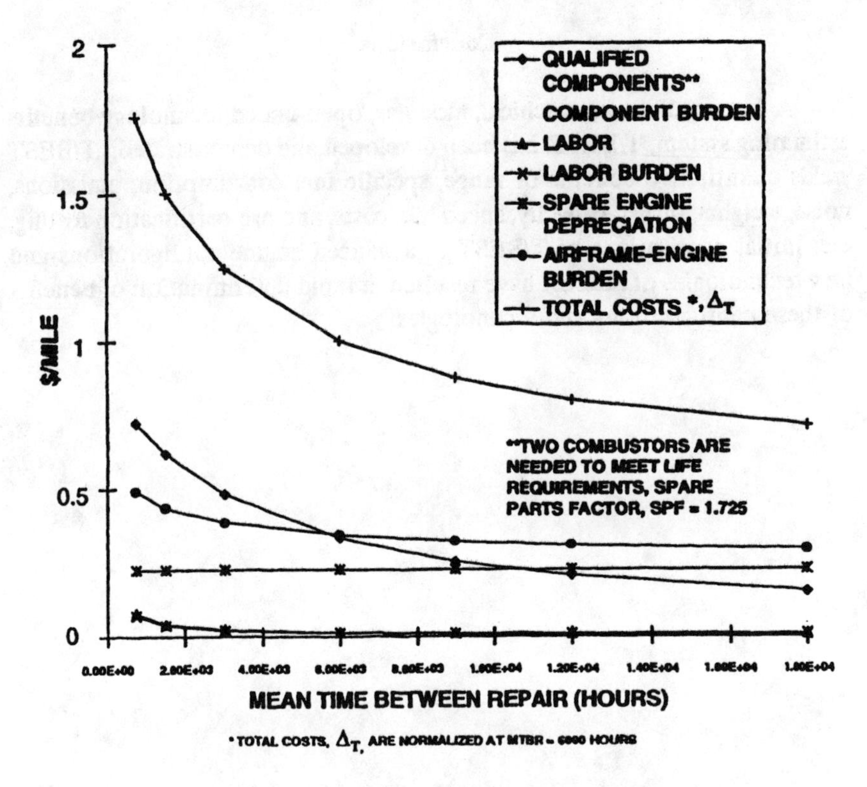

Fig. 18 - Engine maintenance costs.

direct operating cost is shown in Fig. 16. The data is these figures and the constant design data in the next figure are all available from one simulation. Figure 17 indicates typical system level benefits for constant design flow and constant design thrust designs where the combuster efficiency has been reduced by 10%. Parameter changes needed to perform evaluations take only a few seconds. Fig. 18 indicates the increase of engine maintenance costs that occurs if the life of the combuster is modified. A complete simulation analysis that makes use of all T/BEST's modules takes about 20 min. T/BEST may also be used for optimization. For example, optimization of the blade profile of the first fan stage, for minimized weight with tip thickness and resonance margin constraints, takes about 45 min and 80 intracode cycles of the structural analysis code identified as BLASIM in Fig. 4. Verification of the 120,000 lines of T/BEST software code is done as individual modules are attached and by independent evaluation by discipline experts.

Conclusions

A fast, credible, hierarchical, modular, open-ended technology benefits estimating system, T/BEST, has been developed and demonstrated. T/BEST yields quantitative benefits of range, specific fuel consumption, emissions, noise, weights, thrust, capacity, speed life, costs, and pre-certification testing, etc. Initial applications of T/BEST to advanced engine configurations and new technologies of interest have resulted in rapid determination of benefits of these configurations and technologies.

Author Index

PROGRESS IN ASTRONAUTICS AND AERONAUTICS SERIES VOLUMES

***1. Solid Propellant Rocket Research (1960)**
Martin Summerfield
Princeton University

***2. Liquid Rockets and Propellants (1960)**
Loren E. Bollinger
Ohio State University
Martin Goldsmith
The Rand Corp.
Alexis W. Lemmon Jr.
Battelle Memorial Institute

***3. Energy Conversion for Space Power (1961)**
Nathan W. Snyder
Institute for Defense Analyses

***4. Space Power Systems (1961)**
Nathan W. Snyder
Institute for Defense Analyses

***5. Electrostatic Propulsion (1961)**
David B. Langmuir
Space Technology Laboratories, Inc.
Ernst Stuhlinger
NASA George C. Marshall Space Flight Center
J.M. Sellen Jr.
Space Technology Laboratories, Inc.

***6. Detonation and Two-Phase Flow (1962)**
S.S. Penner
California Institute of Technology
F.A. Williams
Harvard University

***7. Hypersonic Flow Research (1962)**
Frederick R. Riddell
AVCO Corp.

***8. Guidance and Control (1962)**
Robert E. Roberson
Consultant
James S. Farrior
Lockheed Missiles and Space Co.

***9. Electric Propulsion Development (1963)**
Ernst Stuhlinger
NASA George C. Marshall Space Flight Center

***10. Technology of Lunar Exploration (1963)**
Clifford I. Cumming
Harold R. Lawrence
Jet Propulsion Laboratory

***11. Power Systems for Space Flight (1963)**
Morris A. Zipkin
Russell N. Edwards
General Electric Co.

***12. Ionization in High-Temperature Gases (1963)**
Kurt E. Shuler, Editor
National Bureau of Standards
John B. Fenn
Associate Editor
Princeton University

***13. Guidance and Control–II (1964)**
Robert C. Langford
General Precision Inc.
Charles J. Mundo
Institute of Naval Studies

***14. Celestial Mechanics and Astrodynamics (1964)**
Victor G. Szebehely
Yale University Observatory

***15. Heterogeneous Combustion (1964)**
Hans G. Wolfhard
Institute for Defense Analyses
Irvin Glassman
Princeton University
Leon Green Jr.
Air Force Systems Command

***16. Space Power Systems Engineering (1966)**
George C. Szego
Institute for Defense Analyses
J. Edward Taylor
TRW Inc.

***17. Methods in Astrodynamics and Celestial Mechanics (1966)**
Raynor L. Duncombe
U.S. Naval Observatory
Victor G. Szebehely
Yale University Observatory

***18. Thermophysics and Temperature Control of Spacecraft and Entry Vehicles (1966)**
Gerhard B. Heller
NASA George C. Marshall Space Flight Center

***19. Communication Satellite Systems Technology (1966)**
Richard B. Marsten
Radio Corporation of America

***20. Thermophysics of Spacecraft and Planetary Bodies: Radiation Properties of Solids and the Electromagnetic Radiation Environment in Space (1967)**
Gerhard B. Heller
NASA George C. Marshall Space Flight Center

*Out of print.

***21. Thermal Design Principles of Spacecraft and Entry Bodies (1969)**
Jerry T. Bevans
TRW Systems

***22. Stratospheric Circulation (1969)**
Willis L. Webb
Atmospheric Sciences Laboratory, White Sands, and University of Texas at El Paso

***23. Thermophysics: Applications to Thermal Design of Spacecraft (1970)**
Jerry T. Bevans
TRW Systems

24. Heat Transfer and Spacecraft Thermal Control (1971)
John W. Lucas
Jet Propulsion Laboratory

25. Communication Satellites for the 70's: Technology (1971)
Nathaniel E. Feldman
The Rand Corp.
Charles M. Kelly
The Aerospace Corp.

26. Communication Satellites for the 70's: Systems (1971)
Nathaniel E. Feldman
The Rand Corp.
Charles M. Kelly
The Aerospace Corp.

27. Thermospheric Circulation (1972)
Willis L. Webb
Atmospheric Sciences Laboratory, White Sands, and University of Texas at El Paso

28. Thermal Characteristics of the Moon (1972)
John W. Lucas
Jet Propulsion Laboratory

***29. Fundamentals of Spacecraft Thermal Design (1972)**
John W. Lucas
Jet Propulsion Laboratory

30. Solar Activity Observations and Predictions (1972)
Patrick S. McIntosh
Murray Dryer
Environmental Research Laboratories, National Oceanic and Atmospheric Administration

31. Thermal Control and Radiation (1973)
Chang-Lin Tien
University of California at Berkeley

32. Communications Satellite Systems (1974)
P.L. Bargellini
COMSAT Laboratories

33. Communications Satellite Technology (1974)
P.L. Bargellini
COMSAT Laboratories

***34. Instrumentation for Airbreathing Propulsion (1974)**
Allen E. Fuhs
Naval Postgraduate School
Marshall Kingery
Arnold Engineering Development Center

35. Thermophysics and Spacecraft Thermal Control (1974)
Robert G. Hering
University of Iowa

36. Thermal Pollution Analysis (1975)
Joseph A. Schetz
Virginia Polytechnic Institute
ISBN 0-915928-00-0

***37. Aeroacoustics: Jet and Combustion Noise; Duct Acoustics (1975)**
Henry T. Nagamatsu, Editor
General Electric Research and Development Center
Jack V. O'Keefe, Associate Editor
The Boeing Co.
Ira R. Schwartz, Associate Editor
NASA Ames Research Center
ISBN 0-915928-01-9

***38. Aeroacoustics: Fan, STOL, and Boundary Layer Noise; Sonic Boom; Aeroacoustics Instrumentation (1975)**
Henry T. Nagamatsu, Editor
General Electric Research and Development Center
Jack V. O'Keefe, Associate Editor
The Boeing Co.
Ira R. Schwartz, Associate Editor
NASA Ames Research Center
ISBN 0-915928-02-7

39. Heat Transfer with Thermal Control Applications (1975)
M. Michael Yovanovich
University of Waterloo
ISBN 0-915928-03-5

***40. Aerodynamics of Base Combustion (1976)**
S.N.B. Murthy, Editor
J.R. Osborn, Associate Editor
Purdue University
A.W. Barrows
J.R. Ward,
Associate Editors
Ballistics Research Laboratories
ISBN 0-915928-04-3

*Out of print.

41. Communications Satellite Developments: Systems (1976)
Gilbert E. LaVean
Defense Communications Agency
William G. Schmidt
CML Satellite Corp.
ISBN 0-915928-05-1

42. Communications Satellite Developments: Technology (1976)
William G. Schmidt
CML Satellite Corp.
Gilbert E. LaVean
Defense Communications Agency
ISBN 0-915928-06-X

***43. Aeroacoustics: Jet Noise, Combustion and Core Engine Noise (1976)**
Ira R. Schwartz, Editor
NASA Ames Research Center
Henry T. Nagamatsu, Associate Editor
General Electric Research and Development Center
Warren C. Strahle, Associate Editor
Georgia Institute of Technology
ISBN 0-915928-07-8

***44. Aeroacoustics: Fan Noise and Control; Duct Acoustics; Rotor Noise (1976)**
Ira R. Schwartz, Editor
NASA Ames Research Center
Henry T. Nagamatsu, Associate Editor
General Electric Research and Development Center
Warren C. Strahle, Associate Editor
Georgia Institute of Technology
ISBN 0-915928-08-6

***45. Aeroacoustics: STOL Noise; Airframe and Airfoil Noise (1976)**
Ira R. Schwartz, Editor
NASA Ames Research Center
Henry T. Nagamatsu, Associate Editor
General Electric Research and Development Center
Warren C. Strahle, Associate Editor
Georgia Institute of Technology
ISBN 0-915928-09-4

***46. Aeroacoustics: Acoustic Wave Propagation; Aircraft Noise Prediction; Aeroacoustic Instrumentation (1976)**
Ira R. Schwartz, Editor
NASA Ames Research Center
Henry T. Nagamatsu, Associate Editor
General Electric Research and Development Center
Warren C. Strahle, Associate Editor
Georgia Institute of Technology
ISBN 0-915928-10-8

***47. Spacecraft Charging by Magnetospheric Plasmas (1976)**
Alan Rosen
TRW Inc.
ISBN 0-915928-11-6

48. Scientific Investigations on the Skylab Satellite (1976)
Marion I. Kent
Ernst Stuhlinger
NASA George C. Marshall Space Flight Center
Shi-Tsan Wu
University of Alabama
ISBN 0-915928-12-4

49. Radiative Transfer and Thermal Control (1976)
Allie M. Smith
ARO Inc.
ISBN 0-915928-13-2

***50. Exploration of the Outer Solar System (1976)**
Eugene W. Greenstadt
TRW Inc.
Murray Dryer
National Oceanic and Atmospheric Administration
Devrie S. Intriligator
University of Southern California
ISBN 0-915928-14-0

51. Rarefied Gas Dynamics, Parts I and II(two volumes) (1977)
J. Leith Potter
ARO Inc.
ISBN 0-915928-15-9

52. Materials Sciences in Space with Application to Space Processing (1977)
Leo Steg
General Electric Co.
ISBN 0-915928-16-7

53. Experimental Diagnostics in Gas Phase Combustion Systems (1977)
Ben T. Zinn, Editor
Georgia Institute of Technology
Craig T. Bowman, Associate Editor
Stanford University
Daniel L. Hartley, Associate Editor
Sandia Laboratories
Edward W. Price, Associate Editor
Georgia Institute of Technology
James G. Skifstad, Associate Editor
Purdue University
ISBN 0-915928-18-3

---
*Out of print.

54. Satellite Communication: Future Systems (1977)
David Jarett
TRW Inc.
ISBN 0-915928-18-3

55. Satellite Communications: Advanced Technologies (1977)
David Jarett
TRW Inc.
ISBN 0-915928-19-1

56. Thermophysics of Spacecraft and Outer Planer Entry Probes (1977)
Allied M. Smith
ARO Inc.
ISBN 0-915928-20-5

57. Space-Based Manufacturing from Nonterrestrial Materials (1977)
Gerald K. O'Neill, Editor
Brian O'Leary,
Assistant Editor
Princeton University
ISBN 0-915928-21-3

***58. Turbulent Combustion (1978)**
Lawrence A. Kennedy
State University of New York at Buffalo
ISBN 0-915928-22-1

***59. Aerodynamic Heating and Thermal Protection Systems (1978)**
Leroy S. Fletcher
University of Virginia
ISBN 0-915928-23-X

60. Heat Transfer and Thermal Control Systems (1978)
Leroy S. Fletcher
University of Virginia
ISBN 0-915928-24-8

61. Radiation Energy Conversion in Space (1978)
Kenneth W. Billman
NASA Ames Research Center
ISBN 0-915928-26-4

62. Alternative Hydrocarbon Fuels: Combustion and Chemical Kinetics (1978)
Craig T. Bowman
Stanford University
Jorgen Birkeland
Department of Energy
ISBN 0-915928-25-6

***63. Experimental Diagnostics in Combustion of Solids (1978)**
Thomas L. Boggs
Naval Weapons Center
Ben T. Zinn
Georgia Institute of Technology
ISBN 0-915928-28-0

64. Outer Planet Entry Heating and Thermal Protection (1979)
Raymond Viskanta
Purdue University
ISBN 0-915928-29-9

65. Thermophysics and Thermal Control (1979)
Raymond Viskanta
Purdue University
ISBN 0-915928-30-2

66. Interior Ballistics of Guns (1979)
Herman Krier
University of Illinois at Urbana-Champaign
Martin Summerfield
New York University
ISBN 0-915928-32-9

***67. Remote Sensing of Earth from Space: Role of "Smart Sensors" (1979)**
Roger A. Breckenridge
NASA Langley Research Center
ISBN 0-915928-33-7

68. Injection and Mixing in Turbulent Flow (1980)
Joseph A. Schetz
Virginia Polytechnic Institute and State University
ISBN 0-915928-35-3

***69. Entry Heating and Thermal Protection (1980)**
Walter B. Olstad
NASA Headquarters
ISBN 0-915928-38-8

***70. Heat Transfer, Thermal Control, and Heat Pipes (1980)**
Walter B. Olstad
NASA Headquarters
ISBN 0-915928-39-6

***71. Space Systems and Their Interactions with Earth's Space Environment (1980)**
Henry B. Garrett
Charles P. Pike
Hanscom Air Force Base
ISBN 0-915928-41-8

***72. Viscous Flow Drag Reduction (1980)**
Gary R. Hough
Vought Advanced Technology Center
ISBN 0-915928-44-2

***73. Combustion Experiments in a Zero-Gravity Laboratory (1981)**
Thomas H. Cochran
NASA Lewis Research Center
ISBN 0-915928-48-5

74. Rarefied Gas Dynamics, Parts I and II (two volumes) (1981)
Sam S. Fisher
University of Virginia
ISBN 0-915928-51-5

*Out of print.

75. Gasdynamics of Detonations and Explosions (1981)
J.R. Bowen
University of Wisconsin at Madison
N. Manson
Universite de Poitiers
A.K. Oppenheim
University of California at Berkeley
R. I. Soloukhin
Institute of Heat and Mass Transfer, BSSR Academy of Sciences
ISBN 0-915928-46-9

76. Combustion in Reactive Systems (1981)
J.R. Bowen
University of Wisconsin at Madison
N. Manson
Universite de Poitiers
A.K. Oppenheim
University of California at Berkeley
R.I. Soloukhin
Institute of Heat and Mass Transfer, BSSR Academy of Sciences
ISBN 0-915928-47-7

***77. Aerothermodynamics and Planetary Entry (1981)**
A.L. Crosbie
University of Missouri-Rolla
ISBN 0-915928-52-3

78. Heat Transfer and Thermal Control (1981)
A.L. Crosbie
University of Missouri-Rolla
ISBN 0-915928-53-1

***79. Electric Propulsion and Its Applications to Space Missions** (1981)
Robert C. Finke
NASA Lewis Research Center
ISBN 0-915928-55-8

***80. Aero-Optical Phenomena (1982)**
Keith G. Gilbert
Leonard J. Otten
Air Force Weapons Laboratory
ISBN 0-915928-60-4

81. Transonic Aerodynamics (1982)
David Nixon
Nielsen Engineering & Research, Inc.
ISBN 0-915928-65-5

82. Thermophysics of Atmospheric Entry (1982)
T.E. Horton
University of Mississippi
ISBN 0-915928-66-3

83. Spacecraft Radiative Transfer and Temperature Control (1982)
T.E. Horton
University of Mississippi
ISBN 0-915928-67-1

84. Liquid-Metal Flows and Magneto-hydrodynamics (1983)
H. Branover
Ben-Gurion University of the Negev
P.S. Lykoudis
Purdue University
A. Yakhot
Ben-Gurion University of the Negev
ISBN 0-915928-70-1

85. Entry Vehicle Heating and Thermal Protection Systems: Space Shuttle, Solar Starprobe, Jupiter Galileo Probe (1983)
Paul E. Bauer
McDonnell Douglas Astronautics Co.
Howard E. Collicott
The Boeing Co.
ISBN 0-915928-74-4

***86. Spacecraft Thermal Control, Design, and Operation (1983)**
Howard E. Collicott
The Boeing Co.
Paul E. Bauer
McDonnell Douglas Astronautics Co.
ISBN 0-915928-75-2

87. Shock Waves, Explosions, and Detonations (1983)
J.R. Bowen
University of Washington
N. Manson
Universite de Poitiers
A.K. Oppenheim
University of California at Berkeley
R.I. Soloukhin
Institute of Heat and Mass Transfer, BSSR Academy of Sciences
ISBN 0-915928-76-0

88. Flames, Lasers, and Reactive Systems (1983)
J.R. Bowen
University of Washington
N. Manson
Universite de Poitiers
A.K. Oppenheim
University of California at Berkeley
R.I. Soloukhin
Institute of Heat and Mass Transfer, BSSR Academy of Sciences
ISBN 0-915928-77-9

***89. Orbit-Raising and Maneuvering Propulsion: Research Status and Needs (1984)**
Leonard H. Caveny
Air Force Office of Scientific Research
ISBN 0-915928-82-5

*Out of print.

90. Fundamental of Solid-Propellant Combustion (1984)
Kenneth K. Kuo
Pennsylvania State University
Martin Summerfield
Princeton Combustion Research Laboratories, Inc.
ISBN 0-915928-84-1

91. Spacecraft Contamination: Sources and Prevention (1984)
J.A. Roux
University of Mississippi
T.D. McCay
NASA Marshall Space Flight Center
ISBN 0-915928-85-X

92. Combustion Diagnostics by Nonintrusive Methods (1984)
T.D. McCay
NASA Marshall Space Flight Center
J.A. Roux
University of Mississippi
ISBN 0-915928-86-8

93. The INTELSAT Global Satellite System (1984)
Joel Alper
COMSAT Corp.
Joseph Pelton
INTELSAT
ISBN 0-915928-90-6

94. Dynamics of Shock Waves, Explosions, and Detonations (1984)
J.R. Bowen
University of Washington
N. Manson
Universite de Poitiers
A. K. Oppenheim
University of California at Berkeley
R.I. Soloukhin
Institute of Heat and Mass Transfer, BSSR Academy of Sciences
ISBN 0-915928-91-4

95. Dynamics of Flames and Reactive Systems (1984)
J.R. Bowen
University of Washington
N. Manson
Universite de Poitiers
A. K. Oppenheim
University of California at Berkeley
R.I. Soloukhin
Institute of Heat and Mass Transfer, BSSR Academy of Sciences
ISBN 0-915928-92-2

96. Thermal Design of Aeroassisted Orbital Transfer Vehicles (1985)
H.F. Nelson
University of Missouri-Rolla
ISBN 0-915928-94-9

97. Monitoring Earth's Ocean, Land, and Atmosphere from Space–Sensors, Systems, and Applications (1985)
Abraham Schnapf
Aerospace Systems Engineering
ISBN 0-915928-98-1

98. Thrust and Drag: Its Prediction and Verification (1985)
Eugene E. Covert
Massachusetts Institute of Technology
C.R. James
Vought Corp.
William F. Kimzey
Sverdrup Technology AEDC Group
George K. Richey
U.S. Air Force
Eugene C. Rooney
U.S. Navy Department of Defense
ISBN 0-930403-00-2

99. Space Stations and Space Platforms – Concepts, Design, Infrastructure, and Uses (1985)
Ivan Bekey
Daniel Herman
NASA Headquarters
ISBN 0-930403-01-0

100. Single- and Multi-Phase Flows in an Electromagnetic Field: Energy, Metallurgical, and Solar Applications (1985)
Herman Branover
Ben-Gurion University of the Negev
Paul S. Lykoudis
Purdue University
Michael Mond
Ben-Gurion University of the Negev
ISBN 0-930403-04-5

101. MHD Energy Conversion: Physiotechnical Problems (1986)
V.A. Kirillin
A.E. Sheyndlin
Soviet Academy of Sciences
ISBN 0-930403-05-3

102. Numerical Methods for Engine-Airframe Integration (1986)
S.N.B. Murthy
Purdue University
Gerald C. Paynter
Boeing Airplane Co.
ISBN 0-930403-09-6

103. Thermophysical Aspects of Re-Entry Flows (1986)
James N. Moss
NASA Langley Research Center
Carl D. Scott
NASA Johnson Space Center
ISBN 0-930430-10-X

*Out of print.

***104. Tactical Missile Aerodynamics (1986)**
M.J. Hemsch
PRC Kentron, Inc.
J.N. Nielson
NASA Ames Research Center
ISBN 0-930403-13-4

105. Dynamics of Reactive Systems Part I: Flames and Configurations; Part II: Modeling and Heterogeneous Combustion (1986)
J.R. Bowen
University of Washington
J.-C. Leyer
Universite de Poitiers
R.I. Soloukhin
Institute of Heat and Mass Transfer, BSSR Academy of Sciences
ISBN 0-930403-14-2

106. Dynamics of Explosions (1986)
J.R. Bowen
University of Washington
J.-C. Leyer
Universite de Poitiers
R.I. Soloukhin
Institute of Heat and Mass Transfer, BSSR Academy of Sciences
ISBN 0-930403-15-0

107. Spacecraft Dielectric Material Properties and Spacecraft Charging (1986)
A.R. Frederickson
U.S. Air Force Rome Air Development Center
D.B. Cotts
SRI International
J.A. Wall
U.S. Air Force Rome Air Development Center
F.L. Bouquet
Jet Propulsion Laboratory, California Institute of Technology
ISBN 0-930403-17-7

108. Opportunities for Academic Research in a Low-Gravity Environment (1986)
George A. Hazelrigg
National Science Foundation
Joseph M. Reynolds
Louisiana State University
ISBN 0-930403-18-5

109. Gun Propulsion Technology (1988)
Ludwig Stiefel
U.S. Army Armament Research, Development and Engineering Center
ISBN 0-930403-20-7

110. Commercial Opportunities in Space (1988)
F. Shahrokhi
K.E. Harwell
University of Tennessee Space Institute
C.C. Chao
National Cheng Kung University
ISBN 0-930403-39-8

111. Liquid-Metal Flows: Magnetohydrodynamics and Application (1988)
Herman Branover, Michael Mond, and Yeshajahu Unger
Ben-Gurion University of the Negev
ISBN 0-930403-43-6

112. Current Trends in Turbulence Research (1988)
Herman Branover, Micheal Mond, and Yeshajahu Unger
Ben-Gurion University of the Negev
ISBN 0-930403-44-4

113. Dynamics of Reactive Systems Part I: Flames; Part II: Heterogeneous Combustion and Applications (1988)
A.L. Kuhl
R & D Associates
J.R. Bowen
University of Washington
J.-C. Leyer
Universite de Poitiers
A. Borisov
USSR Academy of Sciences
ISBN 0-930403-46-0

114. Dynamics of Explosions (1988)
A.L. Kuhl
R & D Associates
J.R. Bowen
University of Washington
J.-C. Leyer
Universite de Poitiers
A. Borisov
USSR Academy of Sciences
ISBN 0-930403-47-9

115. Machine Intelligence and Autonomy for Aerospace (1988)
E. Heer
Heer Associates, Inc.
H. Lum
NASA Ames Research Center
ISBN 0-930403-48-7

116. Rarefied Gas Dynamics: Space Related Studies (1989)
E.P. Muntz
University of Southern California
D.P. Weaver
U.S. Air Force Astronautics Laboratory (AFSC)
D.H. Campbell
University of Dayton Research Institute
ISBN 0-930403-53-3

---
*Out of print.

ISBN 0-930403-68-1

117. Rarefied Gas Dynamics: Physical Phenomena (1989)
E.P. Muntz
University of Southernn California
D.P. Weaver
U.S. Air Force Astronautics Laboratory (AFSC)
D.H. Campbell
University of Dayton Research Institute
ISBN 0-930403-54-1

118. Rarefied Gas Dynamics: Theoretical and Computational Techniques (1989)
E.P. Muntz
University of Southernn California
D.P. Weaver
U.S. Air Force Astronautics Laboratory (AFSC)
D.H. Campbell
University of Dayton Research Institute
ISBN 0-930403-55-X

119. Test and Evaluation of the Tactical Missile (1989)
Emil J. Eichblatt Jr.
Pacific Missile Test Center
ISBN 0-930403-56-8

120. Unsteady transonic Aerodynamics (1989)
David Nixon
Nielsen Engineering & Research, Inc.
ISBN 0-930403-52-5

121. Orbital Debris from Upper-Stage Breakup (1989)
Joseph P. Loftus Jr.
NASA Johnson Space Center
ISBN 0-930403-58-4

122. Thermal-Hydraulics for Space Power, Propulsion and Thermal Management System Design (1989)
William J. Krotiuk
General Electric Co.
ISBN 0-930403-64-9

123. Viscous Drag Reduction in Boundary Layers (1990)
Dennis M. Bushnell
Jerry N. Hefner
NASA Langley Research Center
ISBN 0-930403-66-5

***124. Tactical and Strategic Missile Guidance (1990)**
Paul Zarchan
Charles Stark Draper Laboratory, Inc.

125. Applied Computational Aerodynamics (1990)
P.A. Henne
Douglas Aircraft Company
ISBN 0-930403-69-X

126. Space Commercialization: Launch Vehicles and Programs (1990)
F. Shahrokhi
University of Tennessee Space Institute
J.S. Greenberg
Princeton Synergetics Inc.
T. Al-Saud
Ministry of Defense and Aviation Kingdom of Saudi Arabia
ISBN 0-930403-75-4

127. Space Commercialization: Platforms and Processing (1990)
F. Shahrokhi
University of Tennessee Space Institute
G. Hazelrigg
National Science Foundation
R. Bayuzick
Vanderbilt University
ISBN 0-930403-76-2

128. Space Commercialization: Satellite Technology (1990)
F. Shahrokhi
University of Tennessee Space Institute
N. Jasentuliyana
United Nations
N. Tarabzouni
King Abulaziz City for Science and Technology
ISBN 0-930403-77-0

129. Mechanics and Control of Large Flexible Structures (1990)
John L. Junkins
Texas A&M University
ISBN 0-930403-73-8

130. Low-Gravity Fluid Dynamics and Transport Phenomena (1990)
Jean N. Koster
Robert L. Sani
University of Colorado at Boulder
ISBN 0-930403-74-6

131. Dynamics of Deflagrations and Reactive Systems: Flames (1991)
A.L. Kuhl
Lawrence Livermore National Laboratory
J.-C. Leyer
Universite de Poitiers
A. A. Borisov
USSR Academy of Sciences
W.A. Sirignano
University of California
ISBN 0-930403-95-9

*Out of print.

132. Dynamics of Deflagrations and Reactive Systems: Heterogeneous Combustion (1991)
A.L. Kuhl
Lawrence Livermore National Laboratory
J.-C. Leyer
Universite de Poitiers
A. A. Borisov
USSR Academy of Sciences
W.A. Sirignano
University of California
ISBN 0-930403-96-7

133. Dynamics of Detonations and Explosions: Detonations (1991)
A.L. Kuhl
Lawrence Livermore National Laboratory
J.-C. Leyer
Universite de Poitiers
A. A. Borisov
USSR Academy of Sciences
W.A. Sirignano
University of California
ISBN 0-930403-97-5

134. Dynamics of Detonations and Explosions: Explosion Phenomena (1991)
A.L. Kuhl
Lawrence Livermore National Laboratory
J.-C. Leyer
Universite de Poitiers
A. A. Borisov
USSR Academy of Sciences
W.A. Sirignano
University of California
ISBN 0-930403-98-3

135. Numerical Approaches to Combustion Modeling (1991)
Elaine S. Oran
Jay P. Boris
Naval Research Laboratory
ISBN 1-56347-004-7

136. Aerospace Software Engineering (1991)
Christine Anderson
U.S. Air Force Wright Laboratory
Merlin Dorfman
Lockheed Missiles & Space Company, Inc.
ISBN 1-56347-005-0

137. High-Speed Flight Propulsion Systems (1991)
S.N.B. Murthy
Purdue University
E.T. Curran
Wright Laboratory
ISBN 1-56347-011-X

138. Propagation of Intensive Laser Radiation in Clouds (1992)
O. A. Volkovitsky
Yu. S. Sedenov
L. P. Semenov
Institute of Experimental Meteorology
ISBN 1-56347-020-9

139. Gun Muzzle Blast and Flash (1992)
Gunter Klingenburg
Fraunhofer-Institut fur Kurzzeitdynamik, Ernst-Mach-Institut (EMI)
Joseph M. Heimerl
U.S. Army Ballistic Research Laboratory (BRL)
ISBN 1-56347-012-8

140. Thermal Structures and Materials for High-Speed Flight (1992)
Earl. A. Thornton
University of Virginia
ISBN 1-56347-017-9

141. Tactical Missile Aerodynamics: General Topics (1992)
Michael J. Hemsch
Lockheed Engineering & Sciences Company
ISBN 1-56347-015-2

142. Tactical Missile Aerodynamics: Prediction Methodology (1992)
Michael R. Mendenhall
Nielsen Engineering & Research, Inc.
ISBN 1-56347-016-0

143. Nonsteady Burning and Combustion Stability of Solid Propellants (1992)
Luigi De Luca
Politecnico di Milano
Edward W. Price
Georgia Institute of Technology
Martin Summerfield
Princeton Combustion Research Laboratories, Inc.
ISBN 1-56347-014-4

144. Space Economics (1992)
Joel S. Greenberg
Princeton Synergetics, Inc.
Henry R. Hertzfeld
HRH Associates
ISBN 1-56347-042-X

145. Mars: Past, Present, and Future (1992)
E. Brian Pritchard
NASA Langley Research Center
ISBN 1-56347-043-8

146. Computational Nonlinear Mechanics in Aerospace Engineering (1992)
Satya N. Atluri
Georgia Institute of Technology
ISBN 1-56347-044-6

147. Modern Engineering for Design of Liquid-Propellant Rocket Engines (1992)
Dieter K. Huzel
David H. Huang
ISBN 1-56347-013-6

---
*Out of print.

148. Metallurgical Technologies, Energy Conversion, and Magnetohydrodynamic Flows (1993)
Herman Branover
Yeshajahu Unger
Ben-Gurion University of the Negev
ISBN 1-56347-019-5

149. Advances in Turbulence Studies (1993)
Herman Branover
Yeshajahu Unger
Ben-Gurion University of the Negev
ISBN 1-56347-018-7

150. Structural Optimization: Status and Promise (1993)
Manohar P. Kamat
Georgia Institute of Technology
ISBN 1-56347-56-X

151. Dynamics of Gaseous Combustion (1993)
A.L. Kuhl
Lawrence Livermore National Laboratory
J.-C. Leyer
Universite de Poitiers
A. A. Borisov
USSR Academy of Sciences
W. A. Sirignano

152. Dynamics of Heterogeneous Gaseous Combustion and Reacting Systems (1993)
A.L. Kuhl
Lawrence Livermore National Laboratory
J.-C. Leyer
Universite de Poitiers
A. A. Borisov
USSR Academy of Sciences
W.A. Sirignano
University of California
ISBN 1-56347-058-6

153. Dynamic Aspects of Detonations (1993)
A.L. Kuhl
Lawrence Livermore National Laboratory
J.-C. Leyer
Universite de Poitiers
A. A. Borisov
USSR Academy of Sciences
W.A. Sirignano
University of California
ISBN 1-56347-057-8

154. Dynamic Aspects of Explosion Phenomena (1993)
A.L. Kuhl
Lawrence Livermore National Laboratory
J.-C. Leyer
Universite de Poitiers
A. A. Borisov
USSR Academy of Sciences
W.A. Sirignano
University of California
ISBN 1-56347-059-4

155. Tactical Missile Warheads (1993)
Joseph Carleone
Aerojet General Corporation
ISBN 1-56347-067-5

156. Toward a Science of Command, Control, and Communications (1993)
Carl R. Jones
Naval Postgraduate School
ISBN 1-56347-068-3

157. Tactical and Strategic Missile Guidance Second Edition (1994)
Paul Zarchan
Charles Stark Draper Laboratory, Inc.
ISBN 1-56347-077-2

158. Rarefied Gas Dynamics: Experimental Techniques and Physical Systems (1994)
Bernie D. Shizgal
University of British Columbia
David P. Weaver
Phillips Laboratory
ISBN 1-56347-079-9

159. Rarefied Gas Dynamics: Theory and Simulations (1994)
Bernie D. Shizgal
University of British Columbia
David P. Weaver
Phillips Laboratory
ISBN 1-56347-080-2

160. Rarefied Gas Dynamics: Space Sciences and Engineering (1994)
Bernie D. Shizgal
University of British Columbia
David P. Weaver
Phillips Laboratory
ISBN 1-56347-081-0

161. Teleoperation and Robotics in Space (1994)
Steven B. Skaar
University of Notre Dame
Carl F. Ruoff
Jet Propulsion Laboratory, California Institute of Technology
ISBN 1-56347-095-0

162. Progress in Turbulence Research (1994)
Herman Branover
Yeshajahu Unger
Ben-Gurion University of the Negev
ISBN 1-56347-099-3

*Out of print.

163. Global Positioning System: Theory and Applications Volume I (1995)
Bradford W. Parkinson
Stanford University
James J. Spilker Jr.
Stanford Telecom
Penina Axelrad, Associate Editor
University of Colorado
Per Enge, Associate Editor
Stanford University
ISBN 1-56347-107-8

164. Global Positioning System:Theory and Applications Volume II (1995)
Bradford W. Parkinson
Stanford University
James J. Spilker Jr.
Stanford Telecom
Penina Axelrad, Associate Editor
University of Colorado
Per Enge, Associate Editor
Stanford University
ISBN 1-56347-106-X

165. Developments In High-Speed Vehicle Propulsion Systems (1995)
S.N.B. Murthy
Purdue University
E. T. Curran
Wright Laboratory
ISBN 1-56347-176-0

166. Recent Advances in Spray Combustion (1995)
Kenneth Kuo
Pennsylvania State University
ISBN 1-56347-175-2

167. Fusion Energy in Space Propulsion (1995)
Terry Kammash
University of Michigan
ISBN 1-56347-184-1

168. Aerospace Thermal Structures and Materials for a New Era (1995)
Earl A. Thornton
University of Virginia
ISBN 1-56347-182-5

*Out of print.